# MODELS FOR ENVIRONMENTAL POLLUTION CONTROL

# MODELS FOR ENVIRONMENTAL POLLUTION CONTROL

**ROLF A. DEININGER, editor**

School of Public Health
The University of Michigan
Ann Arbor, Michigan

Second Printing, 1974


P.O. Box 1425, Ann Arbor, Michigan 48106

Library of Congress Catalog Card Number 73-80818
ISBN 0-250-40032-4


Printed in the United States of America

# PREFACE

This book deals with the use of mathematical models and systems analysis techniques for the control of environmental pollution and resources development. It combines two fields of science, namely mathematical modeling, systems analysis, and computer techniques with environmental pollution control technology. The volume is addressed to the scientists and engineers who are involved in the control of environmental pollution.

Environmental pollution control is not cheap. Resources allocated to it have to compete with other needs of society, and it is therefore important that they are used most efficiently. Mathematical modeling and systems analysis techniques offer a unique opportunity to screen the various control technologies and methods and to select those which are most efficient. These new methods by themselves do not solve any problems of environmental degradation, but they aid the decision maker by clearly showing alternate strategies for control, identifying those which are superior to others.

The book was written by an international group of experts from the United States and Europe and represents an excellent collection of papers dealing with the subject matter. Many of the papers are extensively referenced, allowing the reader to further consult the literature. Most of the papers contained in this book were presented at an Advanced Study Institute entitled "Systems Analysis for Environmental Pollution Control" which was held in December of 1972 in Baiersbronn/Tonbach, Germany, and was partly supported through a grant from the Scientific Affairs Division of NATO.

The book is organized in six major parts, each containing one or more individual papers, along the following lines:

Water Pollution Control
Water Supply and Water Resources Development
Air Pollution Control
Solid Waste Disposal
Noise Control
Total Environmental Models

As an introduction, a paper by *Deininger* briefly summarizes the past use of systems analysis and mathematical modeling in the areas of water pollution, air pollution and the disposal of solid wastes. The major studies and publications in these areas are highlighted, as well as an explanation of the utility and importance of using mathematical modeling.

Part I contains six papers on WATER POLLUTION CONTROL and is the largest part of the book, reflecting the fact that indeed the water pollution control area has in the past attracted most of the studies.

A paper by *Pingry* and *Whinston* describes a very interesting model of a river basin and the many alternate abatement strategies available for improving the water quality in the river. The problem is formulated as a nonlinear programming problem, and a new algorithm for solution is developed. The variables considered are the dissolved oxygen, the biochemical oxygen demand, the temperature, and the effects of increasing the flow in the river.

The paper by *Hahn* deals with the problem of regional sewerage systems. The individual parts of a sewerage system--interceptors, pumping stations, treatment plants--show economies of scale indicating that cost savings can be obtained through regional systems. A mathematical formulation of the problem and an algorithm for solution are indicated.

The simulation of the water quality in rivers, lakes, and estuaries is the subject of the next paper authored by *Orlob*. Formal optimization methods are compared with general simulation methods, and the difficulties of direct optimization of complex systems are noted.

*Bender* discusses ecological aspects of water quality models and points to the need for a better modeling of the biological variables of such systems.

A very interesting approach to the modeling of sewerage systems is described in a paper by *Lindholm*. Experiences with such a general simulation system are described, and the general philosophy behind the direct conversational system is explained.

The section closes with a paper by *Chevereau*, which describes in great detail a general water quality model, and points to the difficulty of properly estimating the parameters.

Part II of this book contains four papers concerned with WATER SUPPLY AND WATER RESOURCES DEVELOPMENT. In the first paper *Cembrowicz* surveys the use of systems analysis and mathematical modeling for the design of water supply systems. Major emphasis is placed on the optimal design of the water distribution system.

A paper by *Schmid et al.* describes a systems analysis study of the economic consequences of the construction of weirs and canals in terms of the altered availability of water quality and quantity.

*Emsellem* describes in detail the difficulties with the optimization of water resources for entire river basins in France.

And in the final paper of the section, the use of models and optimization techniques in Italy is described by *Benedini*. The change from the original emphasis on flood control and hydroelectric power to more considerations of water quality is described.

Part III of the book examines the role of mathematical models in AIR POLLUTION CONTROL. In an overview paper by *Burton et al.* the historic development of air quality models is sketched, and the current modeling efforts in the U. S. are described. The theoretical basis of the models and the experiences with their use are discussed. The interplay of different control strategies with the ambient air environment and the costs of the different strategies are presented.

The difficulties of air quality modeling are summarized in a paper by *Gustafson*. The theoretical basis and the type of necessary models are discussed, and several new algorithms for finding optimal and near optimal control strategies are presented.

In the final paper of the section, *Fortak* presents the results of air quality modeling studies of several cities in Germany. The difficulties in these studies are explained, emphasizing the need to know the meteorological variables.

Part IV of the book deals with the safe DISPOSAL OF SOLID WASTES. In a comprehensive review paper *Clark* describes the many studies which were undertaken in the U. S. Both simulation and optimization models are described, and the need for data collection and analysis is emphasized.

*Coyle* describes his experiences with vehicle routing models in the United Kingdom and cautions not to rely solely on mathematical modeling. He argues for the proper mixture of formal optimization and practical engineering.

In a paper by *Kühner* the problem of the proper location of waste disposal facilities is described, and algorithms for a solution of mathematically difficult problems are described.

*Liebling*'s paper closes this section with a discussion of the routing of street cleaning and snow removal vehicles. Experiences with the algorithms and computer programs are presented and a case example for the city of Zurich is discussed.

Part V contains only one paper on NOISE CONTROL. Comparatively little has been done in this area, and the paper by *Loucks* presents the first effort to introduce mathematical modeling and formal optimization techniques into this badly neglected area. While there has been little application of these models until now, the paper clearly points to the future and will be a classic before long.

The final part of the book, Part VI, looks at TOTAL ENVIRONMENTAL MODELS and contains two papers. *Spofford* describes the efforts to integrate the models of air and water pollution and solid waste disposal into an overall environmental quality model. These types of models have only recently been studied, but again they point toward the future.

The final paper by *Häfele* draws the analogy between the safeguard of nuclear materials and environmental pollution control. Many of the techniques and surveillance strategies of the former should be directly applicable to environmental control, and the author clearly identifies these.

Overall, the book summarizes the present state-of-the-art in mathematical modeling for environmental pollution control. The original articles in the book, and the many literature references cited, present an extensive review of past studies, present efforts, and the future in this rapidly expanding field.

Rolf A. Deininger
Ann Arbor, Michigan
August, 1973

## TABLE OF CONTENTS

### INTRODUCTION



### PART I
### WATER POLLUTION CONTROL

## PART II
## WATER SUPPLY AND WATER RESOURCES DEVELOPMENT



## PART III
## AIR POLLUTION CONTROL



## PART IV
## SOLID WASTE DISPOSAL

## PART V
## NOISE CONTROL



## PART VI
## TOTAL ENVIRONMENTAL MODELS

# INTRODUCTION

MODELS FOR ENVIRONMENTAL POLLUTION CONTROL


# SYSTEMS ANALYSIS FOR ENVIRONMENTAL POLLUTION CONTROL

Rolf A. Deininger*

## I. INTRODUCTION

The use of systems analysis and mathematical modeling for formulating and solving problems of environmental pollution is only of very recent origin and has been used for only a little more than ten years. To define exactly what systems analysis means or involves is somewhat difficult, and there is no general agreement on a definition except to say that it is an approach to problem solving that focuses on the entire problem or system, rather than on each of its separate components. It is an approach that tries to quantitatively define alternate courses of action, and provides a means of evaluating them in an orderly fashion with the aim of selecting those that are better than others. This idea of a comprehensive analysis is not new, but it is only in recent years that the mathematical and computational tools have become available that permit the analysis of many more alternatives than have heretofore been possible. And, it is hoped, by being able to compare many alternatives, the decision-makers should be able to define those combinations of system components that best meet management objectives.

There are a number of more or less standard steps in a systems analysis study:

---

*Dr. Rolf A. Deininger is Professor of Environmental Health, School of Public Health, The University of Michigan, Ann Arbor, Michigan 48104, U.S.A.

1. formulation of the problem
2. construction of a mathematical model that describes the most important variables of the system
3. definition of a criterion function or measure of merit
4. collection of data to allow an estimation of the various parameters of the model
5. derivation of optimal solutions through formal algorithms
6. testing of the model, the solutions, and the sensitivity of the parameters
7. implementation of the best solution.

The first step involves the formulation of the problem and a definition of the objectives of the study. Alternate courses of action must be determined and the relation between the courses of action and the major variables of the problem under study must be delineated. This leads to step 2, namely the construction of a mathematical model. The quantitative relationships between the variables of the problem must be established. The third step is the definition of a criterion function, or measure of merit. Since the system under investigation may be in many different states, or the system to be constructed may consist of many alternate solutions, it is necessary to have a criterion which compares one system against another. Such criteria might be a minimum cost system or one having the greatest benefit-cost ratio.

The fourth step is the collection of data to allow an estimate of the parameters of the model. This may be the most time-consuming step, and depending on its outcome the entire model may have to be modified due to a lack of data. Thus step 4 should be carried out simultaneously with steps 1 to 3.

Step 5 then seeks to determine an optimal solution to the problem, if it is possible at all, through the application of a number of formal algorithms that will be described later on.

The most important step is a testing of the model, the solutions, and the sensitivity of the parameters. To make the model mathematically tractable usually some simplifications are introduced. These and the influence certain parameters have on the optimal solution must be studied.

The techniques and tools of systems analysis are shown below.

*Systems Analysis*

| | |
|---|---|
| Linear programming | Queuing theory |
| Nonlinear programming | Simulation techniques |
| Dynamic programming | Network techniques |
| Integer programming | PERT, CPM |
| Stochastic programming | Gradient methods |

Digital computer

The techniques of linear and nonlinear programming deal with optimization problems where the maximum or minimum of linear or nonlinear functions is to be found subject to equality or inequality constraints. Several examples of its use will be given below. Dynamic programming deals with the optimization of multistage decision processes, whereas integer programming attempts to find solutions to problems where the variables may take on only integer values. Stochastic programming is concerned with problems in which some of the variables and parameters are random variables. Queuing theory, which was originally developed to analyze the queues in telephone systems, waiting lines at supermarkets and toll booths of turnpikes, has found new use in the analysis of reservoirs and their operation. Network techniques, PERT, which stands for program evaluation and review technique, and CPM, which stands for critical path method, are management techniques for the control of large scale research and development projects or for large construction jobs. Simulation techniques are used for the simulation of large systems without the explicit attempt to find the optimum but rather just to observe what happens when certain basic parameters are changed.

All of these techniques would be mostly of academic interest were it not for the concurrent development of high speed digital computers. It is well to remember that the first commercially available computer came on the market only in 1951 with UNIVAC I. Now, only two decades later, the digital computer is having an unbelievable impact on our societies. Computing speeds have increased more than 100-fold, and storage capacities a thousand-fold. And costs have decreased to about 1/1000 of the early models. But most of all, computers have become so widespread that today no engineering firm or university is either without such a system or without easy access to it.

All these mathematical modeling and computational techniques cannot by themselves solve environmental pollution problems. The mathematical algorithms and computers cannot provide solutions to all the current problems of society, especially those requiring public policy decisions. They merely increase the capability of generating information that can be used in the decision-making process.

One should also be aware that optimal solutions to *models* of environmental pollution are often not the optimal solutions to the actual *problems* themselves. Perhaps the most significant reason for this is the inability of the policy-makers to define their objectives. This is due in part to the inability of the engineers, ecologists and economists to define and quantify many of the important physical, biological, economic and social interrelations between the many components of any environmental pollution control system. Partly because of these political technological and economic uncertainties one should not expect the results from any quantitative systems analysis study to be implemented without modifications, no matter how impressive the results may appear on the computer printout.

This paper will first examine water pollution problems, with special attention paid to the historic development of the use of systems analysis, then solid waste disposal problems, then air pollution problems, and finally look at total environmental models.

## II. WATER POLLUTION CONTROL MODELS

Probably because water pollution was recognized early and partly because research funds were available for this area, the use of systems analysis and mathematical models is more developed in this area than in any other. Models have been constructed that start with the collection of wastewaters, their treatment, and their final disposal in a body of water and the quality variations in this body of water.

Despite the fact that the expenditures for wastewater collection systems are very large, relatively little work has been conducted on these aspects. Only recently have efforts been undertaken to develop general mathematical models for simulating the entire rainfall-runoff process from the onset of the precipitation, through collection, conveyance, storage, treatment and final disposal to the water course. Basically these models aim at describing the entire, unsteady flow in the canal network to study its functioning and behavior; to see if it is over- or underdesigned, and to get traces of the water quantity and quality at the various outlet points as a function of time.

Other efforts have been to determine the least costly collection systems for sanitary wastes alone under steady-state conditions. These models try to determine networks that will collect all the wastes at the required points, convey them to a central collection point, and minimize the cost of the system, consisting of the costs of the pipes and the costs of excavation. From a strict mathematical point of view, many of these problems are unsolved; from a practical point of view, these models allow the designer to find very good solutions, even if they may not be optimal in the strict mathematical sense.

The design of wastewater treatment plants has found more attention. Typically, the models view the treatment plant consisting of a number of unit processes and the objective is to select that combination of processes that will attain the required effluent quality at minimum overall cost. Both linear and dynamic programming models have been formulated and solved. In general, the mathematics are ahead of the data requirements regarding the costs of the individual units and their efficiency. Other current research focuses on the actual operation of the treatment plant, attempting to simulate the behavior of the plant under varying inflows in quantity and quality. Other research and development efforts are aimed at the automatic design of treatment plants on computers, and the automatic drawing of construction plans.

The planning of regional and intercommunity wastewater collection and treatment systems has also been given attention. The aim of these models is to determine the most economical location of treatment plants and the interconnecting sewerage system so that the regional costs are minimized. Expansion of capacity and possibly changing degrees of treatment over time have been given consideration.

By far the most attention has been given to models of entire river basins. Under the assumption that a regional water authority exists, models have been constructed for the most economical location of treatment plants and their required efficiencies.

The earlier models considered in essence only the degree of treatment as a variable, whereas the later models also include effluent charges, storing of wastewaters in lagoons, and artificial stream aeration. Newer models consider the effects of low flow augmentation from reservoirs and the problems of heat discharges.

The frontier in systems modeling is the study of lakes. One of the most pressing problems of today is a complete ecological model of a lake; that is, to predict the quality changes over time as a function of the nutrient inputs to the lake. There are several research studies going on in this area.

## III. MODELS OF THE DISPOSAL OF SOLID WASTES

The entire solid-waste system encompasses everything from the manufacturing process through the generation of wastes and their collection to the ultimate disposal and recycling. Similar to the modeling situation in water pollution control, attempts at constructing models of the entire system tend to be simply conceptual, while more detailed, practical models are constructed of smaller subsystems. Few models have been built of the overall problem and most attention has been given to two specific problem areas: the routing of collecting and transfer vehicles, and the location of treatment and disposal facilities.

The first models focused on the collection system and were simulation models aimed at generating information on workloads along vehicle routes and the queuing problems at disposal and transfer sites. Attempts have been made to develop a completely general simulation system of the solid-waste collection, transfer and disposal systems. Options included various lengths of work day, alternate locations of transfer stations and disposal sites, and other system variables. Current work in this area attempts to find patterns of vehicle routing that will minimize the total distance traveled by the vehicles, taking into consideration the existence of one- and two-way streets. The mathematics for solving these problems becomes quite intricate, and some of the problem formulations from Operations Research such as the "Traveling Salesman Problem" and the "Chinese Postman Problem" are being used. Again, mathematically speaking, optimal solutions are not available at the moment, but most of the techniques lead to better routings.

The problem of selecting the optimal location for transfer stations, treatment facilities, and ultimate disposal sites is similar to general facility location problems that have appeared in the literature in many different contexts. Details of the various methods are not important; the differences of formulation are usually in the form of cost functions assumed and whether the facility may be located anywhere on the plane or only at selected points of a network. But basically the models try to find locations of transfer stations, treatment facilities and ultimate disposal sites so that the total cost of transportation, treatment and disposal is minimized.

In general the development of solid-waste models passed through three stages. In the first stage, extremely simple models were produced. They were followed in the second stage by more refined models, which reflect the physical system better. As these approaches become more sophisticated the social (nonphysical) aspects of the problem come to be recognized and appear slowly in the models. The better understood the feature of the problem, the sooner it is likely to be incorporated in the model. Thus financing alternatives are attached rather early, while models that include human behavior and political preference appear rather late.

## IV. AIR POLLUTION CONTROL MODELS

By comparison to the previously mentioned areas the use of mathematical models and optimization techniques in air pollution control is of more recent origin and fewer studies are available. It may also be that air pollution is more difficult to model. Water pollution is essentially a one-dimensional problem--along a river; solid-waste disposal is two-dimensional--location of facilities in a plane; but air pollution is a three-dimensional problem, namely modeling the distribution and ultimate deposition of pollutants in all three dimensions.

Thus at the beginning of modeling the focus was primarily on obtaining the data and coefficients of the pollutants in the atmosphere for a diffusion model. There are several of these models available and while there is some difference in the mathematics, the greatest uncertainties lie with the coefficients and parameters that enter the equations. The selection of one model over the other is based on the analyst's trust of the various parameters.

With a basic dispersion model available, the first steps again were simple simulation models. Given a number of line, area or point sources, the task was to simulate how the pollutants would distribute over an area of interest. The simulation was usually limited to two parameters, sulfur-dioxide and particulates. The typical types of questions asked and answered in these simulation studies were: what will happen if one switches from 2% sulfur coal to 1% sulfur coal? Or what will happen to the particulate matter concentration if the major sources will be required to install precipitators? Or, if one switches to oil or gas of certain qualities, how will this affect the pollution levels?

In analyzing these many alternatives and combinations thereof, it became evident that some formal optimization procedures were necessary to screen the alternatives; thus now some models have appeared that aim at finding that combination of control processes and energy sources that will attain the air-quality standards and minimize the economic costs for the region. Most of the models are still in a crude form, and none of the political and social aspects have been introduced yet, but these undoubtedly will follow.

## V. TOTAL ENVIRONMENTAL MODELS

It has long been recognized that the problems of environmental pollution should not be attacked piecemeal and that one should look at the total environment, taking into consideration the air, water and the soil. Some of the interdependence has been known: the treating of wastewaters will create sludge disposal problems. Should they be disposed on land, or should they be incinerated, shifting part of the load temporarily to the air environment?

There have been a few attempts at modeling the total environment but all of them are still in the conceptual stage and only prototype models have been forthcoming. Probably the best work in this area was done by a group of scientists at Resources for the Future who have set down a framework for a large scale, regional environmental quality model dealing with the air, water and land. They divide their overall management model into three parts:

1. an interindustry input-output model reflecting alternate production processes and a final vector of waste discharges
2. a set of environmental diffusion models describing the fate and ultimate disposal of the pollutants in the environment
3. a set of damage functions at various receptor points relating the ambient concentrations of various pollutants to costs and damages at these points.

The model aims then at an optimal selection of production and discharge of pollutants for the given region. At this time the model is still largely conceptual, since the waste coefficients for the interindustry model are known only for a few industries, and the damage functions at the receptor points are virtually nonexistent.

There is no doubt that one should move in this direction, but at the moment our knowledge does not permit us to have such fully operational models and it will require much more research effort.

## VI. GLOBAL ENVIRONMENTAL MODELS

Probably the most ambitious models are those which try to model the complete economic, sociologic and environmental factors of the world. Typical of these are the studies by Forrester and Meadows. These models tie environmental pollution to models of population, industry, land use, capital and general resources. The models are simulation models trying to predict what would happen under different assumptions regarding pollution resources, and population growth. Much more study of the parameters and assumptions built into these models will be necessary before the predictions of these models are reliable.

## VII. SUMMARY OF CURRENT MODELING EFFORTS

From the brief discussion of the modeling efforts it should be evident that much work remains to be done. Some of the major difficulties are defining and quantifying environmental quality and other public policy objectives. There is often a great difference between the optimal solution to a mathematical model and the eventual political solution to the problem that the model was structured to solve. Clearly problem definition, modeling and analysis involves considerable judgment in addition to some mathematical and computational skills. Just how this is accomplished in any particular situation depends not only on the problem itself, but also on the skill of the system analyst, the available data, programming algorithms, and computational facilities. Despite these caveats the use of systems analysis and mathematical models will aid in the decision-making process in four major ways:

1. The use of these methods leads to an increased capability for defining and evaluating possible alternatives and provides for a wider range of options at every level of decision-making.
2. There is an improved capacity for testing assumptions and data to estimate the effects of economic, hydrologic, political and technological uncertainties.
3. The use of systems analysis forces us to make explicit all assumptions and judgments, the consequences of which are available for all to see and question.
4. Systems analysis is a means of communication between all the participants such as planners, engineers, ecologists, hydrologists and economists, helping in understanding what each has to do.

But above all, systems analysis is never completed since the problems are continually changing requiring a continuous updating of information and techniques.

## REFERENCES

Following is a list of books, pamphlets and papers that in one way or another use systems analysis or mathematical models for the formulation of environmental problems. The list is by no means complete, but could serve as an introduction to the studies to date.

### 1. *Literature on Water Pollution Control*

Anderson, M. W. and H. J. Day. "Regional Management of Water Quality--a Systems Approach," *J. Water Pollut. Cont. Fed., 40(10),* (1968).

Berthouex, P. M. and L. B. Polkowski. "Optimum Waste Treatment Plant Design Under Uncertainty," *J. Water Pollut. Cont. Fed., 42(9),* (1970).

Berthouex, P. M. and L. B. Polkowski. "Design Capacities to Accommodate Forecast Uncertainties," *J. San. Eng. Div., ASCE, 96,* No. SA5, Proc. Paper (1970).
Chen, C. "Computer Simulation of Urban Storm Water Runoff," *Proc. Am. Soc. of Civil Eng., Hydraulics Div., 97,* No. HY2 (1971).
DeCicco, P. R. *et al.* "Use of Computers in Design of Sanitary Sewer Systems," *J. Water Pollut. Cont. Fed., 40(2),* 269 (1968).
Deininger, R. A. "Computer Aided Design of Waste Collection and Treatment Systems," *Proc. of 2nd Annual Am. Water Res. Conf.,* Chicago, Illinois, (November, 1966).
Deininger, R. A. "The Economics of Regional Pollution Control Systems," *Proc. 21st Industrial Waste Conf.,* Purdue University, Ext. Ser. 121 (1966).
Deininger, R. A. "Water Quality: Management--the Planning of Economically Optimal Control Systems," *Proc. of the 1st Annual Meeting of the Am. Water Res. Assoc.* (December, 1965).
Deininger, R. A., R. Parrott, and H. Akfirat. "Computer Aided Design of Waste Water Treatment Plants," *Water Research, 3* (1969).
Deininger, R. A. and D. P. Loucks. "Systems Approaches to Problems of Water Pollution Control," *Proc. Annual Meeting, Am. Assoc. for the Advance. of Science,* Chicago, Illinois (December, 1970).
Dysart, B. C. "Water Quality Planning in the Presence of Interacting Pollutants," *J. Water Pollut. Cont. Fed., 42 (8),* (1970).
Evenson, D. E., G. T. Orlob, and J. R. Monser. "Preliminary Selection of Waste Treatment Systems," *J. Water Pollut. Cont. Fed., 41(11),* (1969).
Fisher, J. M., *et al.* "Design of Sewer Systems," *Water Res. Bull., 7(2),* 294 (1971).
Frankel, R. J. "Water Quality Management: Engineering Economic Factors in Municipal Waste Disposal," *Water Resources Res., 1(2),* (1965).
Galler, W. S. "The Use of Operations Research Techniques in Wastewater Treatment Plant Design," *Proc. National Symp. on Anal. Water Resources Systems,* Amer . Water Res. Assoc., Denver, Colo., 225 (1968).
Goodman, A. S. and W. E. Dobbins. "Mathematical Model for Water Pollution Control Studies," *J. San. Eng. Div., Proc. Am. Soc. of Civil Eng., 92(SA6),* (1966).
Grantham, G. R., *et al.* Model for Flow Augmentation Analysis--an Overview," *J. San. Eng. Div., ASCE, 96(SA5),* Proc. Paper 7578 (1970).
Graves, G. W., C. B. Hatfield, and A. Whinston. "Water Pollution Control Using By-Pass Piping," *Water Resources Res., 5(1),* (1969).

Harrington, J. J. "The Role of Computers in Planning and Managing Water Utilities," *J. New England Water Works Assoc., 81(3)*, (1967).

Harrington, J. J. "Systems Analysis for Water Supply and Pollution Control," *Proc. AWRA National Symposium on the Analysis of Water Resource Systems*, Denver, Colorado (July, 1968).

Hass, J. E. "Optimal Taxing for the Abatement of Water Pollution," *Water Resources Res., 6(2)*, (1970).

Heaney, J. P., R. S. Gemmell, A. Charnes, and H. B. Gotaas. "Impact of Institutional Arrangements on the Available Alternative Development Paths for Water Allocation and Pollution Control in the Colorado River Basin," *Proc. 3rd Annual Am. Water Resources Conf.*, San Francisco (November, 1967).

Hoover, T. E. and R. A. Arnoldi. "Computer Model of Connecticut River Pollution," *J. Water Pollut. Cont. Fed., 42(2)*, (1970).

Jaworski, N. A., W. J. Weber, and R. A. Deininger. "Optimal Reservoir Releases for Water Quality Control," *J. San. Eng. Div., Proc. Amer. Soc. of Civil Eng., 96(SA6)*, (1970).

Johnson, E. L. "A Study in the Economics of Water Quality Management," *Water Resources Res., 3(2)*, (1967).

Kerri, K. D. "A Dynamic Model for Water Quality Control," *J. Water Pollut. Cont. Fed., 39(5)*, (1967).

Kerri, K. D. "An Economic Approach to Water Quality Control," *J. Water Pollut. Cont. Fed., 38(12)*, (1966).

Kneese, A. V. *The Economics of Regional Water Quality Management*. (Baltimore, Md.: Johns Hopkins University Press, 1964).

Kneese, A. V. and B. T. Bower. *Managing Water Quality: Economics, Technology, Institutions*. (Baltimore, Md.: Johns Hopkins University Press, 1968).

Liebman, J. C. "A Heuristic Aid for the Design of Sewer Networks," *J. San. Div., Proc. Am. Soc. of Civil Eng., 93(SA4)*, (1967).

Liebman, J. C. and W. R. Lynn. "The Optimal Allocation of Stream Dissolved Oxygen," *Water Resources Res., 2(3)*, (1966).

Loucks, D. P. "Risk Evaluation in Sewage Treatment Plant Design," *J. San. Eng. Div., ASCE, 93(SA1)*, (1967).

Loucks, B. P. and W. R. Lynn. "Probabilistic Models for Predicting Stream Quality," *Water Resources Res., 2(3)*, (1966).

Loucks, D. P., C. W. ReVelle, and W. R. Lynn. "Linear Programming Models for Water Pollution Control," *Management Science, 14(4)*, (1967).

Lynn, W. R. "Application Systems Analysis to Problems in Water and Waste Treatment," *J. Am. Water Works Assoc., 58(6)*, (1966).

Lynn, W. R. "Stage Development of Wastewater Treatment Works," *J. Water Pollut. Cont. Fed., 36(6),* (1964).

Lynn, W. R., J. A. Logan, and A. Charnes. "Systems Analysis for Planning Wastewater Treatment Plants," *J. Water Pollut. Cont. Fed., 34(6),* (1962).

Montgomery, M. M. and W. R. Lynn. "Analysis of Sewage Treatment Systems by Simulation," *J. San. Eng. Div., ASCE 90(SA1),* (1964).

O'Connor, D. J. and J. A. Mueller. "A Water Quality Model of Chlorides in Great Lakes," *J. San. Eng. Div., ASCE 96(SA4),* (1970).

Quirk, T. P. and L. J. Eder. "Evaluation of Alternative Solutions for Achievement of River Standards," *J. Water Pollut. Cont. Fed., 42(2),* (1970).

ReVelle, C., G. Dietrich, and D. Stensel. "The Improvement of Water Quality Under a Financial Constraint," *Water Resources Res., 5(2),* (1969).

ReVelle, C. S., D. P. Loucks, and W. R. Lynn. "Linear Programming Applied to Water Quality Management," *Water Resources Res., 4(1),* (1968).

Russell, C. S. and W. O. Spofford, Jr. "A Quantitative Framework for Residuals--Environmental Quality Management." In: *Natural Resource Systems Models in Decision Making,* G. H. Toebes, ed. (Lafayette, Indiana: Water Resources Research Center, Purdue University, 1969).

Shih, C. S. and J. A. DeFilippi. "System Optimization of Waste Treatment Plant Process Design," *J. San. Eng. Div., ASCE 96(SA2),* (1970).

Shih, C. S. and P. Kirshan. "Dynamic Optimization for Industrial Wastewater Treatment," *J. Water Pollut. Cont. Fed., 41(10),* (1969).

Shih, C. S. and J. A. DeFilippi. "Optimization Models for River Basin Water Quality Management and Waste Treatment Plant Design," *Proc. 4th Amer. Water Resources Conf.,* New York, 754 (1968).

Shih, C. S. "System Optimization for River Basin Water Quality Management," *J. Water Pollut. Cont. Fed., 42(10),* (1970).

Thayer, R. and R. G. Krutchkoff. "Stochastic Model for BOD and DO in Streams," *J. San. Eng. Div., ASCE, 93(SA3),* (1967).

Thomann, R. V. "Variability of Waste Treatment Plant Performance," *J. San. Eng. Div., ASCE, 96(SA3),* (1970).

Thomann, R. V. and M. J. Sobel. "Estuarine Water Quality Management and Forecasting," *J. San. Eng. Div., ASCE, 90(SA5),* (1964).

Upton, C. "Optimal Taxing of Water Pollution," *Water Resources Res., 4(5),* (1968).

Upton, C. "A Model of Water Quality Management Under Uncertainty," *Water Resources Res., 6(3),* (1970).

Worley, J. L., F. J. Burgess, and W. W. Towne. "Identification of Low-Flow Augmentation Requirements for Water Quality Control by Computer Techniques," *J. Water Pollut. Cont. Fed., 37(5),* (1965).

Young, G. K. and M. A. Pisano. "Nonlinear Programming Applied to Regional Water Resources Planning," *Water Resources Res., 6(1),* (1970).

2. *Literature on Solid Waste Disposal Problems*

American Public Works Association. *Proceedings of the National Conference on Solid Wastes Research.* (Chicago: American Public Works Association, 1963).

Clark, Robert M. and Billy P. Helms. "Decentralized Solid Waste Collection Facilities," *J. San. Eng. Div., ASCE, 96(SA5),* Proc. Paper 7594, (October, 1970), p. 1035.

Clark, Robert M. and Billy P. Helms. "Fleet Selection for Solid Waste Collection Systems," *J. San. Eng. Div., ASCE, 97(SA1),* Proc. Paper 8720, (February, 1972), p. 71.

Golueke, C. G. and P. H. McGauhey. *Comprehensive Studies of Solid Waste Management.* Public Health Service Publication No. 2039 (Washington, D.C.: U.S. Government Printing Office, 1970).

Golueke, C. G. and P. H. McGauhey. "Comprehensive Studies of Solid Wastes Management," First Annual Report, Sanitary Engineering Research Laboratory, University of California, Berkeley, SERL Report No. 67-7 (1967).

Helms, B. P. and R. M. Clark. "Locational Models for Solid Waste Management," *J. Urban Planning and Development Div., ASCE, 97(UPI),* 1 (1971).

Marks, D. H. "Facility Location and Routing Models in Solid Waste Collection Systems," Ph.D. dissertation, Johns Hopkins University (1969).

Owen, E. H. "Computer Program Cuts Costs of Urban Solid Waste Collection," *Public Works, 101(1),* (1970).

Quon, J. E., A. Charnes, and S. J. Wersan. "Simulation and Analysis of a Refuse Collection System," *J. San. Eng. Div., ASCE, 91(5),* 17 (1965).

Quon, J. E., R. R. Martens, and M. Tanaka. "Efficiency of Refuse Collection Crews," *J. San. Eng. Div., ASCE, 94(SA2),* (1970).

Quon, J. E., M. Tanaka, and S. J. Wersan. "Simulation Model of Refuse Collection Policies," *J. San. Eng. Div., ASCE, 95(SA3),* Proc. Paper 6626, (June, 1969), p. 575.

ReVelle, C. S., D. H. Marks, and J. C. Liebman. "An Analysis of Private and Public Sector Location Models," *Management Science, 16(11),* (1970).

Skelly, M. J. "Planning for Regional Refuse Systems," Ph.D. Dissertation, Cornell University, Ithaca, New York (1968).

Truitt, M. M., J. C. Liebman, and C. W. Kruse. "Simulation Model of Urban Refuse Collection," *J. San. Eng. Div., ASCE, 95(SA5)*, (1969).

Wolfe, H. and R. Zinn. "Systems Analysis of Solid Waste Disposal Problems," *Public Works, 98(9)*, 99 (1967).

3. *Literature on Air Pollution Problems*

Bowne, N. E. "Simulation Model for Air Pollution Over Connecticut," *J. Air Pollut. Cont. Assoc., 19*, 570 (1969).

"Development of a Training Exercise on Benefit-Cost Evaluation of Air Pollution Control Strategies," Final Report to U.S. Department of Health, Education and Welfare, Resources Research, Inc., Virginia (December, 1969).

Gorr, W., S. A. Gustafson, and K. O. Kortanek. "Optimal Control Strategies for Air Quality Standards and Regulatory Policy," Research Report No. 18, Carnegie-Mellon University, Pittsburgh, Pa. (August, 1971).

Hamburg, F. C. and F. L. Cross. "A Training Exercise on Cost-Effectiveness Evaluation of Air Pollution Control Strategies," *J. Air Pollut. Cont. Assoc., 21(2)*, 66 (1971).

Hickey, H. R. and W. D. Rowe. "A Cost Model for Air Quality Monitoring Systems," Paper presented at 63rd Annual Meeting of Air Pollution Control Assoc., St. Louis, Missouri (June 14-18, 1970).

"Human Investment Programs: An Economic Analysis of the Control of Sulphur Oxides Air Pollution," U.S. Dept. of Health, Education and Welfare (December, 1967).

Kohn, R. E. *Capital Intensiveness of Air Pollution Control*, Reprint (Pittsburgh: Air Pollution Control Assoc., 1970).

Kohn, R. E. "Linear Programming Model for Air Pollution Control: A Pilot Study of the St. Louis Airshed," *J. Air Pollut. Cont. Assoc., 21(2)*, 78 (1970).

Martin, D. O. and J. A. Tikvart. "A General Atmospheric Diffusion Model for Estimating the Effects of Air Quality of One or More Sources," *J. Air Pollut. Cont. Assoc., 21 (1)*, 16 (1971).

McElroy, J. L. "A Comparative Study of Urban and Rural Dispersion," *J. Appl. Meteorol., 8(19)*, (1969).

Milford, S. N., *et al.* "Developing a Practical Dispersion Model for an Air Quality Region," *J. Air Pollut. Cont. Assoc., 21(9)*, 549 (1971).

Moses, H. "Mathematical Urban Air Pollution Models," Research Report No. ANL/ES-RPY-001, Argonne National Laboratory, Illinois (April, 1969).

Sklarew, R. C. "A New Approach: The Grid Model of Urban Air Pollution." Paper presented at 63rd Annual Meeting of Air Pollution Control Assoc., St. Louis, Missouri (June 14-18, 1970).

Stern, A. C. "The Systems Approach to Air Pollution Control," Clean Air Conf., Sydney, Australia (May 21, 1969).

Turner, D. B. "A Diffusion Model for an Urban Area," *J. Appl. Meteorol.*, *3*, 83 (1964).

Turner, D. B. *Workbook of Atmospheric Dispersion Estimates.* (U.S. Dept. of Health, Education and Welfare, Public Health Service, 1969).

U.S. Environmental Protection Agency, Air Pollution Control Office. *Proc. Symposium on Multiple-Source Urban Diffusion Models.* (1970).

Wolozin, H. ed. *The Economics of Air Pollution.* (New York: W. W. Norton, 1966).

4. *Global Environmental Models*

Forrester, J. W. *World Dynamics.* (Cambridge, Mass.: Wright Allen Press, 1971).

Meadows, D. H., *et al.* *The Limits To Growth.* (Washington, D.C.: Potomac Assoc., Inc., 1972).

# PART I

# WATER POLLUTION CONTROL

MODELS FOR ENVIRONMENTAL POLLUTION CONTROL


# CHAPTER 1

# MODELING OF WASTEWATER DISPOSAL SYSTEMS

Oddvar Georg Lindholm*

## I. INTRODUCTION

The Norwegian Institute for Water Research (NIVA) has developed four computer programs for a total analysis and simulation of sewerage systems. The main objectives of these programs, written in FORTRAN, were to characterize the flow of wastewaters through the various components of the sewerage system, to minimize the "leakage" of wastewater from the system, and to determine economically optimal systems. Four specific models were developed:

1. a waste- and stormwater run-off model
2. a model of the wastewater treatment plant
3. a sludge treatment plant model
4. an overall model for the economics of the system.

The models may be linked together but at present most are independently operated. Their general interrelation is shown in Figure 1.1.

The work on these models was started in July, 1971, and was financed mainly by the Norwegian government. The actual coding of the programs and the computer implementation was developed in close cooperation with a private firm, A/S Computas.

---

*Oddvar G. Lindholm is an environmental engineer with the Norwegian Institute for Water Research, P.O. Box 260,Blindern, Oslo 3, Norway.

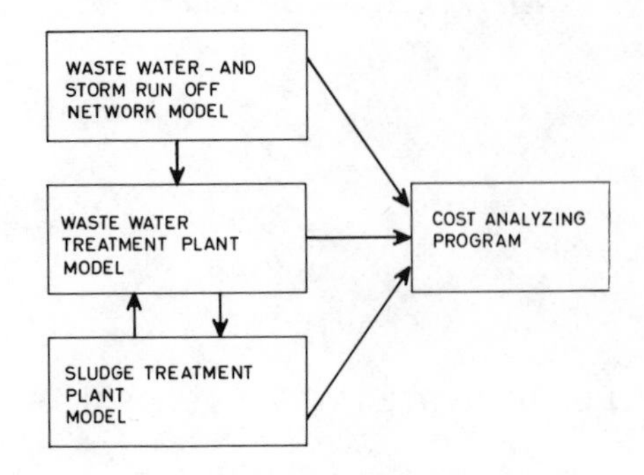


*Figure 1.1. Data flows between the programs.*

## II. THE SEWER NETWORK MODEL

The main objective of this model was to describe the pattern of stormwater runoff in the sewer network, determining the variation in discharge at any point in the network for each minute of a given rainfall. A hydrograph method that considered the effects of storage capacity in the sewer network and the true water velocity when the pipes were partly filled was sought. On the other hand the method could not be too complicated since the cost of computation should be kept at a reasonable level.

### *Features of the Network Model*

The following features were built into the model:

1. The rainfall intensity can vary with time, *i.e.*, for each minute of rainfall, different intensities may be given.
2. The runoff coefficient can vary with time, *i.e.*, for each minute of rainfall, different coefficients may be given.
3. Time of entry *vs.* contributing area functions may be given. In most methods only the time of entry is given to indicate the runoff behavior, which means a linear relationship between time of entry and contributing area for the specified sewer line. Figure 1.2 shows the five standard relations which can be chosen. The default option here is the linear function. Each sewer line can have its own relation, and if no standard relation is found suitable, the user can specify additional relations he wishes to use on separate data sheets.

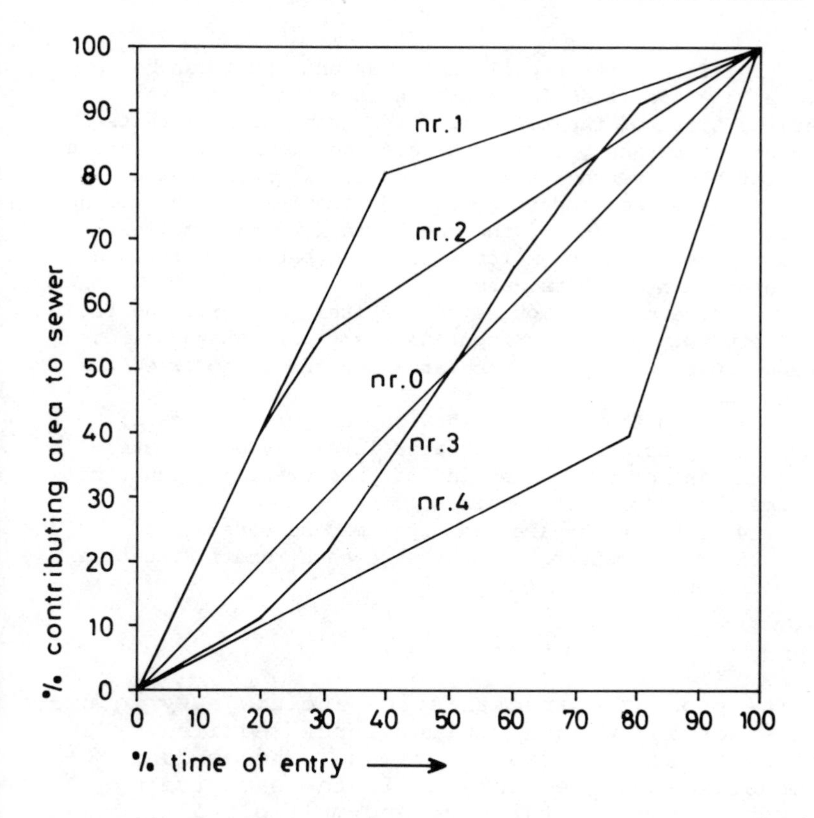


*Figure 1.2. Standard time of entry functions.*

4. The storage capacity of each of the sewer lines is considered.

5. The velocity of flow is made a function of the water depth in the partially filled sewer.

6. Storage tanks may be built in at any point of the system. The necessary tank volume for a given rainfall can be computed when a maximum outlet discharge from the tank is specified.

7. Storm overflows can be considered at any point of the system. The total bypassed and diverted volumes of water are computed.

8. The sum of industrial and domestic wastewater flow together with infiltration water is considered as a constant discharge in time, and each sewer line may have its own value for wastewater "production."

9. Pumping stations may be given at any point of the system.

10. Transport of pollutants per unit time can be computed at any specified location and as a function of time after the start of the rainfall. The grams of pollutants produced per person per day must also be given. Total amount of pollutants diverted from storm overflows is computed.

11. Besides computing water discharge (l/s) and transport of pollutants (g/s), the model can find the smallest standard pipe dimension which will avoid backwater for the particular rainfall considered.

12. The model computes the capital costs for the total sewer network. Necessary input data are percentage of rock in each cross-sectional trench area and the diameter of the sewer when this is not computed.

13. When the sewer network capacity is too small, backwater may occur. The backwater level may be computed for each point of the system and is presented as a function *vs.* time after start of a rainfall.

14. Tunnels, canals and pipes may be considered.

15. The computing step between each runoff situation may be chosen.

### *Program Characteristics*

The program characteristics include easy to use data stacking, with a minimum input required by extensive use of default options. Independent alternatives may be analyzed in the same run by altering changed data only, and multiple datasets may be entered during the same execution. There is an extensive error analysis of input data, with error messages and warnings printed. Furthermore, attention was paid to modular program construction that will simplify future expansions and/or modifications; the coding allows for easy implementation on different computers and machine configurations.

### *Theory of the Model*

The following section briefly describes the computation of the hydrographs for each pipe.

From the input data a rain intensity (l/sec ha) and a runoff coefficient are used for each sewer line and each minute of the rainfall. The runoff for each sewer line is computed for every minute according to Equation 1.

$$Q_n = \phi_N \cdot A \cdot I_N \qquad N = 1, 2, \ldots, M \tag{1}$$

where

$Q_N$ = runoff (l/sec) at minute N

$\phi_N$ = runoff coefficient at minute N

A = area (ha) draining to sewer line

$I_N$ = rainfall intensity (l/sec ha) at minute N

M = rainfall duration in minutes (see Figure 1.3)

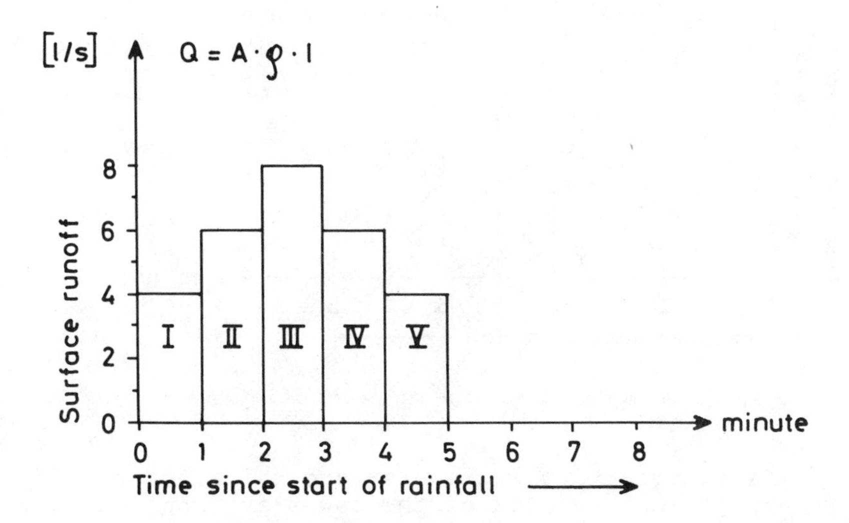


*Figure 1.3. Rainwater runoff for each minute of a rainfall.*

The delayed water discharge into the sewer inlet is computed from Equation 2.

$$D_N = Q_1 \cdot t_E + Q_2 \cdot t_{E-1} + Q_3 \cdot t_{E-2} + \ldots + Q_M \cdot t_1 \qquad (2)$$

$$N = 1, 2, \ldots, M + E$$

where

$D_N$ = discharge into sewer inlet (l/sec) at minute N after start of rainfall

E = time of entry (min)

$t_{E-1}$ = per cent of a sewer's total runoff area contributing at E-1 minutes after start of rainfall (see Figure 1.3).

Equations 1 and 2 lead to the inlet hydrograph shown in Figure 1.4. This hydrograph consists of superimposed "minute-runoffs."

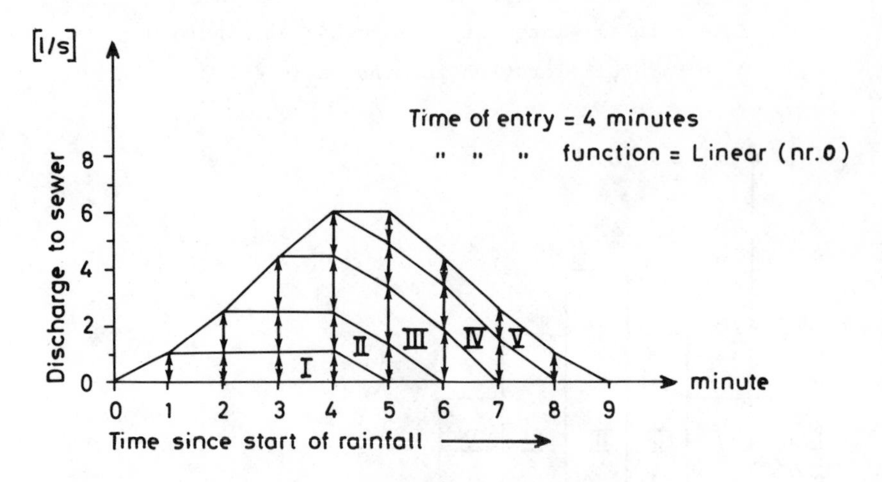


*Figure 1.4. Rainwater discharge at sewer inlet.*

The contributing wastewater and rainwater runoff from the tributary area of the sewer are considered to adjoin at the lower end of the sewer line. The discharge from the upstream sewer adjoining the sewer line, considered at its inlet, goes through two additional analyses before the total outlet hydrograph of the sewer line is found.

The hydrograph from the upstream sewer is manipulated in a pipestorage procedure. The method is very similar to the one described by Watkins,[1] which was developed by the British Road Research Laboratory. However, the RRL-method considers the whole network system in one operation, which may lead to inaccuracies. The method described in this paper analyzes each sewer line independently. In Figure 1.5 the inflow and outflow hydrographs to a sewer line are shown and symbols used in the following equation are defined.

$$S_2 - S_1 = [(P_2 - Q_2) + (P_1 - Q_1)]\ t/2$$

$$S_2 + Q_2 \cdot t/2 = (P_1 + P_2 - Q_1)\ t/2 + S_1$$

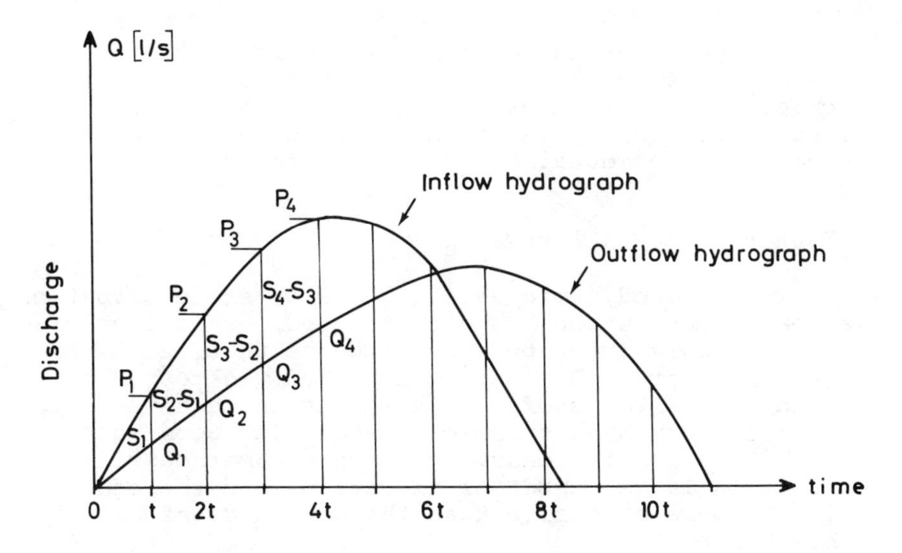


*Figure 1.5. Effect of network storage volume.*

In general the equation will be:

$$S_n + \frac{Q_n \cdot t}{2} = (P_n + P_{n-1} - Q_{n-1}) \frac{t}{2} + S_{n-1} \qquad (3)$$

In Equation 3 the right-hand side is always known from the previous computation (previous min) and from the inflow hydrograph. Since the relationship between the stored volume $S_n$ at minute n and the discharge $Q_n$ is known from curves of partial pipe filling, both $Q_n$ and $S_n$ can be found. In this way the outflow hydrograph for a sewer line is found step by step.

The velocity of flow is computed in each sewer line for each minute. The corresponding discharge used is the average value from the inflow and outflow hydrograph. Curves of partial pipe filling that are stored in the computer are used to compute velocities at partial filling. Since length of the sewer line is known, the time of flow through the sewer can be computed. Each value of the outflow hydrograph (Figure 1.5) is translated the amount of computed time of flow through the sewer line, along the horizontal time-axis. If the sewer is almost empty, the translations are much greater than those for a nearly filled sewer.

A superimposing routine takes care of the branching sewers so that their hydrographs will fit correctly into the total hydrograph. A detailed listing of the input data and the logic of the program may be obtained by writing to the author.

### *Limitations of the Model*

In a hydrodynamic aspect a steady state situation at each computation has been assumed. This inaccuracy is negligible in comparison with the uncertainty of the input data. Besides, the method was tested against on-site runoff measurements and against nonstationary hydrodynamic methods with good agreement. The nonstationary hydrodynamic method will require a large number of iterations and therefore will be more expensive than the method described in this paper.

## III. THE SEWAGE TREATMENT PLANT MODEL

The main objective of this model was to study the performance of a sewage treatment plant receiving both sewage and rainwater runoff over a given period or a whole year. The total program configuration, *i.e.*, the combination of network model and treatment plant model, can be used to analyze the performance of a combined or a separate sewerage system in a given area. On the basis of rain intensity *vs.* duration, total precipitation-duration and frequencies for a year's rain activity, usually 5 to 10 representative base rainfalls will be chosen.

These base rainfalls will be given as input to the sewer network model for a given area. The results, in terms of flow (l/sec) and BOD-load (g/sec) as functions of time, may be used further as input to the sewage treatment plant model.

### *General Description of the Sewage Treatment Plant Model*

Figure 1.6 shows the basic units in the model. As indicated, the flow may pass the chemical stage before the biological, and vice versa. It is also possible to let part of the flow bypass the biological stage, flowing directly for chemical treatment. Any unit may be left out, if desired.

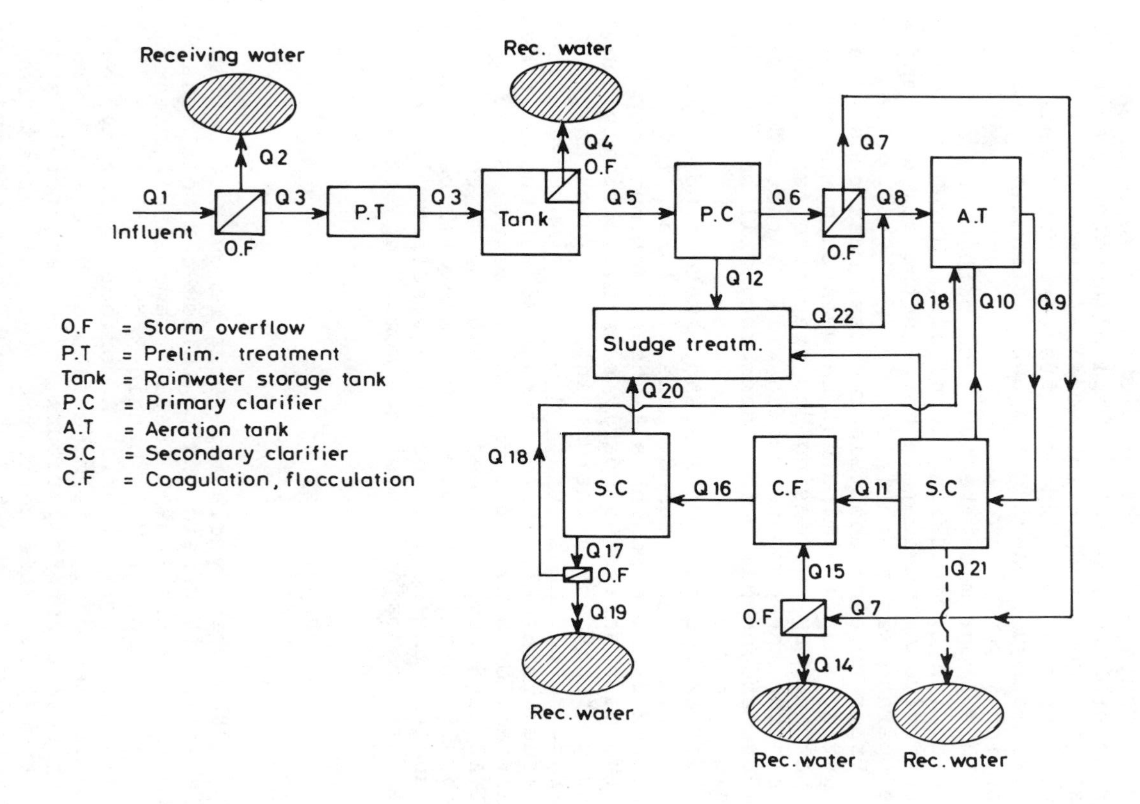


*Figure 1.6. Sewage treatment plant model.*

It should therefore be possible to simulate the actual treatment plants.

To start with, BOD was the only loading parameter considered; however, incorporation of phosphorus in the model is now realized. The removal efficiency for each unit is based mostly on empirical relationships since a more theoretical approach with extensive use of process kinetics was found too deficient and inaccurate for the prediction of removal rates for an arbitrary plant. Capital and total annual costs are computed in the model. The capital cost functions are based on nonlinear empirical unit costs, each unit having its own distinctive cost function.

This program may be run from both batch and interactive time-sharing terminals. However, the interactive communication between the computer and the operator is essential.

Prior to each analysis the user must chose the computing step (in minutes) between each runoff situation. By increasing the step size from 1 to 10 minutes the computer costs will be reduced by 80-90%, and normally the inaccuracy will be insignificant.

For each situation a steady-state condition in the plant was assumed. The final result was obtained by superposing all the situations computed. If the main part of a yearly rainfall activity can be represented by 5 basic rainfalls, which on an average result in a duration of 150 minutes runoff for each rainfall, 751 different situations must be analyzed for a specific treatment plant. Since a UNIVAC 1108 computer needs 3/1000 sec to compute each situation, a total of 0.25 sec will be needed to analyze the performance of a specific plant for one year, choosing ten-minute computing steps.

### *Specific Description of the Sewage Treatment Plant Model*

In the primary clarifier the removal of settleable solids is considered to follow the relations described by Husmann.[2] The BOD removal efficiency in the aeration tank is considered to follow Wuhrmann's observations.[2] A correction for hydraulic load in the aerator is made according to Munz,[2] and a correction for temperature in the aerator is also incorporated. This correction value is taken from Eckenfelder and adjusted to unpublished observational data from Norway.[3] Amount of excess sludge from the biological stage is calculated on the basis of the sludge load factor and the temperature.[4]

The performance of the secondary clarifier will vary from plant to plant. For the present, observational data by Eye[5] have been implemented, representing a relationship between concentration of suspended solids (SS) in the supernatant and the overflow rate. The amount of BOD in the SS is a function of the sludge load factor in the aerator. This function is taken from Eckenfelder.[6]

Formulation of performance for the chemical treatment stage is mostly a result of observations made in an extensive research project for physical/chemical treatment under way at our institution in Oslo. Details of the logic and the optional command words are available from the author.

*The Sludge Treatment Plant Model*

Essentially, the sludge treatment model is not developed enough to be of practical use. What has been accomplished so far will be briefly described.

For the time being the following units are considered: thickener, anaerobic digestion, aerobic stabilization, sludge centrifuge, sieve band press, sludge drying beds, vacuum filters, and final disposals. Figure 1.7 shows how one unit fits into the model.

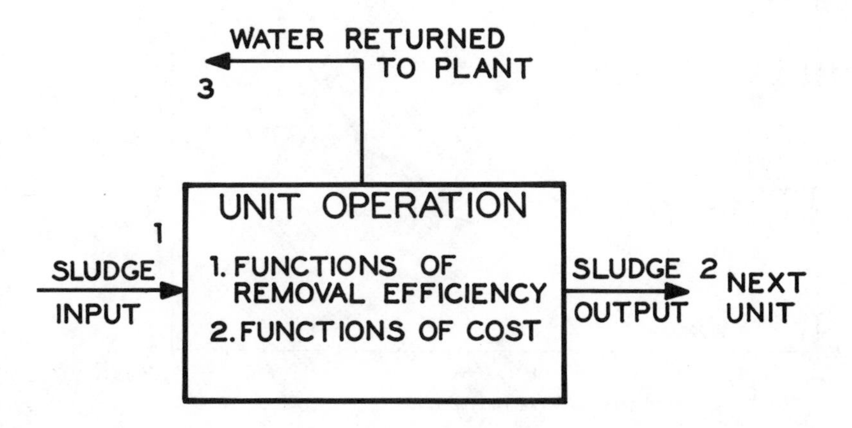


*Figure 1.7. Unit operations in sludge model.*

The combinations of units may be chosen by the user (process alternative). Each stream flow (vector) contains water discharge (l/sec), suspended solids (mg/l), BOD particulate (mg/l), BOD solved (mg/l), phosporus total (mg/l), and nitrogen (mg/l).

*Optimization Technique*

In the sewage treatment plant model there is implemented an automatic optimization subroutine. The optimization technique used is a combination of a trial-and-error method and a gradient method. Figure 1.8 shows how the VAR command and testings of

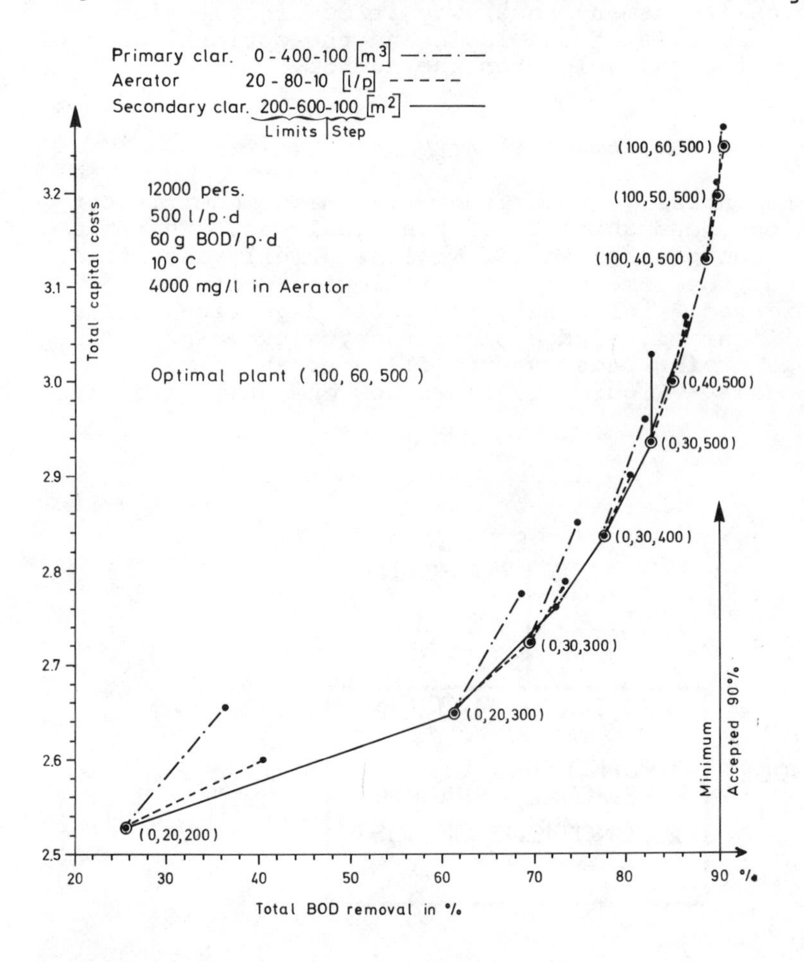


*Figure 1.8. Example of an optimization.*

the steepness of each variable's cost gradient may result in an optimal plant. Each curve represents variation in one variable, while the others are held constant.

The computer selects the variable with the least cost gradient and increases only this variable with a given step, while the others are held constant. Then, from this new base-situation the computer selects the variable again with the least cost gradient. It may now be different from the previous one, providing an increase in the size with a given step. In this way the optimal plant is found step by step. The computer (UNIVAC 1108) needs one second to compute the example shown in Figure 1.8.

*Simplifying Assumptions*

In each computation steady state was assumed. The time aspect is considered by superimposing several situations. The removal efficiency functions are based on empirical data presented in the literature. In addition, the following assumptions are made:

1. homogeneous mixture at the storm overflows
2. discharge into a unit is equal to discharge out (except in storage tank)
3. BOD-inflow is completely mixed in storage tank
4. no removal efficiency in storage tank
5. of the BOD coming to primary clarifier, 28% is settleable, 36% colloidal, and 36% dissolved matter
6. oxygen, pH, and biocides are not included in the model
7. return sludge discharge does not influence the overflow rate in the secondary clarifier
8. the ratio between suspended solids and BOD in stream Q15 is 1.4
9. flocculation, pH, and chemical dosage are optimal in the flocculation unit.

*Future Improvements*

The BOD and phosphorus removal functions will be improved continuously as the knowledge in this area develops.

In the sludge treatment plant model the cost functions will be improved. The removal functions will also be corrected as knowledge in this area develops.

## IV. EXAMPLES AND DEMONSTRATIONS

### *Sewer Network Model*

To illustrate the special features of the model, different flow theories on eight sewer lines linked together have been tested. Only one inflow hydrograph in the upper end of sewer line No. 1 was generated. This configuration allows a study of the damping and transport effects. Figure 1.9 shows the damping effects in the existing version of the model. The maximum discharge in sewer line No. 8 is one-third of the maximum inflow to sewer line No. 1.

Figure 1.10 shows the effect of using different flow theories in the model. The outflow hydrograph with the lowest maximum discharge represents the existing version of the model.

The conclusion is that in short intensive rainfalls the storage capacity in sewer lines must be considered, and that the partly filled velocity procedure also may have an influence.

### *Sewage Treatment Plant Model*

An example has been shown in Section III. In addition an example showing the computation for a treatment plant in a combined sewerage area is presented here. The main data from the area are:

| | |
|---|---|
| number of person equiv. | 10,000 |
| area | 1,000 ha |
| runoff coeff. | 0.7 |
| dry weather flow | 500 l/p.d. |
| BOD per person equiv. | 60 g/p.d. |
| precipitation | 0.832 m per year |
| total duration of precipitation | 792 hours per year |

Five basic rainfalls were chosen from the rainfall statistics to represent the yearly rainfall activity. The number of each rainfall per year is shown in Figure 1.11. The hydrographs in Figure 1.11 are the inflow hydrographs to the treatment plant caused by the five basic rainfalls.

In Figure 1.12 the "step function" is used to compute the BOD in runoff. These input data are used by the treatment plant model, and Figure 1.13 shows the results of using a storage tank in the sewage treatment plant serving the combined system area.

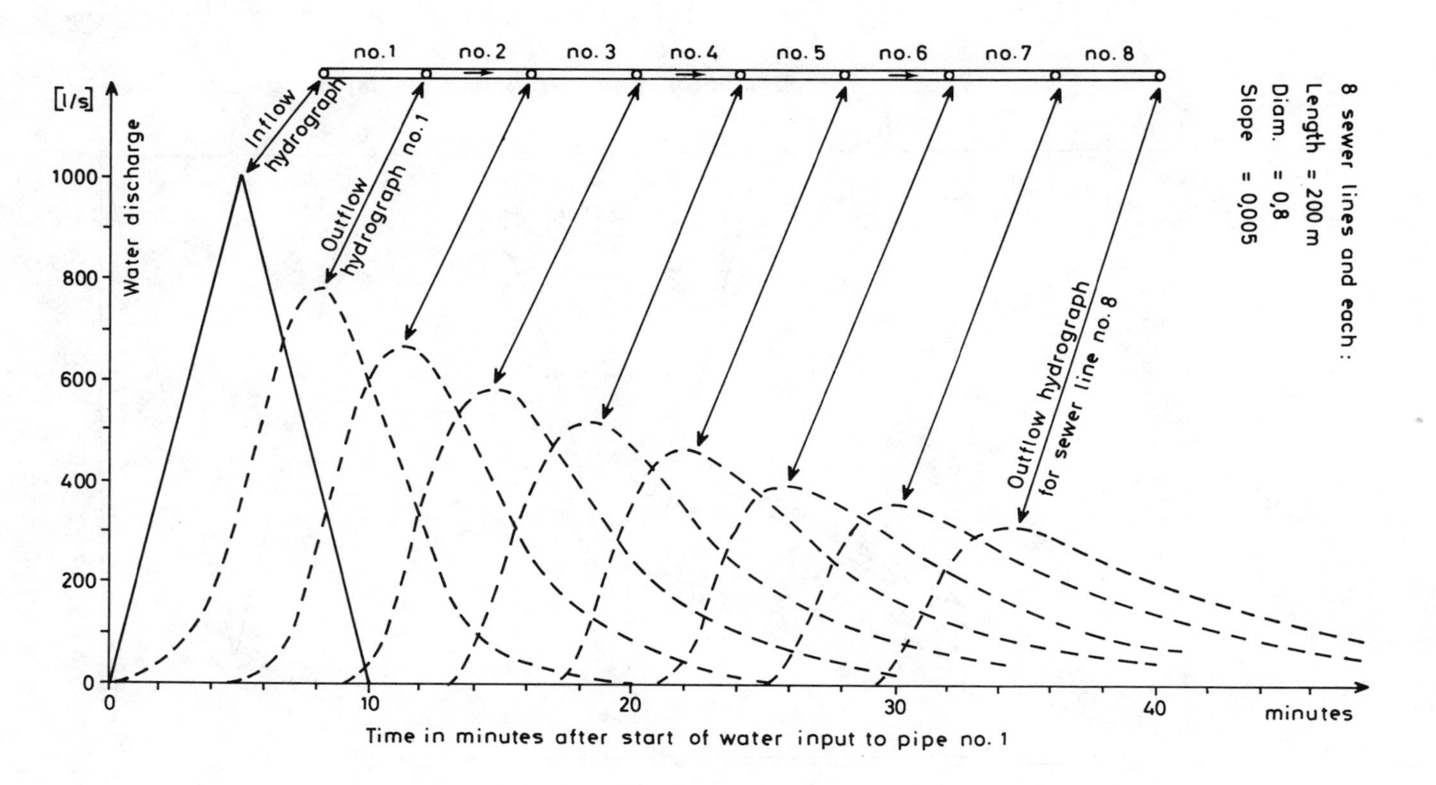


*Figure 1.9. Damping effect in a sewer line.*

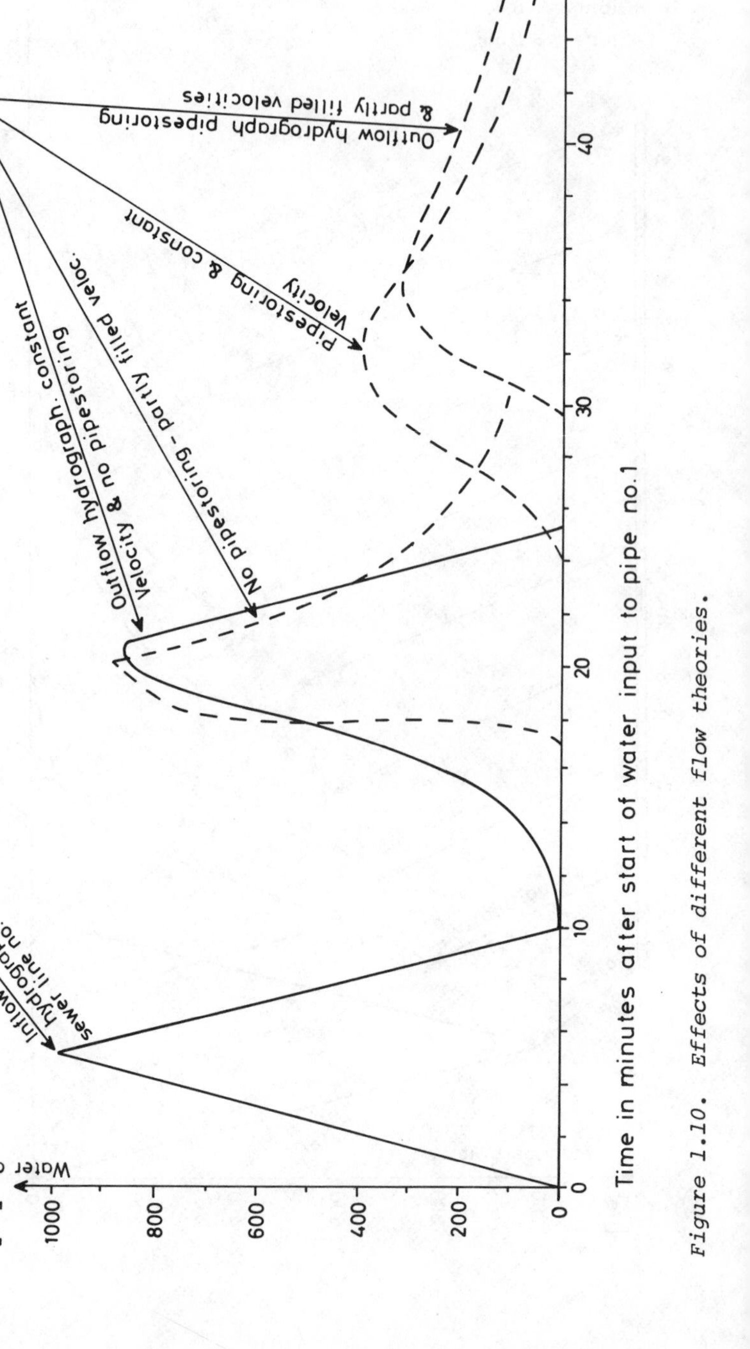


*Figure 1.10. Effects of different flow theories.*

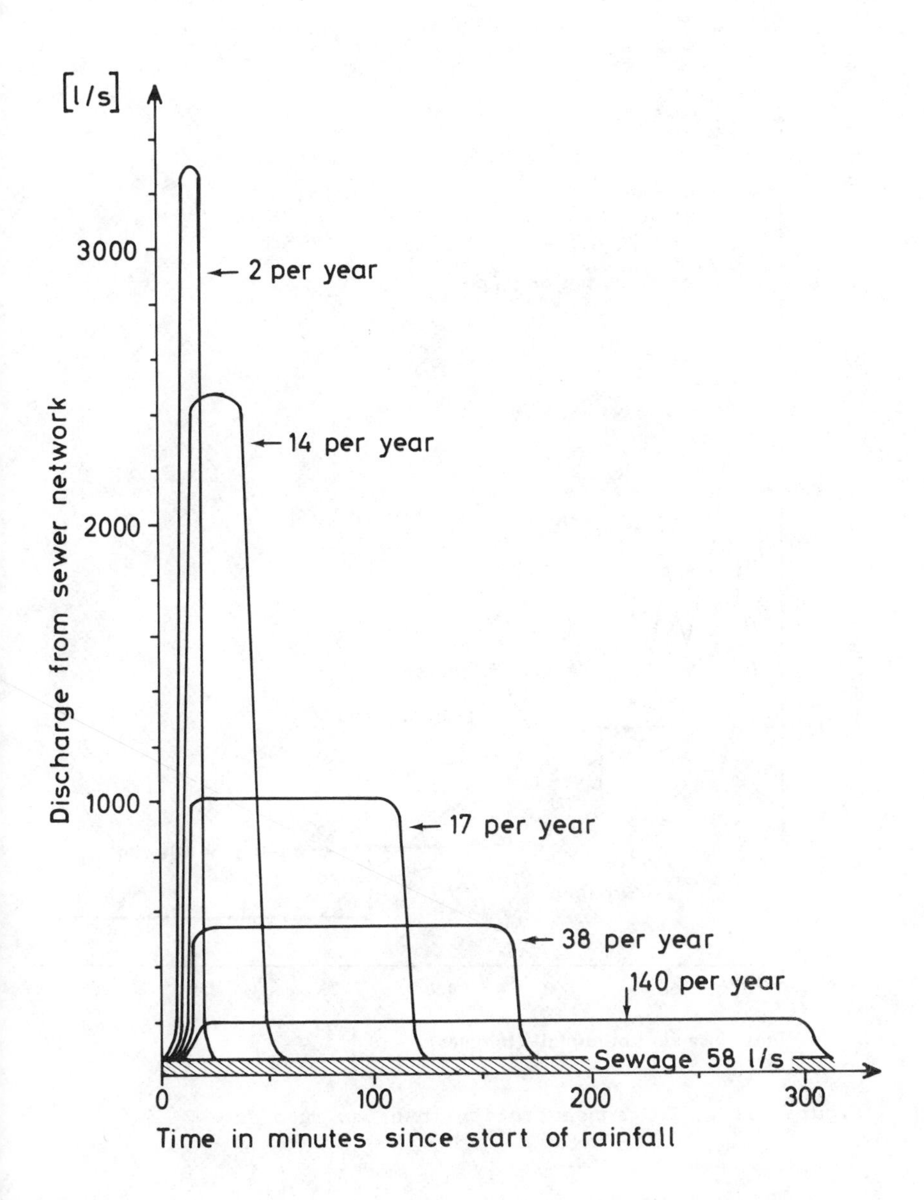


*Figure 1.11. Hydrographs caused by 5 base-rainfalls.*

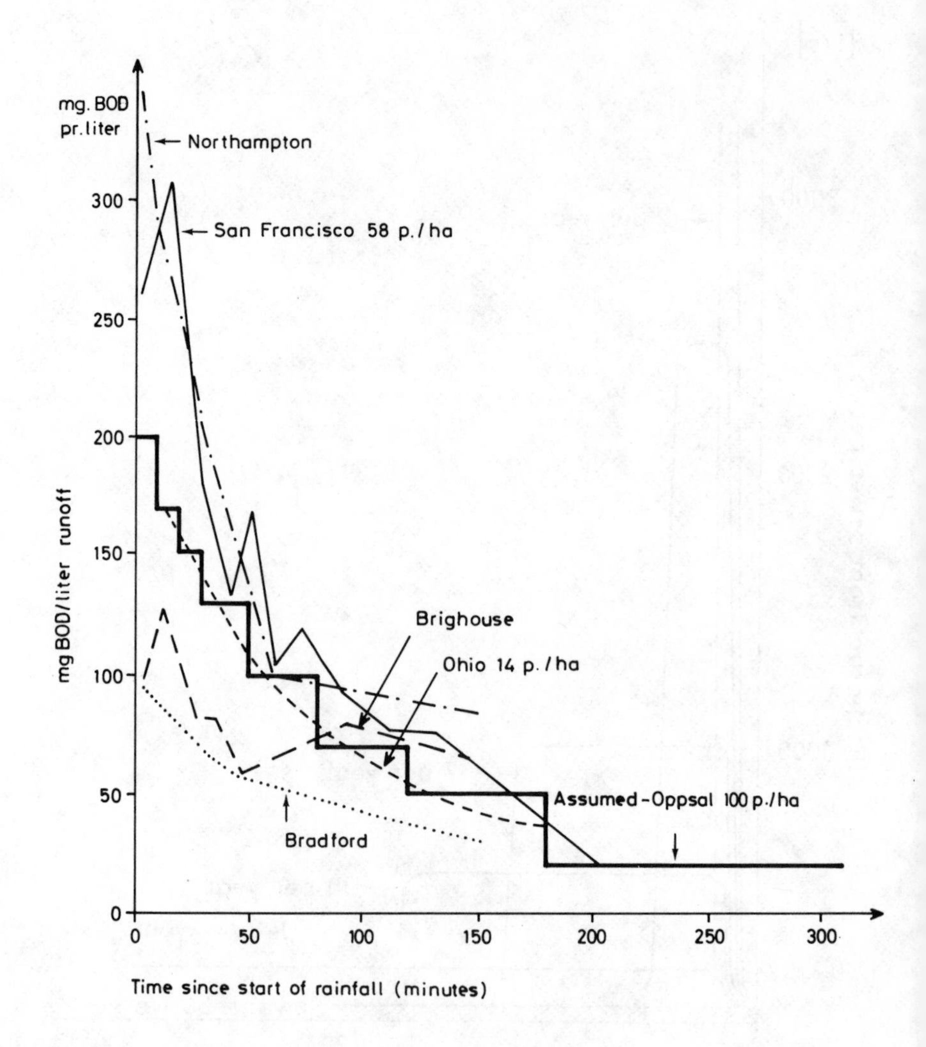


*Figure 1.12. BOD concentration in urban runoff.*

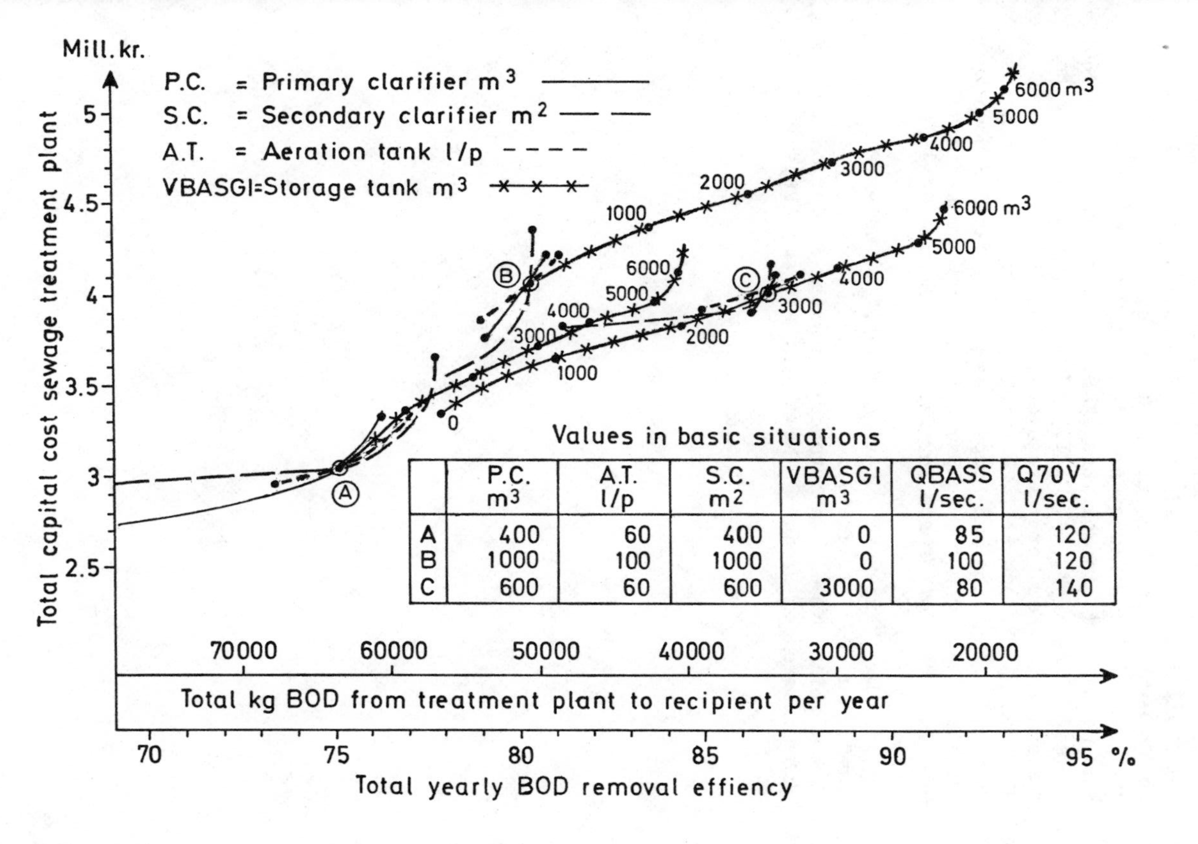


Values in basic situations

| | P.C. m3 | A.T. l/p | S.C. m2 | VBASGI m3 | QBASS l/sec. | Q70V l/sec. |
|---|---|---|---|---|---|---|
| A | 400 | 60 | 400 | 0 | 85 | 120 |
| B | 1000 | 100 | 1000 | 0 | 100 | 120 |
| C | 600 | 60 | 600 | 3000 | 80 | 140 |

*Figure 1.13. Example of a treatment plant.*

REFERENCES


1. Watkins, L. H. "The Design of Urban Sewers Systems," *Road Research Technical Papers No. 55*. (London: Dept. of Scientific and Industrial Research, 1962).
2. Munz, W. "Die Wirkung verschiedener Gewässerschutzmassnahmen auf den Vorfluter," *Hydrologie 28(2)*:184 (1966).
3. Eckenfelder, W. W. and D. I. O'Connor. *Biological Waste Treatment*. (New York: Pergamon Press, 1961).
4. Hopwood, A. P. and A. L. Downing. "Factors Affecting the Rate of Production and Properties of Activated Sludge in Plants Treating Domestic Sewage," *J. and Proceedings. The Institute of Sewage Purification, London 64*:435 (1965).
5. Eye, I. D. "Extended Aeration Plant," *J. Water Poll. Cont. Fed. 41*:1313 (1969).
6. Eckenfelder, W. W. In *Advances in Water Pollution Research, 4th Conference, Prague*. (New York: Pergamon Press, 1969), p. 592.

MODELS FOR ENVIRONMENTAL POLLUTION CONTROL


# CHAPTER 2

# REGIONAL WASTEWATER MANAGEMENT SYSTEMS

Hermann H. Hahn, Peter M. Meier and Hermann Orth*

## I. THE PROBLEM

The tendency today is toward a regionalization in water supply and wastewater disposal. In the Netherlands, for instance, 96% of the whole population is served by regional water supply systems.[1] This tendency towards regionalization may be explained by the following:

1. a decrease in the unit costs of construction and operation of centralized plants within a certain plant size range
2. an increase of cost in the sewerage and canal system as a consequence of the longer distances to be covered and the larger wastewater quantities to be transported; these increased costs are compensated for by reductions in cost mentioned under (1) and (3)
3. definite advantages and quantitative benefits from planning, constructing and operating larger plants
4. pressures from administrations to form regional water and wastewater management systems
5. an increased use of surface water
6. a stricter control and more stringent supervision of larger plants and thus a more reliable operation.

Items 4, 5, and 6 cannot be described quantitatively; nevertheless, they are considered in the decision

*Hermann H. Hahn is Professor of Siedlungswasserwirtschaft (Environmental Engineering) and Director of the Institut für Siedlungswasserwirtschaft an der Universität Karlsruhe, Karlsruhe, West Germany. Peter M. Meier and Hermann Orth are Research Associates at the same institute.

making process when regional management systems are planned.

At present this tendency toward a regional solution of problems of water supply and wastewater disposal is seen both in the planning of *new* systems as well as the renewal and reconstruction of *old* plants which are in part overloaded or which function improperly. In the future, however, the enlargement and renewal of existing plants will gain in importance The increase in the per-capita water demand and wastewater production, as well as population increases, will be characteristic for areas of high population concentration and are the prime reasons for this development. It is therefore absolutely necessary, when developing new methods for the planning of regional water and wastewater systems, to consider the aspects of updating and renewing existing individual plants for the purpose of forming regional systems.

In order to describe the problem in more detail, a typical situation, as illustrated in Figure 2.1, shall be discussed. In general the planning engineer will receive a topographical map containing additional information on the location of wastewater sources and the quantities produced. In cooperation with the respective administrative offices the planning engineer may gather additional information on the following elements:

1. all possible locations of wastewater treatment plants
2. all possible locations for storm water outlets and storm water treatment facilities
3. all possible locations for pumping stations, etc.

In addition, the engineer must define all possible connections between these points; topographical, geological and administrative aspects may be of utmost importance in this phase. It is necessary to define the costs, in this instance as unit cost for the transport and treatment of one unit of wastewater along the connections between different points. In addition to these more technical aspects there are also other considerations. For example, the engineer must take into account the aversion or propensity of people towards certain treatment plant locations as well as certain transport elements and the possibilities of changing wastewater quantities and qualities by means of defining additional points, connecting transport elements, fictive costs, etc.

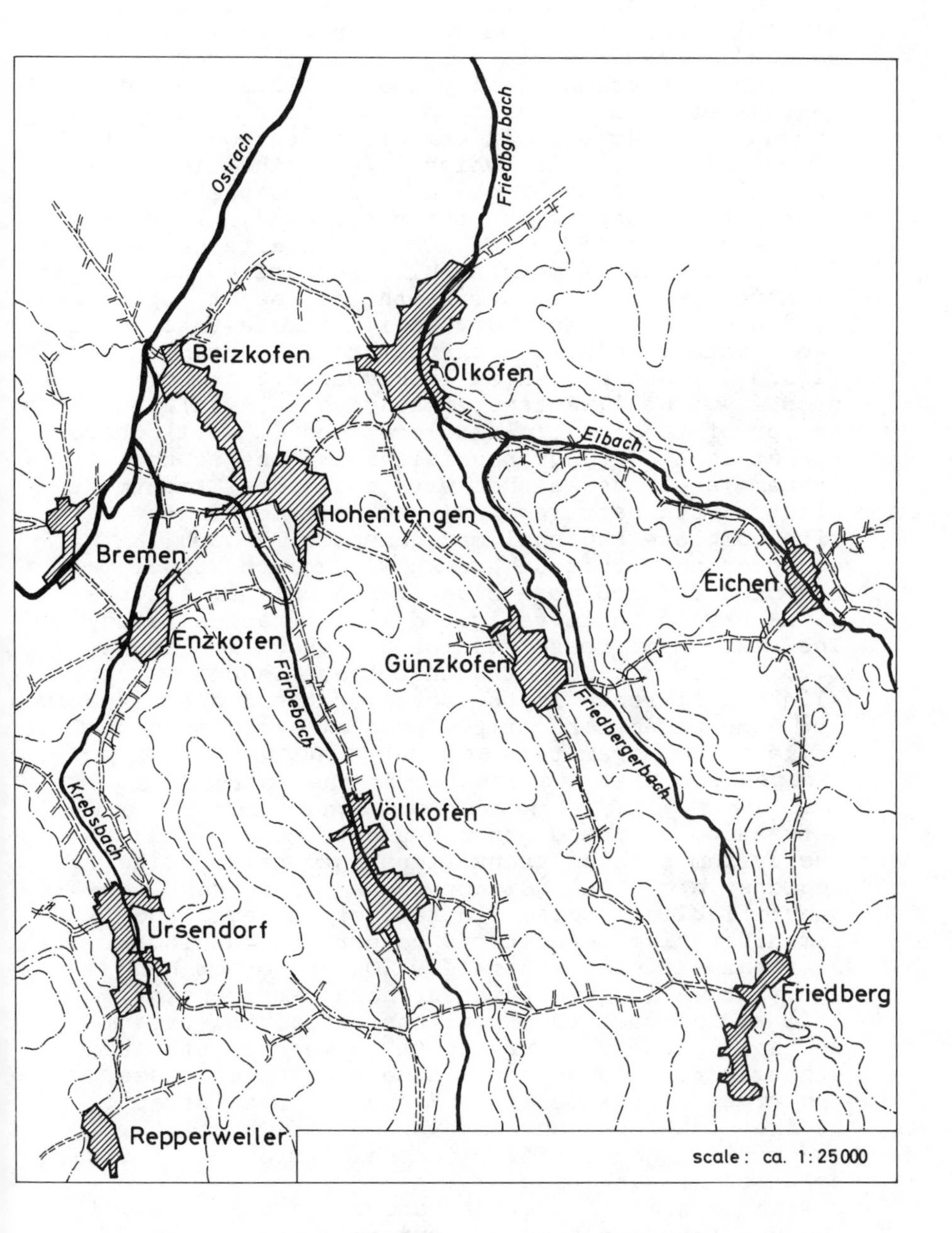


*Figure 2.1. Topographical map of a proposed regional waste-water system.*

All this information may be displayed in a matrix, which is the basis for the regional planning of a wastewater authority. If the gradients of all possible connections between matrix points as well as the increase/decrease in wastewater production and the population density and possibilities of wastewater treatment all show a certain tendency of increase or decrease in one direction, one speaks of an ordered matrix, which defines the limits of the solution to the problem. In a completely unstructured matrix, on the other hand, there are innumerable possible and meaningful alternatives.

The number of meaningful alternatives which must be investigated will define the degree of "openness of the matrix." In order to illustrate this point two possible solutions of the problem shown in Figure 2.1 are presented in Figure 2.2.[2] For this problem with 11 wastewater sources (n = 11) and the number of treatment plant locations (m = 11) there are at least 500,000 possible solutions if the system is planned either with individual treatment plants, with part regionally operating plants or with complete regional management. Through engineering intuition and knowledge, the problem has been defined more closely and only 17 alternatives have been investigated in this actual study. The results of these studies are shown in Figure 2.2. It should be noted that there is only a very slight difference in the costs of the two alternatives.

Some parameters of the problem are time dependent while other parameters are independent. The time independent elements consist of the coordinates of the matrix points, the various connecting elements between points, and other boundary conditions of the system such as transitions into neighboring regions, etc. In addition there may be some non-quantifiable factors with respect to legal and administrative regulations concerning the location of treatment plant outfalls. On the other hand the wastewater quantities at each matrix point, as well as the cost functions characterizing treatment at these points, or transport between points are strongly time dependent. A separation of these two types of parameters and special consideration of the time dependent development of wastewater quantities and cost functions are important for the solution of such problems.

The arguments sketched above are valid only if the participants are of comparable size. The treatment costs at each point must compare in order

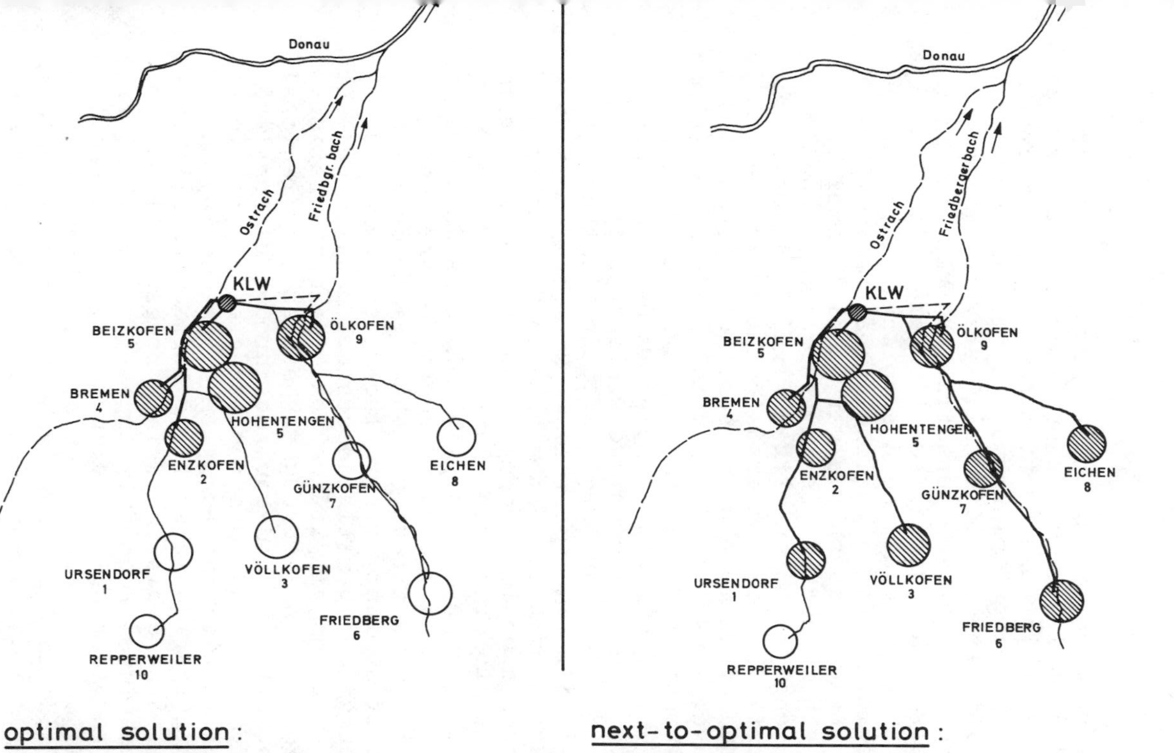


*Figure 2.2. Comparison of optimal and next-to-optimal solution.*

of size to the transport costs for wastewater from one matrix point to another. In addition these discussions are valid only if the economies of scale for transport costs are comparable to those for treatment costs.

## II. TRADITIONAL SOLUTION OF THE PROBLEM

The considerations underlying decisions in practical engineering can be summarized as follows: "The savings in construction costs by designing a regional treatment plant may be offset by the increase in cost of additional wastewater transport elements. If this compensation is fully given or the costs of a regional solution are only slightly higher than those of the nonregional solution, then one should aim for a regional management system."[3] In all these cases only a very small number of possible alternatives is investigated. Usually only one regional solution with a fixed location of the treatment plant is compared to the nonregional, *i.e.*, individual treatment plant solution. Occasionally a third alternative of a partial regional management may be considered. In the problem presented in Figure 2.1, the regional solution of a treatment plant between Beizkofen and Ölkofen would be compared with the nonregional solution of treatment plants in all municipalities. Possibly the additional alternative of treatment plants in Ölkofen as well as Beizkofen would be compared with one treatment plant operator for both plants.

In this type of analysis one considers only construction and operating costs for treatment plants in a quantitative fashion. The degree of loading of a plant, its influence on operating efficiency, and operating costs are usually neglected. Savings through simplification of administration, reduction of costs through subsidies from regional planning offices, and savings through rationalization in design and construction are only considered in a qualitative fashion. In a similar way the self-purification in small to large rivers is included in the decision-making process. All these latter aspects frequently appear in a contradictive manner in practical decision making: "The decision is based on a comparison of costs. This comparison can be made in different fashions; it will, however, always juxtapose construction and operating costs of the regional solution to those of the nonregional solution. The way in which operating costs are added to treatment costs and thus enter the considerations depends upon the respective way of financing the object."[3]

In a different, very detailed and extensive method of investigation one studies a "partly predefined" matrix with several alternative solutions.[2] Engineering knowledge and intuition reduce the large number of possible alternative solutions of the problem given in Figure 2.1 to a number of alternatives that can be investigated with reasonable effort. In this actual study the municipalities Beizkofen and Ölkofen as well as a matrix point between these two municipalities, were investigated as locations for treatment plants. There was, however, no treatment plant location south of Beizkofen or Ölkofen. The various investigated alternatives were found by adding one or more or all towns to each of the three possible treatment plants.

The next step in solving this problem was to use a type of hydraulic tabulation whereby the necessary transport capacities between the different matrix points for each alternative were determined. By means of an estimation of the necessary construction efforts and the connected costs, one could define the transport costs along the transport elements for each alternative. So far generalized cost functions are not used in this planning method even though the simplifications which might be necessary for such cost functions are comparable to those simplifications necessary for the next steps of the planning process.

Next, all possible alternatives are enumerated and the costs of each individual project are determined. Thus all possible solutions are tested and then the optimal solution is selected on the basis of transport and treatment costs. To find the solution, one must define all intermediate solutions explicitly. A systematic searching of the solution space which is defined by all possible alternatives, leaving out the consideration of some intermediary alternatives, is not possible in this planning method.

A summary of the construction costs on one hand and the total costs (including OMR costs) on the other hand for all different alternatives serves as the basis for the decision. If it were made on the basis of the comparison of true costs, the total costs of each alternative must be considered. Frequently, however, the decision is strongly influenced by considering only construction costs; this practical consideration is based on the bias that construction costs can be determined more precisely than OMR costs.

Such detailed determination of costs of each alternative shows that in most cases the economies of scale in wastewater treatment do influence the solution to a large degree and must be considered.

Although only a rather small fraction of all possible alternatives are considered, the effort in this study is very large. In order to investigate the matrix given in Figure 2.1 and its predefined 17 alternatives, 50-60 working days were necessary. The determination of treatment costs took 2-3 weeks, the hydraulic tabulation for the determination of transport costs 3-4 weeks and the summary of total costs about 4 weeks. Not included in this list of expanded efforts are all those preliminary studies which deal with the formulation of generally valid cost functions for the construction and operation of treatment plants as well as for the construction of wastewater transporting elements. In some cases such information can be transferred from one project to another with the necessary modification and then compared to values given in the respective literature.[4,5]

When planning regional wastewater management systems one first determines which combinations of wastewater producers and wastewater treatment plants are technically possible. For example, it may be decided that wastewater pumps will lift wastewater only to a certain height; otherwise costs will increase too greatly. Furthermore technically feasible or infeasible solutions are determined by considering maximum flow times. If wastewater is to be transported longer than this maximum flow time, certain treatment facilities must be provided, which again cause undue costs. After the network has been predefined in such a technical way, the economic factors characterizing this predefined solution are considered.

Looking at the given problem in this way, however, one can rarely distinguish between technical feasibility and economic feasibility. If the problem is to be solved correctly, one must consider both technical and economic variables. A simultaneous technical and economic optimization is necessary. The following question defines the problem: which is the optimal allocation of wastewater to different wastewater treatment plants, located at different points, so that total costs minus total benefits are minimized? This somewhat detailed and more specified question leads to a classic problem in Operations Research.

## III. A NEW METHOD FOR DETERMINATION OF THE OPTIMAL SOLUTION

The necessity of technical and economic optimization in the planning of regional wastewater management systems results from the following two facts:

1. It is desirable to investigate the largest possible number of meaningful alternatives without excluding some solutions on the basis of engineering intuition. In most cases this greatly enlarged solution space leads to a better solution than the original strongly reduced number of alternatives.
2. The best utilization of large quantities of information (generalized cost functions, topographic, demographic and other parameters) should be warranted. In addition it would be meaningful if the method of investigation could test the influence of the quality of available information upon planning (sensitivity analysis). This would indicate how far one should go in increasing quality and quantity of input parameters.

In order to demonstrate the necessary steps of this new method, a problem with three municipalities (shown in Figure 2.3) is discussed. To solve the problem, one needs the following information: (1) quantity and quality of wastewater produced at each point, (2) all costs and possible limitations for each individual connection between the various matrix points, (3) costs for treatment of wastewater at each individual point under consideration. In this instance it was assumed for simplicity's sake that all sources in this region have a comparable size; particularly large or particularly small sources are not included. Furthermore one assumes that all participants can be represented as point sources, *i.e.*, local wastewater sewerage systems are neglected.

The final structure of the overall network is determined with an optimization procedure, whereby optimization in this context means a "selection of one solution from a group of several alternatives in such a way that a predefined purpose is realized as best as possible."[6] The mathematical formulation

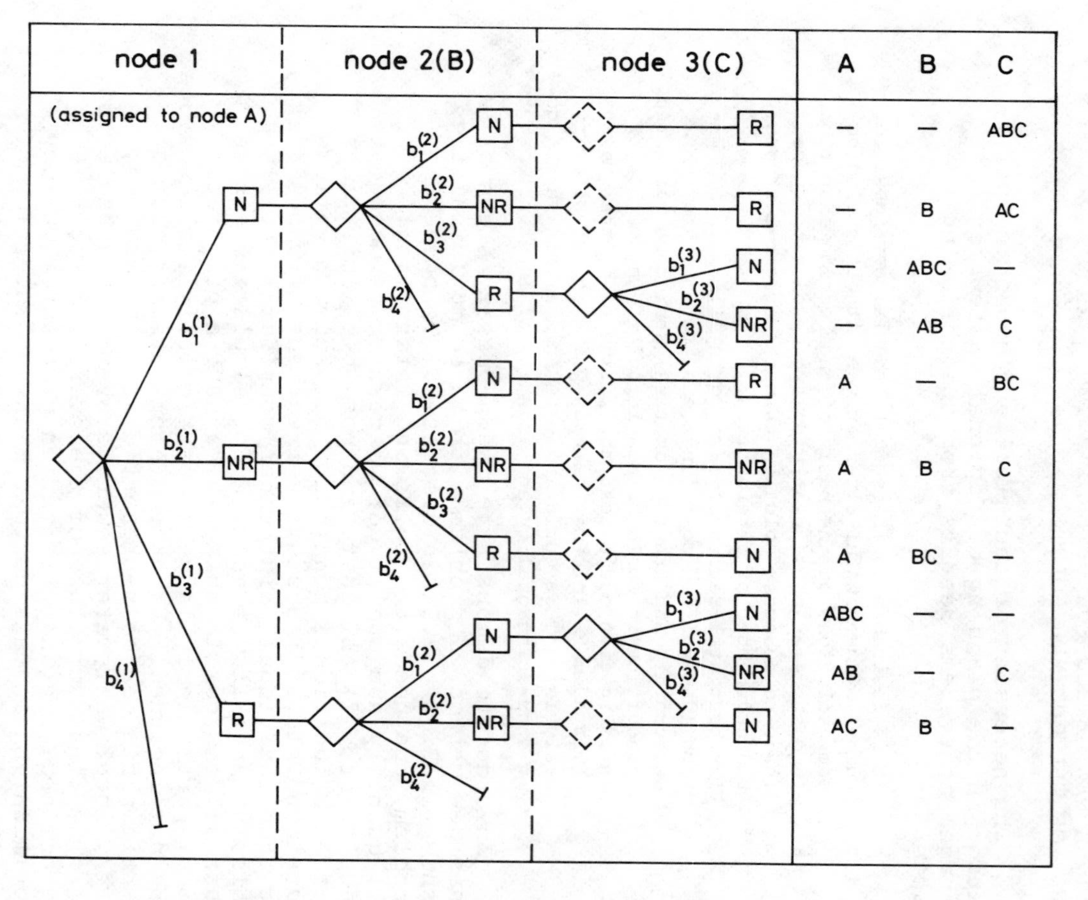


*Figure 2.3. Decision tree for the regionalization problem.*

of this regionalization problem is as follows:

$$\text{Minimize:} \quad \sum_{\substack{i \\ (i,j)\varepsilon U}} \sum_{j} f_{ij}(x_{ij}) + \sum_{\substack{j \\ j\varepsilon T}} g_j(Y_j) \qquad (1)$$

$$\text{Subject to:} \quad - \sum_{\substack{i \\ (i,j)\varepsilon U}} x_{ij} + \sum_{\substack{i \\ (j,i)\varepsilon U}} x_{ji} + \underset{j\varepsilon T}{Y_j} = \underset{j\varepsilon S}{D_j} \qquad j = 1,2,\ldots,m \qquad (2)$$

$$x_{ij} \geq 0 \qquad i,j = 1,2,\ldots,m$$

$$Y_j \geq 0$$

where:

$f_{ij}(x_{ij})$ = transport costs from i to j
$x_{ij}$ = wastewater transported from i to j
$Y_i$ = wastewater at j
U = sum of all possible canals
T = sum of all possible locations for treatment plants
S = sum of all wastewater sources
$g_j(Y_j)$ = treatment costs at throughput $Y_j$
$D_j$ = wastewater quantity at source j

All cost functions have been simplified. It is understood that treatment costs are dependent not only upon the wastewater throughput but also upon the degree of treatment. In a similar fashion transport costs are rarely explicit functions of an average throughput. Yet even with such simplified cost functions one encounters already large difficulties in solving the problem.

At first one should note a certain similarity of this problem to the transshipment problem known in Operations Research literature.[7] If economies of scale in construction and operation of treatment plants as well as in the construction of wastewater transport elements would be neglected, *i.e.*, if all costs would be represented by linear functions through the origin, well-tested solution algorithms[8] that lead to the so-called global optimum are known.

However, it is impossible in this case to assume that the objective function is linear. As mentioned earlier, the increase in costs of transportation is mainly compensated for by savings in treatment costs due to the economies of scale. If, for instance, two small treatment plants each for 10,000 inhabitants would cost together as much as one treatment plant for 20,000 inhabitants, then there would be no savings due to regional wastewater management. Thus the known and tested solutions for the linear transport problem cannot be used directly in this case.

Another proposal[7] should be mentioned in this context. Here the nonlinear cost functions are approximated by various linear functions. In this case modifications of the above-mentioned algorithms do not necessarily yield the global optimum. One will arrive at a solution that is better than a certain number of other solutions, but one cannot prove that it is better than *all* possible solutions. If one tries to solve nontrivial problems, one encounters computational difficulties, which are mentioned in the Operation Research literature.[9,10]

If economies of scale are to be considered also, the problem must be reformulated in such a way that it corresponds to the fixed costs transport model. The typical cost functions here are linear functions that do not go through the origin. They may be stated as follows:

$$f_{ij}(x_{ij}) = \alpha_{ij} \cdot \delta_{ij} + \Pi_{ij} x_{ij} \tag{4}$$

where

$$\delta_{ij} = 0 \quad \text{if} \quad x_{ij} = 0$$

$$\delta_{ij} = 1 \quad \text{if} \quad x_{ij} > 0$$

and

$$g_j(Y_j) = \beta_j \cdot \gamma_j + \lambda_j Y_j \tag{5}$$

where

$$\gamma_j = 0 \quad \text{if} \quad Y_j = 0$$

$$\gamma_j = 1 \quad \text{if} \quad Y_j > 0$$

This formulation of the problem approximates better the real situation. It leads, however, to a mixed integer programming problem for which some solution algorithms are known, but which have been tested only to a small extent and in any case require large computational efforts and machine capacity.

However, the regionalization problem resembles more the warehouse location problem than the transshipment problem. This problem is solved today mainly by heuristic algorithms that use the special structure of the problem.[9,11] This allows inclusion of cost functions that represent cost degression. Nevertheless there is still an essential difference between this warehouse location problem and the here-discussed project: the nonlinear transport costs.

Therefore, a solution is proposed that takes advantage of the specialty of this given regional system: if there are no restrictions on the capacity of the transport network and the treatment plant, then the system can be transformed into a discrete one because wastewater produced at location "i" will be distributed to only one treatment plant, corresponding to the single assignment property of the warehouse location problem. In the example with three participants (Figure 2.3) there are exactly ten possible alternatives. As mentioned already, the number of possible solutions increases rapidly: with 4 possible locations or participants 41 combinatorial possibilities are obtained; with 5 participants, more than 200 alternatives. Even with today's electronic computers, an investigation of all combinations of a "ten-locations-problem" would be impossible. Through the use of an "intelligent" search strategy it is, however, possible to determine the optimum by analyzing only a certain fraction of all solutions. A suitable search strategy is the Branch-and-Bound method. The main advantage of this solution method is the possibility of including all nonlinear functions and nonlinear constraints. Figure 2.4 shows an example of a nonlinear cost function for treatment plants where the stepwise increase is caused by the different treatment methods. In addition, in each step of this cost function the cost degression for increased throughput are shown along the y axis.

Another advantage of the Branch-and-Bound method is the possibility of introducing necessary suboptimization problems. One can, for instance, include directly a mathematical model of the receiving river into this system, introducing the required degree of treatment at each treatment plant location.

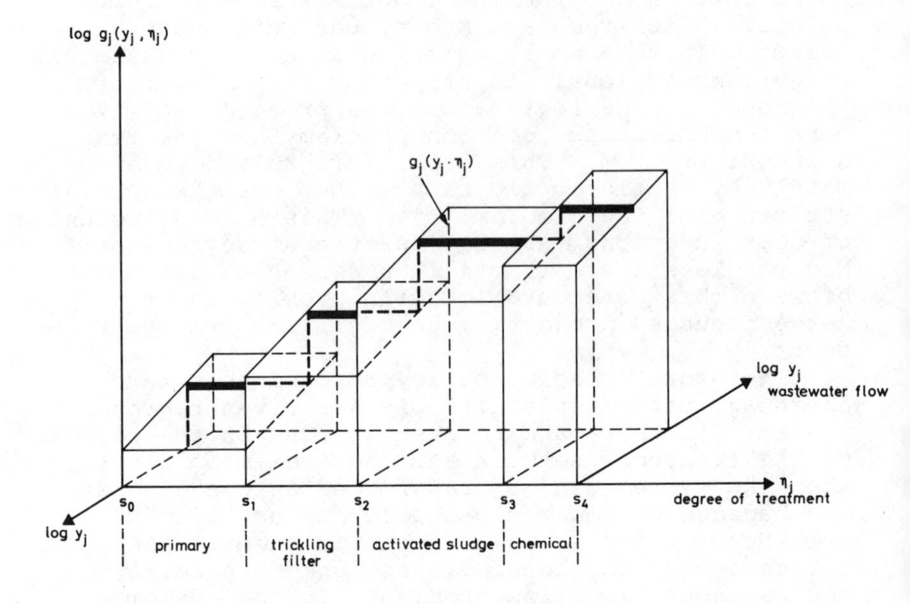


*Figure 2.4. Treatment plant cost functions.*

## IV. THE BRANCH-AND-BOUND STRATEGY

In the following paragraph the main ideas of the Branch-and-Bound search strategy are discussed briefly. The interested reader is referred to the exact formulation in Meier.[12] Due to the nonlinear character of the problem the algorithm consists of two interconnected Branch-and-Bound procedures. At each possible treatment plant location one is presented with three alternatives (see Figure 2.3):

1. do not plan a treatment plan in this location (solution "N")
2. plan a treatment plant for the wastewater of only the considered municipality, *i.e.*, arrive at the nonregional solution (solution "NR")
3. plan a regional treatment plant that also treats wastewater arriving from other points (solution "R").

If all possible treatment plant locations are investigated in this manner, one obtains a decision tree as shown in Figure 2.3. Following all the branches of this tree one obtains the above-mentioned ten alternatives. The assignment of the different locations to the nodes of the tree is arbitrary and would only change the sequence of the solutions shown in the right column of Figure 2.3. The fact that all branches of this tree need not be pursued if one follows at a node that particular branch for which minimal costs are expected is the basis of the Branch-and-Bound algorithm. It is thus necessary to compute minimal bounds $b_j^{(k)}$ in each node. For example, in Figure 2.3 these bounds are for node 1 according to their size: $b_2^{(1)} < b_1^{(1)} < b_3^{(1)}$. Thus one would plan a nonregional treatment plant in location "A" and go on considering node 2. Here the bounds are: $b_2^{(2)} < b_1^{(2)} < b_3^{(2)}$. Thus one would plan a regional treatment plant in location "B" treating also the wastewaters arriving from "C." Since all points or municipalities are already included in the optimal solution, no further decisions are necessary. Only one branch and two nodes have been investigated instead of studying all ten alternatives.

In order to use this strategy one has to compute the bounds $b_j^{(k)}$. The value $b_1^{(1)}$ represents the minimal costs that occur if one does not build a treatment plant in "A" and therefore must provide one in location "B" or "C." Whether a treatment plant will in fact be built in "B" or "C" can only be decided in subsequent nodes. In the first node only the expected optimal possibilities are decisive. The exact calculation of the bounds $b_j^{(k)}$ is complex and requires a second Branch-and-Bound search strategy. This second algorithm determines at each point the optimal transport network and its costs.[12]

To date the computer program consists of about 4,000 FORTRAN instructions and is written for computers of the third generation (IBM 360/85, UNIVAC 1108, etc.). The program was tested in two different ways: for small artificial problems the optimal solution was checked by complete enumeration; for larger problems the solution was checked by rearranging the assignment of locations to various nodes in the decision tree. If the logic of the computer program is correct, the same solution should always be reached.

## V. DISCUSSION OF THE RESULTS OF THIS NEW PLANNING METHOD

From its very first conception the proposed planning method was tested against real world problems. Therefore results of computer runs were checked against practical solutions even while the program was under development. By comparing solutions arrived at by today's engineering methods and the computer solution the program was improved step by step. Furthermore it was possible to estimate the degree of precision necessary for the input data.

The particular characteristics of this new planning method are summarized in Table 2.1, showing all the necessary assumptions.

*Table 2.1*

*Summary of Assumptions Used in the Presented Optimization Procedure*

| | |
|---|---|
| transportation network | local network neglected ("point sources")<br>wastewater can reach "i" from "j" via one and only one connection<br>matrix elements: "sources" - wastewater producers; "sinks" - treatment plants; "connecting elements" - canals |
| flow of wastewater | no splitting of wastewater at any matrix point<br>only one type of water to be transported<br>no time dependent variation in wastewater production and purification |
| cost functions for waste treatment and transport | treatment costs can be formulated as: tables, polynomials, exponential functions; for example<br>$K = g(Q) = \alpha \cdot Q^{\beta}$<br>transport costs can be approximated by linear functions; for example<br>$K = f(Q) = \alpha \cdot Q$ |

A comparison of the planning of regional wastewater systems with present day engineering methods and with the Branch-and-Bound strategy is shown in Table 2.2. In particular this summary draws

*Table 2.2*

*Comparison of Steps in Planning Regional Wastewater Management Systems by Traditional Methods and the New Search Strategy*

| | *Methods in practice today* | *Proposed Branch-and-Bound strategy* |
|---|---|---|
| definition of matrix | wastewater sources "i,"<br>limited number of locations for treatment plants "j,"<br>connections between "i" and "j" that are apparent from engineering intuition<br>limited number of alternatives | wastewater sources "i,"<br>all possible locations for treatment plants "j,"<br>all possible connections between "i" and "j"<br>maximum number of meaningful alternatives |
| determination of cost functions | treatment costs: based on own estimates and literature<br>transport costs: either neglected or evaluated through detailed tabulation | treatment costs: functions as reported in the literature[5]<br>transport costs: simplified and generalized functions or detailed tabulation (if present program is modified slightly) |
| usefulness of result | design is useful within the framework of today's requirements from practice<br>frequently only local optimum | computer design is as good as input data; with sufficient sensitivity analysis yields further insight in system;<br>if large number of geological, topographical, etc. details are included, then result better than with conventional method<br>always global optimum |
| expanded effort | for local optimum about 50-60 days;<br>for global optimum complete enumeration | for global optimum (implicit enumeration)<br>data preparation: about 5 days<br>computation time: about 1 minute |

attention to the fact that in practice only a very small number of alternatives are investigated. The new search strategy, on the other hand, studies a maximum of meaningful alternatives. The cost functions for wastewater transport and treatment are comparable in both planning methods. The result of today's engineering planning frequently represents only a local optimum due to the very early fixation of the structure of the system, while the Branch-and-Bound search strategy arrives at the global optimum. The usefulness of the computer solution depends on the quality of the input data. Furthermore Table 2.2 shows clearly the significantly reduced effort if the planning method is used. The main difference between the two procedures is that the Branch-and-Bound strategy uses implicit enumeration in order to find the global optimum while the engineering method must rely on explicit or total enumeration of all possible alternatives if the global optimum is the object.

In the next phase of the investigations the restrictions on the form of the cost functions were eased. Basically, the Branch-and-Bound algorithm allows the use of any type of cost function. Thus the usefulness of the computer-aided design as a basis for the detailed engineering planning is increased significantly. The degree of precision of the computer solution can be increased at liberty if one concentrates on information gathering, especially for the cost functions. Furthermore it is possible to check the sensitivity of the model by repeated simulations with systematically changed input data. From such investigations the precision of input data necessary for realistic solutions can be defined quantitatively. For the planning of wastewater management systems of smaller regions a solution algorithm was proposed that is somewhat simpler than the version discussed so far.[13,14] To also apply this algorithm for large-scale problems might prove unwise due to the more elaborate Branch-and-Bound strategy being more economical if many alternatives must be investigated.

## VI. SUMMARY

In continuing these investigations the focus will be on relaxing the constraints introduced by all the assumptions listed in Table 2.1. At present additional engineering aspects are introduced into the matrix that to date contains only three elements: (a) source-wastewater producer, (b) sink-wastewater treatment plant, and (c) vectors connecting sinks and sources-wastewater canals. This should make the matrix more realistic. Thus it is intended to describe, for instance, pumps that are used partially to increase transport capacities of wastewater canals or to lift wastewater at certain points in such a way that their hydraulic functions, their cost functions, etc., resemble the already existent matrix elements. In this way they could enter the computational procedure without any difficulties.

A further desirable extension of this work aims at a more precise description of wastewater production and its time dependent distribution. These tasks will present mathematical as well as computational difficulties. Beyond that one will also have to take into account that sometimes one deals with combined sewerage systems while on other occasions with separate storm and sewage canals. Finally it must be noted that in the case of the multiperiod model one deals with wastewater quantities that are expected in the future and thus are of stochastic nature. The extension of the existing algorithm to accommodate stochastic input parameters is a logical next step.

In summary, then, the optimization procedure will give the practicing engineer a method to find the true optimal solution by means of mathematical optimization algorithms and in connection with large-scale computation centers. It is understood, of course, that the costs of writing and testing such programs are considerable and should be expended only if the program will be used more than once. The optimization techniques demand of the practicing engineer a knowledge of existing solution algorithms as well as an understanding of the possibilities and limitations of such methods.

REFERENCES

1. Fair, G. M. Report for *Resources for the Future* (1969).
2. Ingenieurbüro, Gesellschaft für Kläranlagen und Wasserversorgung Mannheim mbH, (GKW) *Erläuterungsbericht, Technische und wirschaftliche Untersuchung über die Bildung von Abwasserzweckverbänden Raum Göge*, Mannheim (1970).
3. Kassner, W. and C. H. Buchard. "Gruppenkläranlage oder Einzelkläranlagen," *Württembergische Gemeindezeitung*, 23 (1968).

4. Kehr, D. and H. Teichmann. "Bau- und Betriebskosten öffentlicher Kläranlagen in der Bundesrepublik," *Institut für Siedlungswasserwirtschaft TH Hannover, Veröffentlichungen,* 8 (1961).
5. Schmidt, U. "Über die Kosten der biologischen Abwasserreinigung," *Institut für Siedlungswasserwirtschaft TH Hannover, Veröffentlichungen,* 13 (1964).
6. Münnich, F. E. "Das Prinzip der Optimierung," *Stadtbauwelt 51:*275 (1969).
7. Deininger, R. "Über die Planung von interkommunalen Systemen von Kläranlagen," *Gas and Wasserfach 110(8):* 1443 (1969).
8. Berge, B. and A. Ghouila-Houri. *Programme, Spiele und Transportnetze, Teil B* (Leipzig: Teubner Verlag, 1967).
9. Feldmann, E., F. Lehrer, and B. Ray. "Warehouse Location under Continuous Economies of Scale," *Management Science 12:*670 (1966).
10. Landis, W. "Optimale Dimensionierung eines Bewässerungsprojektes in Nievergelt," *Praktische Studien zur Unternehmensforschung* (Heidelberg: Springer Verlag, 1970).
11. Kuehn, A. and M. Hamburger. "A Heuristic Program for Location Warehouses," *Management Science 10:*643 (1963).
12. Meier, P. "Möglichkeiten zur technischen und wirtschaftlichen Optimierung von Zweckverbänden," *Wasser und Abwasser in Forschung und Praxis* (Bielefeld: E. Schmidt Verlag, 1972).
13. Orth, H. "Die binäre Optimierung als Hilfsmittel bei der Planung regionaler Abwasserbeseitigungssysteme, I." *Technischer Bericht Nr. 7,* Institut für Siedlungswasserwirtschaft, Universität Karlsruhe. (1972).
14. Ahrens, W. "Die binäre Optimierung als Hilfsmittel bei der Planung regionaler Abwasserbeseitigungssysteme, II." *Technischer Bericht Nr. 8,* Institut für Siedlungswasserwirtschaft der Universität Karlsruhe.(1972).
15. Orth, H. and H. H. Hahn. "Die mathematische Optimierung als Hilfsmittel bei der Planung regionaler Abwasserbeseitigungssysteme," *gwf-wasser/abwasser 114* (1973).
16. Orth, H. "Dekompositionsmethoden - Ein Hilfsmittel bei der Planung regionaler Abwasserbeseitigungssysteme," *gwf-wasser/abwasser 114* (1973).

MODELS FOR ENVIRONMENTAL POLLUTION CONTROL


# CHAPTER 3

# A REGIONAL PLANNING MODEL FOR WATER QUALITY CONTROL

David E. Pingry and Andrew B. Whinston*

## I. INTRODUCTION--PLANNING MODELS AND EXTERNALITIES

With increased public interest in the quality of the environment, and water quality in particular, economists have increased their work in the analysis of the more general problem of external effects or externalities. A good deal of this work has been the theoretical exploration of optimal conditions that would characterize an efficient allocation of resources. Very often it is then presumed that if an appropriate decentralized mechanism or centralized authority with perfect information were instituted then the optimal allocation could be obtained.

Unfortunately, the information problem when dealing with external effects of water pollution is immense. For comparison, consider the traditional resource allocations problem with no production or consumption external effects. In this case it is clear that allocation of a unit of resource to this use has the effect of eliminating all other uses of that resource, and all the benefits and costs accrue to the consumer of the resource. In the case of external effects the problem becomes more complicated. Usually the external effect (*i.e.*, cost or benefit) is transferred to the affected party through some complex series of physical, chemical or biological

---

*David E. Pingry is a Visiting Assistant Professor of Economics at Purdue University and Andrew B. Whinston is Professor of Economics at Purdue University, West Lafayette, Indiana.

processes. For example, in the water pollution case the initial effluent discharged by the polluter may affect a downstream user through some complex chemical reaction. In these cases the opportunity costs of a specific resource allocation are not at all clear.

The existence of this information problem has led to attempts to produce the effects of resource allocation under external effects by modeling the appropriate physical system. It is in this spirit that various planning models have been developed in the area of water quality control. The core of these models has been a simulation model of a river that is used to determine the water quality effects of given effluent patterns. These models have been used to form constraints for a mathematical programming problem of the form

minimize: treatment costs
subject to: quality goals satisfied.

Using these models a considerable case has been made for treating water pollution as a regional or basin wide problem instead of imposing discharge limitations and deciding on the construction of treatment plants on an *ad hoc* basis. This approach is advocated because the effect on water quality of waste discharge in a particular river varies greatly with the actual point of discharge. If the discharge is at a point where the assimilative capacity is great, then the added value of advanced waste treatment facilities could be small. On the other hand, in an area with a large concentration of water users the addition of further treatment, flow augmentation, or other means of treatment can have a significant effect on improving water quality. Thus for any given set of quality goals for a river, significantly different policies for each polluter may be appropriate. This can be contrasted with current regulations which impose various uniform rules such as secondary treatment regardless of the effect on the quality goals. Once a regional approach is considered, other treatment strategies may also be evaluated alongside treatment at polluters. Combining effluents from various sources and treating them in a large regional plant may, because of possible economies of scale, lead to a lower per-unit cost. Low flow augmentation, by-pass piping and instream aeration are other possibilities that, if used judiciously, can achieve further economies.

While it is understood that in view of the large variation in stream assimilative capability, uniform rules introduce significant inefficiencies, the problems of organizing and implementing a scheme that takes advantage of the difference may be extremely difficult. In order to determine effectively the most economical way of achieving the water quality standards basin-wide, enormous amounts of information must be available in an analyzable form. This problem is twofold:

1. Data must be collected and organized to determine water quality (and implicitly water use) implied by alternative treatment strategies.
2. Efficient solutions must be isolated. In other words the least cost combinations of treatment strategies given the water quality (water use) goals must be found.

The planning model presented in this paper attacks both of these problems. A simulation model of the river basin water quality serves as the constraint set for an optimization model that searches for the efficient solutions given the quality goals. The role this model plays in the planning process is outlined in Figure 3.1. Basically the planning procedure is iterative. As the implications for groups and individuals of specific quality levels are determined, these groups would demand that alternative goals be examined, some of which may require additional data or expansion of the scope of the model. The planning could take place in a basin commission with representation for "important" groups. These groups could represent the various conflicting interests such as boating, fishing, swimming, waste disposal, water supply and irrigation. Presumably the planning process would continue until an "acceptable" allocation was reached. The speed of convergence to a solution would depend on the type of decision rule used.

The question of resolving the conflicting interests of the various "important" groups (*i.e.*, the decision rule) is avoided in this paper. However, the technique that we propose will make available to the affected parties information on the alternative allocations of the river water resource and the costs associated with each. This information does not make the selection between alternatives a trivial problem. However, it does allow the participants to reach decisions using some level of knowledge of the

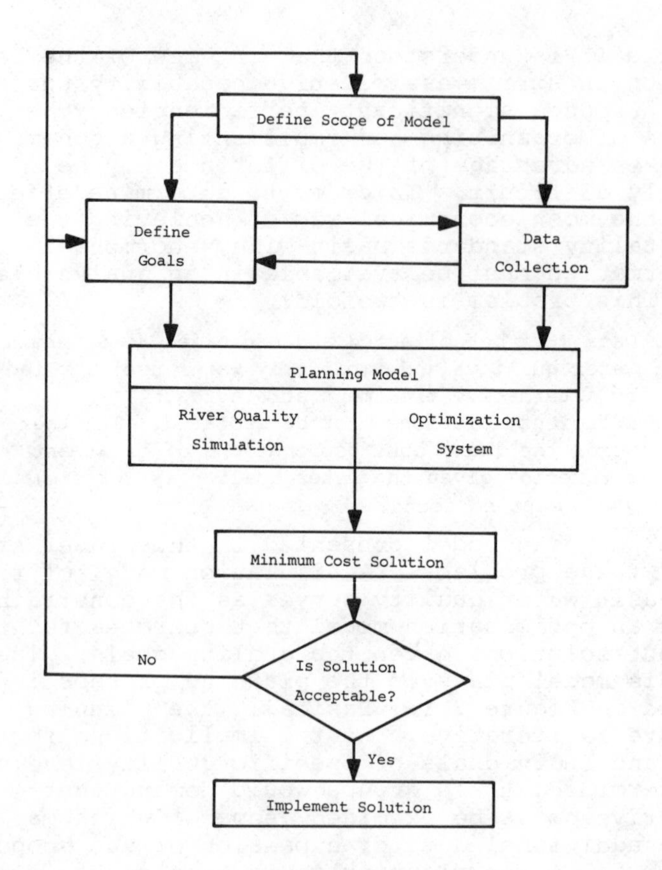


*Figure 3.1. Role of planning model.*

alternatives and their implications, material not ordinarily available for externality problems.

The water quality basin planning model presented in this paper is a significant advance over the previous models. The major contributions of this model are the explicit consideration of the water quality measure, temperature and its relationship with the other major water quality parameter, the dissolved oxygen level, and the simultaneous consideration of several treatment strategies and their costs. In particular, the model presented will be able to answer the question: what is the least-cost combination of treatment facilities that will meet given temperature and dissolved oxygen goals? Using such a model a river basin authority could present an assortment of river use and cost combinations for consideration by the affected parties.

## II. WATER QUALITY MEASURES AND TREATMENT STRATEGIES

Water quality is usually "measured" by the deviation of the concentration levels of certain chemicals or materials from "desired" levels. Some of the typical concentrations that are considered important are dissolved oxygen level, dissolved oxygen deficit (DOD), heat concentration (temperature level) and concentrations of various chemicals. All of these river concentrations can be altered by mixing the river flow with influent flows of various concentrations, the new concentrations being a weighted average of the influent and mainstream concentrations.

Many of these concentration levels are interrelated, and change for a period of time after initial mixing. For example, the discharge of organic material into a river will create an additional demand on the existing supply of oxygen in the river (called Biochemical Oxygen Demand or BOD). The reaction using the oxygen takes place over a period of time during which oxygen enters at the surface of the river, making available new oxygen. At any time after the effluent has been discharged the oxygen concentration level is a function of the initial concentrations of oxygen and BOD and the rates of supply (reaeration) and demand (deoxygenation), both of which are functions of the heat concentration or temperature. Given enough time the DOD level should go to zero and the temperature to equilibrium temperature determined by meteorological and other conditions. Heat and organic material are considered degradable materials because these effects are eliminated with time. Other materials (classed as nondegradable) do not converge to a natural equilibrium, at least not in a relevant time horizon.

Historically the dissolved oxygen level has been the common measure of water quality. Recently, however, thermal pollution has become increasingly important not only because of its direct effects on aquatic life but because of its indirect effects on the dissolved oxygen level.

The dissolved oxygen level is affected in two ways by the temperature. First, the higher the temperature is, the lower the saturation level of dissolved oxygen. This means at high temperatures less oxygen will be available to oxidize the organic waste material. And second the temperature level affects the rate of the oxidation reaction and the rate of reaeration of the water. The combination of these effects has never been explicitly handled in previous basin planning models. The addition of the temperature in the model allows for the consideration of the treatment strategy of cooling towers to control either the dissolved oxygen or temperature or both.

The treatment strategies open to a river basin authority can be classified into three major categories

1. more efficient use of existing assimilative capacity
2. increase assimilative capacity
3. treatment in a closed system outside of river.

Under the first category is the treatment strategy of by-pass piping. This system attempts to take advantage of the natural ability of the river to oxidize organic material and to transfer heat to the atmosphere. The goal of this technique is to disperse the waste material enough so as to maintain a desired level of quality.

The second category includes the technique of flow augmentation. The effectiveness of this technique depends critically on the availability of large quantities of clean cool water from reservoirs or wells during critically low flow periods. While in this respect we have looked at flow augmentation in terms of its effect on the dissolved oxygen levels in the stream, there are other benefits that occur. The reservoir that would be constructed for this purpose can be used for recreational and irrigation purposes. Thus any evaluation of the value of flow augmentation must be coupled with some attempt at including other uses.*

The third category consists of regional and on-site waste-treatment plants and cooling towers. This procedure simply amounts to performing the oxidation of organic material or transfer of heat to the atmosphere in a controlled system outside of the river.

The treatment strategies can be combined in an infinite variety of combinations by constructing various structures such as dams, pipes, treatment plants and reservoirs. Each of these strategies has an associated cost given a predetermined set of quality goals.

---

*These areas may sometimes be conflicting. For example, a greatly fluctuating water level may not be conducive to recreation but may be necessary to maintain a flow augmentation program.

## III. DISSOLVED OXYGEN-TEMPERATURE MODEL

A first attempt to describe the relationship between DO, atmospheric reaeration, and bacterial respiration was formulated by Streeter and Phelps in 1925.[1] The relations were expanded by the addition of further sources and sinks by O'Conner,[2] Dobbins,[3] and others.

The model presented by Streeter and Phelps (see Appendix A for notation) can be described as follows:

Assume

$$dB/dt = -K_1 B \tag{1}$$

and

$$dD/dt = K_1 B - K_2 D \tag{2}$$

Equations 1 and 2 integrate to yield

$$B = B^\circ C_1 \tag{3}$$

and

$$D = KB^\circ [C_1 - C_2] + D^\circ C_2 \tag{4}$$

and

$$C_1 = \exp(-K_1 t)$$

$$C_2 = \exp(-K_2 t)$$

$$K = K_1/(K_2 - K_1)$$

In the context of a river basin, if the volumetric flow and velocity of the flow are assumed to be constant over some river segment of length a, then time can be interpreted as

$$t = a/v \tag{5}$$

Ordinarily the velocity of flow can be written as

$$v = F/A \tag{6}$$

where A is the cross-sectional area of the stream. However, as the volumetric flow varies in a fixed river bed, the cross-sectional area also varies. In other words, Equation 6 can be rewritten as

$$v = F/A(F) \tag{7}$$

For the purpose of this paper the relationship in Equation 7 has been assumed to have the form

$$v = \delta F^{\sigma} \tag{8}$$

where $\delta$ and $\sigma$ are parameters estimated for each river segment.

It is also assumed that the rate of reaeration depends on the velocity of flow and therefore on the volumetric flow rate. This relationship is assumed to have the form

$$K_2^{20} = \emptyset F^{\mu} \tag{9}$$

where $\emptyset$ and $\mu$ are estimated parameters which are determined at 20°C.

BOD is a measure of degradable organic waste material. If oxygen is available, the organic material will be oxidized. At the point when the oxygen from the sources enters at a faster rate than it is being used, the river will begin to recover.

Heat can also be considered as waste. The waste heat from power plants and other industries is usually carried off by water that is discharged in rivers. The heat is then transferred to the atmosphere by evaporation and heat conduction. After a period of time the river returns to its normal temperature.

The behavior of temperature in rivers has been described using an exponential decay model similar to the one presented above for BOD.

Assume:

$$d(T - T_E)/dt = -K_3(T - T_E) \tag{10}$$

Equation 6 integrates to yield

$$T - T_E = (T^{\circ} - T_E)c_3 \tag{11}$$

where

$$c_3 = \exp(-K_3 t)$$

As noted in the previous section, temperature affects the rate of deoxygenation and reaeration. These relations have been assumed to have the form

$$K_1 = K_1^{20}\ \theta_1^{T-20} \tag{12}$$

$$K_2 = \emptyset F^{\mu}\ \theta_2^{T-20} \tag{13}$$

where $K_1^{20}$ and $K_2^{20}$ are the experimentally determined values of $K_1$ and $K_2$ at 20°C and $\theta_1$ and $\theta_2$ are estimated parameters. Also the level of saturation of dissolved oxygen in water is affected by the temperature. This directly affects the level of DOD. The relation between temperature and the dissolved oxygen saturation level is usually assumed to be

$$D_s(T) = \zeta + \psi T + \zeta T^2 + \omega T^3 \tag{14}$$

where $\zeta$, $\psi$, $\zeta$ and $\omega$ are estimated parameters. Applications of Equations 12-14 can be found in References 4, 5, and 6. Equations 12-14 can easily be substituted into Equations 3 and 4 to determine dissolved oxygen sag curves for each different level of temperature.

The procedure described above to relate temperature and dissolved oxygen is correct as long as the basin temperature is fixed. However, if some of the effluent sources have effluent of a high temperature (*i.e.*, above the equilibrium value, given the meteorological conditions) then a model as in Equation 10 is appropriate for the temperature. This implies that a three-equation differential equation model including Equations 1, 2, and 10 should be solved to predict the level of dissolved oxygen. This was noted by Dysart and Hines.[4] However, they did not solve the new systems of equations but used the temperature decay model to fix the temperature at a point and assume that temperature constant for a small river section. In this paper the three-equation system is solved directly and applied to a river basin model. This allows the addition of thermal quality constraints to the basin model and consideration of cooling tower as a waste treatment system not only for heat pollution but also for increasing dissolved oxygen.

The new model, which will include the decay of temperature, can be written:

$$d(T - T_E)/dt = - K_3(T - T_E) \tag{15}$$

$$dB/dt = - K_1 B \tag{16}$$

$$dD/dt = K_1 B - K_2 D \tag{17}$$

$$K_1 = K_1^{20}\ \theta_1^{T-20} \tag{18}$$

$$K_2 = \phi F^{\mu}\ \theta_2^{T-20} \tag{19}$$

$$D_s = \zeta + \psi T + \zeta T^2 + \omega T^3 \tag{20}$$

The implicit form of the solution to this model is

$$T = T(F,\ T°,\ |\ T_E,\ K_3,\ a,\ \delta,\ \sigma) \tag{21}$$

$$B = B(F,\ T°,\ b°\ |\ T_E,\ K_3,\ a,\ \delta,\ \sigma,\ \theta_1,\ K_1^{20}) \tag{22}$$

$$D = D(F,\ T°,\ b°,\ D°\ |\ T_E,\ K_3,\ a,\ \delta,\ \sigma,\ \theta_1,\ K_1^{20},\ \theta_2,\ \mu,\ \phi) \tag{23}$$

This set of equations describes the water quality at a point on a river as a function of the "state" parameters of that river and the "control" variables. (The explicit form of the equations is developed in Appendix B.) The planning model presented in the next section concentrates on the costs of altering the control variables to achieve various quality goals.

## IV. NONLINEAR PROGRAMMING RIVER BASIN PLANNING MODEL

Using the water quality relationships described in the previous section, a programming model of the following form can be constructed:

minimize: total cost of abatement structures
subject to: water quality goals satisfied

The constraints of this model are constructed by dividing the river into sections and constraining the water quality to be met at the end of each section. A new section begins where one of the following occurs:

1. effluent flow enters the river
2. incremental flow enters the river (ground water, tributary flow, etc.)
3. the flow in the main channel is augmented or diverted
4. "state" parameters are altered.

Although new sections must be introduced where one of the above occurs, another consideration is that the section not be so long as to allow the dissolved oxygen sag curve to be completely contained in its length. This may require additional sections not needed because of entering flows or parameter changes. The river sections are numbered sequentially starting with the headwaters section including possible tributaries of a maximum length of one section (see Figure 3.2). The values of F, T°, B° and D° must

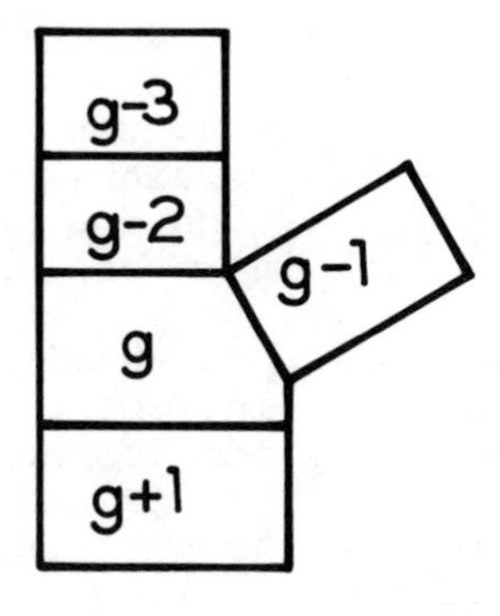


*Figure 3.2. Typical river sections.*

be recalculated at the headwaters of each section and are a function of the various quality control strategies employed. The cost of the basin system or objective function of the nonlinear model is also dependent on the alternative control strategies.

For this model we will assume the following alternative control strategies:

1. by-pass piping
2. flow augmentation
3. regional and on-site treatment plants
4. cooling towers.

The control variables associated with these strategies are:

$f_{gi}$ flow from polluter i to section g

$p_{mi}$ flow from polluter i to treatment plant m

$t_{gm}$ flow from treatment plant m to section g

$r_m$ per cent removal of BOD at treatment plant m

$F_{g_3}$ flow augmentation flow in section g

$s_g$ per cent removal of heat in cooling tower in section g.

The particular pattern of piping flows determines the regional and on-site plants operating.

There is an implicit assumption that the two groups of polluters are mutually exclusive. In other words heat polluters are not BOD polluters and vise versa. Operationally the assumption is that the temperature of all BOD flow is $T_E$ and the BOD and DOD of the heat effluent are assumed to have some constant value. The cooling alternative consists of simply building an on-site cooling tower or discharging directly. This limiting assumption does not appear to be serious since the major thermal polluters are power plants which do not directly affect BOD levels to a great degree. The major problem caused by the removal of this assumption would be the modeling of the cooling, which would take place in a BOD treatment plant, and the BOD oxidation, which would take place in a cooling tower.

To write the explicit nonlinear programming planning model one must write the values of F, T°, B° and D° at the head of each river section as functions of the control variables. These relations are conditional on the location of the particular section relative to the tributaries.

If g is the headwaters section or a tributary,

$$F_g = \sum_{j=1}^{4} F_{gj} \tag{24}$$

$$T^\circ_g = \sum_{j=1}^{4} T_{gj} F_{gj}/F_g \tag{25}$$

$$B^\circ_g = \sum_{j=1}^{4} B_{gj} F_{gj}/F_g \tag{26}$$

$$D^\circ_g = \sum_{j=1}^{4} D_{gj} F_{gj}/F_g \tag{27}$$

If section g is neither a tributary nor directly preceded by a tributary,

$$F_g = \sum_{j=1}^{4} F_{gj} + F_{g-1} \tag{28}$$

$$T^\circ_g = \left[\sum_{j=1}^{4} T_{gj} F_{gj} + T_{g-1} F_{g-1}\right]/F_g \tag{29}$$

$$B^\circ_g = \left[\sum_{j=1}^{4} B_{gj} F_{gj} + B_{g-1} F_{g-1}\right]/F_g \tag{30}$$

$$D^\circ_g = \left[\sum_{j=1}^{4} D_{gj} F_{gj} + D_{g-1} F_{g-1}\right]/F_g \tag{31}$$

If section g is directly preceded by a tributary,

$$F_g = \sum_{j=1}^{4} F_{gj} + F_{g-1} + F_{g-2} \tag{32}$$

$$T^\circ_g = \left[\sum_{j=1}^{4} T_{gj}F_{gj} + T_{g-1}F_{g-1} + T_{g-2}F_{g-2}\right]/F_g \tag{33}$$

$$B^\circ_g = \left[\sum_{j=1}^{4} B_{gj}F_{gj} + B_{g-1}F_{g-1} + B_{g-2}F_{g-2}\right]/F_g \tag{34}$$

$$D^\circ_g = \left[\sum_{j=1}^{4} D_{gj}F_{gj} + D_{g-1}F_{g-1} + D_{g-2}F_{g-2}\right]/F_g \tag{35}$$

where

$$F_{g2} = \sum_i f_{gi} + \sum_m t_{gm} \tag{36}$$

$$F_{g2}B_{g2} = \sum_i \overline{B}_i f_{gi} + \sum_m (1-r_m)[\sum_i \overline{B}_i p_{mi}/\sum_i p_{mi}]t_{gm} \tag{37}$$

$$F_{g2}D_{g2} = {}_i\overline{D}_i fg_i + \sum_m\left[[\sum_i \overline{D}_i P_{mi}/\sum_i P_{mi}]t_{gm}\right] \quad 38)$$

and

$$T_{g4} = (1-s_g)(T^*_g - T_E) + T_E \tag{39}$$

The planning model can now be explicitly written in terms of the control variables as follows:

minimize: $TC = C^L + C^R + C^P + C^T$

where (see Appendix C for discussion of cost functions)

$$C^L = \sum_{gi}\sum C^L_{gi} + \sum_{mi}\sum C^L_{mi} + \sum_{gm}\sum C^L_{gm}$$

$$C^L_{gi} = 1.865\ a_{gi}(f_{gi})\ .598$$

$$C^L_{hi} = 1.865\ a_{mi}(p_{mi})\ .598$$

$$C^1_{gh} = 1.865\ a_{gm}\ (t_{gm})\ .598$$

$$C^R = \sum_g C^R_g F_{g3}$$

$$C^P = \sum_h 49.22\ (\sum_i p_{mi})\ 3/4\ [8.0\ (r_m - .5)^3 + 1]$$

$$C^T = \sum_g C^T_g s_g\ (T^*_g - T_E)\ F_{g4}$$

subject to:

$$T_g = T_g\ (F_g,\ T^\circ_g \mid T_{Eg},\ a_g,\ \delta_g,\ \sigma_g) \leq \overline{\overline{T}}_g \qquad G = i,\ N \tag{40}$$

$$B_g = B_g\ (F_g,\ T^\circ_g,\ B^\circ_g \mid T_{Eg},\ K_{3g},\ a_g,\ \delta_g,\ \sigma_g,\ \theta_{1g},\ K^{20}_{1g}) \leq \overline{\overline{B}}_g \tag{41}$$

$$D_g \leq D_g\ (F_g, T^\circ_g, B^\circ_g, D^\circ_g \mid T_{Eg}, K_{3g}, a_g, \delta_g, \sigma_g, \theta_{1g}, K^{20}_{1g}, \theta_{2g}, \mu_g) \leq \overline{\overline{D}}_g \tag{42}$$

where $F_g$, $T^\circ_g$, $B^\circ_g$, and $D^\circ_g$ are described as functions of the control variables in Equations 24-39. The following flow balance equations are also needed:

$$\sum_{gg} f_{gi} + \sum_m p_{mi} - \overline{f}_i = 0\ (i=1,\ P,\ m=1,\ L) \tag{43}$$

$$\sum_m p_{mi} - \sum_m t_{gm} = 0 \qquad (i=1,\ P,\ g=1,\ N) \tag{44}$$

This explicitly states the planning model in terms of the control variables available to the planner.

The nature of this model is that the constraints are sequentially dependent on the concentrations of all the upstream flows. The temperature level affects the BOD and DOD levels directly and the DOD levels indirectly through the BOD. The BOD level affects the DOD level directly. Since the level of BOD only affects the water quality indirectly through the DOD, the constraint set

$$B_g = B_g\ (F_g,\ T^\circ_g,\ B^\circ_g \mid T_{Eg},\ K_{3g},\ a_g,\ \delta_g,\ \sigma_g,\ \theta_{1g},\ K^{20}_{1g}) \leq \overline{\overline{B}}_g$$

$$g = 1,\ N$$

can be eliminated from direct consideration. The BOD levels must still be calculated in order to determine the DOD levels. However, the constraints on DOD implicitly constrain the BOD levels.

The major problem of adapting this model for solution by the nonlinear algorithm described in Equations 8 and 9 is the calculation of the partial derivative of the constraints and objective function. These calculations are needed to set up a local linear programming to determine a direction of search. If all the constraints were of a different functional form, the computer coding necessary for partial derivative evaluation would be immense for any large-scale problem. However, as in most large-scale problems, the constraints of the water-quality model can be classified into a relatively few functionally homogeneous groupings.

The function groupings for the water-quality model are represented in Table 3.1.

*Table 3.1*

*Function Groupings for Water-Quality Model*

| | $f_{gi}$ | $P_{mi}$ | $t_{gm}$ | $r_m$ | $F_{g3}$ | $s_g$ |
|---|---|---|---|---|---|---|
| Temperature constraints | $\Lambda_{11}$ | $\Lambda_{12}$ | $\Lambda_{13}$ | $\Lambda_{14}$ | $\Lambda_{15}$ | $\Lambda_{16}$ |
| Quality constraints | $\Lambda_{21}$ | $\Lambda_{22}$ | $\Lambda_{23}$ | $\Lambda_{24}$ | $\Lambda_{25}$ | $\Lambda_{26}$ |
| Flow conservation (polluter) | $\Lambda_{31}$ | $\Lambda_{32}$ | $\Lambda_{33}$ | $\Lambda_{34}$ | $\Lambda_{35}$ | $\Lambda_{36}$ |
| Flow conservation (treatment plants) | $\Lambda_{41}$ | $\Lambda_{42}$ | $\Lambda_{43}$ | $\Lambda_{44}$ | $\Lambda_{45}$ | $\Lambda_{46}$ |
| Objective function | $\Lambda_{51}$ | $\Lambda_{52}$ | $\Lambda_{53}$ | $\Lambda_{54}$ | $\Lambda_{55}$ | $\Lambda_{56}$ |

Each $\Lambda_{ij}$ represents a matrix of partial derivatives between the constraints and the appropriate variables. For example, if the quality constraints are labeled $Q_1, \ldots, Q_N$, then the submatrix $\Lambda_{11}$ would be

$$\Lambda_{11} = \begin{pmatrix} \dfrac{\partial Q_1}{\partial f_{11}} & \cdots & \dfrac{\partial Q}{\partial f_{Np}} \\ \vdots & & \vdots \\ \dfrac{\partial Q_N}{\partial f_{11}} & & \dfrac{\partial Q_N}{\partial f_{Np}} \end{pmatrix}$$

Each block in $\Lambda$ can be treated as a unit for the computer coding of the problem. Some of the blocks require little calculation. For example, the blocks such as $\Lambda_{21}$ and $\Lambda_{22}$ are simply linear functions with coefficients of 1 requiring only that a 1 be placed in the appropriate place. On the other hand, other blocks such as $\Lambda_{14}$ and $\Lambda_{15}$ require considerable sorting and calculation. The elements of $\Lambda$ are generated a column at a time and only the columns associated with the basis variables of the linear programming problem are stored. The other columns are generated and updated as needed.

Notice that even with the simplication of the problem from the treatment of homogeneous function blocks there are still 25 different blocks to deal with. This means 25 different partial derivative forms must be precalculated and programmed. In the water-quality case the "not-quite sequential nature" of the constraints further complicates the problem.

Because of the vast number of variables in the water quality problem the following solution technique was adopted to search the feasible set. A number of piping patterns which seemed reasonable from knowing the problem were read in as the initial solution and the present removal variables of the appropriate treatment cooling plants and towers the level of flow augmentation were given a high priority level. This technique saved considerable computer time and still allowed the feasible set to be adequately searched.

## V. APPLICATION OF PLANNING MODEL

The programming model as proposed has been applied to the West Fork White River in Indiana. The West Fork White River has its source near the Indiana-Ohio border and flows southwesterly for 371 miles through the state of Indiana. At this point it joins the East Fork White River and flows to the Ohio.

The major city on the West Fork is Indianapolis, a city of over 600,000 which is 234 miles from the mouth. Two other cities, Anderson and Muncie, are upstream from Indianapolis. The concentration of population and industry around these three cities causes the major portion of the pollution problem in the West Fork White.

For the purpose of this paper a length of the West Fork White which runs from the headwaters above

Muncie to Spencer below Indianapolis was chosen. This section described is 172.8 miles long and was divided into 62 sections based on information about polluters, incremental flow, and river parameters. The sections range in length from 0.1 miles to 6.2 miles.

Thirteen major polluters and three wasteheat dischargers are in this part of the river basin (see Map 3.1). Detailed listings of the parameters and pollution sources are available from the authors.

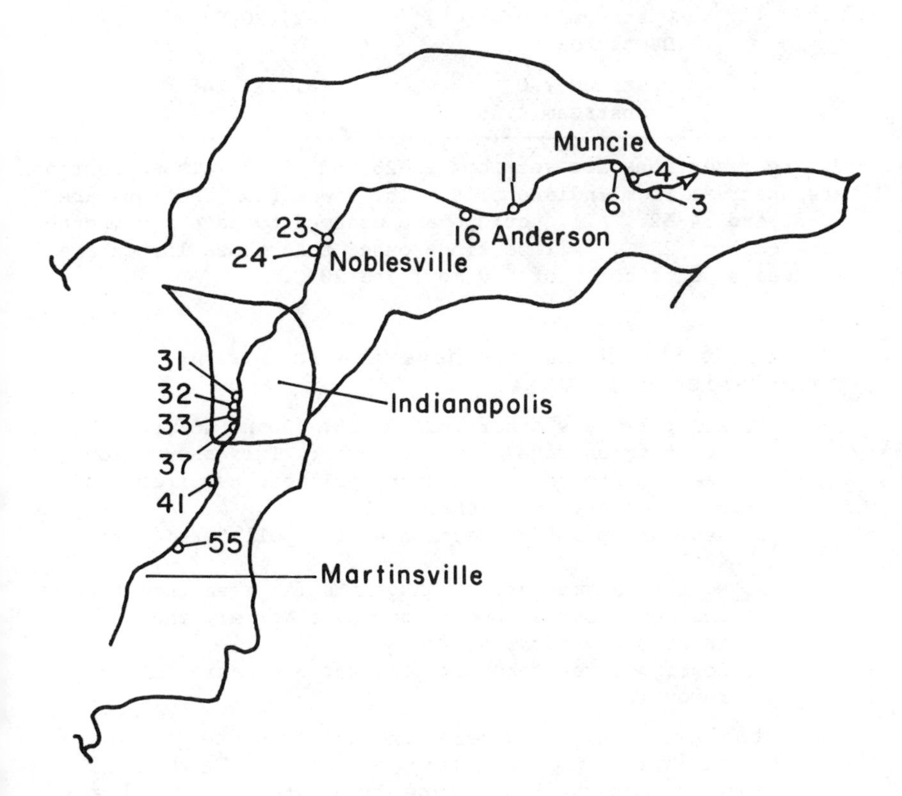


*Map 3.1. West Fork White River, location of polluters.*

In order to demonstrate the model, solutions were found for four different sets of quality goals which are described in Table 3.2 along with the cost associated with the particular solution.

*Table 3.2*

*River Section DO Goals and Costs*

| *Solution* | *Goals* (mg/l DO)* | *Cost* |
|---|---|---|
| #1 | Upstream 5.0<br>Downstream 5.0 | $2,484,269 |
| #2 | Upstream 7.0<br>Downstream 5.0 | $2,641,646 |
| #3 | Upstream 5.0<br>Downstream 4.0 | $2,320,881 |
| #4 | Upstream 7.0<br>Downstream 4.0 | $2,477,248 |

*Upstream sections are sections 21-26 and 28-33. These sections are upstream from Indianapolis. The downstream sections are 50-52 and 54-62. All sections are assumed to have a temperature goal of 22°C. All sections other than those listed are assumed to have goals of 5.0 mg/l and 20°C.

All of the solutions have the following characteristics:

1. All polluters other than 6, 16, 40 and 46 are treating on-site at 80% removal. This assumption was made to reflect current polluter requirements of secondary treatment.
2. Regional plant at section 46 for polluters 40 and 46.
3. No flow augmentation. Augmentation even from the cheapest source appears to be relatively ineffective in reducing total cost.
4. Cooling tower for heat polluter 3 treating at 82% removal.

The actual treatment levels at the on-site plants 6 and 16 and the regional plant 46 are in Table 3.3.

It is of interest to note that solutions #1 and #4 differ by only $6,000 in annual costs. The river use implied by each set of goals, however, may be considerably different. Solution #4 represents a gain of a river segment with 7.0 mg/l DO upstream at the cost of a 1.0 mg/l DO decrease downstream. This could yield excellent game fishing at little or no loss in boating or fishing opportunities, which could be moved downstream. In fact, of the 11 downstream sections with the 4.0 mg/l DO constraint only one of the sections (52) has a binding

*Table 3.3*

*Treatment Levels at Treatment Plants*

| *Solution* | *Plant* | *Treatment Level* |
|---|---|---|
| #1 | 6 | 0.94636 |
| | 16 | 0.92600 |
| | 46 | 0.97227 |
| #2 | 6 | 0.96950 |
| | 16 | 0.98762 |
| | 46 | 0.97206 |
| #3 | 6 | 0.94681 |
| | 16 | 0.92566 |
| | 46 | 0.96184 |
| #4 | 6 | 0.96950 |
| | 16 | 0.98762 |
| | 46 | 0.96163 |

constraint. The DO levels in the other ten sections range from 4.08 (55) to 6.40 (62). The same is true of the 12 upstream sections whose DO levels range up to 8.09, which is near the saturation level.

The heat removal solution indicates on this river that the thermal level is relatively independent of the BOD solution. The location of the heat polluters does not correspond to the location of the major BOD polluters. The lowest DO levels occur in sections 22 and 52, both of which are just upstream from thermal polluters (24 and 55).

Of course the temperature goals can also be varied. Another run was made with 7.0 mg/l upstream 4.0 mg/l downstream DO goals along with the temperature goals listed in Table 3.4.

*Table 3.4*

*Temperature Goals for Solution #5*

| *Sections* | *Goal (°C)* |
|---|---|
| 1-10 | 25.0 |
| 11-20 | 22.0 |
| 12-62 | 21.6 |

The temperature goals just below the major heat polluter were relaxed and the rest were left the same or tightened. The resulting cost of solution #5 was $2,493,624, an addition of $16,000 over solution #4. This cost yields 50 sections of almost equilibrium temperature at the cost of 10 sections which are 4°C above the equilibrium temperature.

It is of interest that as the nonlinear algorithm reduced the heat removal level in section 3 from 90% to 58%, the DOD level varied by 0.5 mg/l. This indicates that the heat could have a sustained effect on DO and, therefore, on basin treatment costs if introduced near the bottom of the DO sag curve.

## VI. SUMMARY

An outline of a planning model that could be used in resource allocation problems characterized by strong interactions among the various users has been presented. The apparent philosophy underlying recent federal legislation has been to emphasize treatment at the source and imposition of uniform standards. The results of this analysis and those of other reports[7,8,9] indicate that the possibility of enormous quality gains at a much lower cost than incurred under the current approach does exist. Although the use of joint facilities such as flow augmentation, regional plants and by-pass piping cause some cost allocation problems, they are not by any means insurmountable as evidenced by the cost allocation schemes in References 8 and 10. In fact the planning models are utilized in these reports to determine equitable cost shares. It is possible that in some regions the cost of attaining the federally required uniform quality goals may be prohibitive.

## APPENDIX A. NOTATION

All Greek letters in water-quality model are experimentally determined parameters.

$a_g$ length of river section g (miles)

$a_{gi}$ length of pipe from polluter i to section g (miles)

$a_{mi}$ length of pipe from polluter i to treatment plant m (miles)

$a_{gm}$ length of pipe from treatment plant in to section g (miles)

$B_g$ biochemical oxygen demand at end of river section g (mg/l)

$B^\circ_g$ biochemical oxygen demand at headwater of river section g (mg/l)

$\overline{B}_i$ biochemical oxygen demand of effluent from polluter i (mg/l)

$\overline{\overline{B}}_g$ biochemical oxygen demand goal in river section g (mg/l)

$D_g$ dissolved oxygen concentration at end of river section g (mg/l)

$D^\circ_g$ dissolved oxygen concentration at headwaters of river section g (mg/l)

$\overline{D}_i$ dissolved oxygen concentration of effluent from polluter i (mg/l)

$\overline{\overline{D}}_g$ dissolved oxygen concentration level goal in river section g (mg/l)

$D_s$ dissolved oxygen saturation level (mg/l)

$\overline{f}_i$ effluent flow from polluter i (c.f.s.)

$F_g$ river flow in section g (c.f.s.)

$F_{g1}$ incremental flow entering river section g (c.f.s.)

$F_{g2}$ effluent flow entering river section g (c.f.s.)

$F_{g3}$ augmentation flow entering river section g (c.f.s.)

$F_{g4}$ effluent flow from heat polluter g (c.f.s.)

$K^{20}_{g1}$ deoxygenation rate constant at 20°C (days $^{-1}$)

$K^{20}_{g2}$ reaeration rate constant at 20°C (days $^{-1}$)

$K_3$ temperature decay rate constant (days $^{-1}$)

$T$ time (days)

$T_g$ temperature at end of river section g (°C)

$T^\circ_g$ temperature at headwaters of river section g (°C)

$T^*_g$ temperature of effluent from heat polluter g (°C)

$T_E$ equilibrium basin temperature (°C)

$\overline{\overline{T}}_g$ dissolved oxygen concentration goal in river section g (mg/l)

## APPENDIX B. SOLUTION OF DISSOLVED OXYGEN-TEMPERATURE MODEL

The new model as described in Section III of this paper is:*

$$\frac{d(T-T_E)}{dt} = -K_3(T-T_E) \tag{1}$$

$$\frac{dB}{dt} = -K_1 B \tag{2}$$

$$\frac{dD}{dt} = K_1 B - K_2 D \tag{3}$$

$$K_1 = K_1^{20}\ \theta_1^{T-20} \tag{4}$$

$$K_2 = \phi F^{\mu}\ \theta_2^{T-20} \tag{5}$$

$$D_s = \zeta + \psi T + \zeta T^2 + \omega T^3 \tag{6}$$

For the purpose of solving this system it is useful to note the causality relations assumed. Temperature affects the dissolved oxygen level both directly and through biological oxygen demand. The biological oxygen demand affects the dissolved oxygen deficit directly. These relations imply that we can solve Equation 1 independently and substitute the solution into Equations 2 and 3. Equation 2 can then be solved and substituted into 3, which can then be solved to complete the solution of the system.

Proceeding in the manner prescribed, one notes that the solution to Equation 1 is the same as the one presented in Section I.

$$T = T_E + (T° - T_E)\ c_3 \tag{7}$$

The value for T can be substituted into Equation 2 to yield

$$\frac{dB}{dt} = (-\ A_1 \theta_1^{\overline{T}c_3})B \tag{8}$$

---

*Section subscripts g do not appear in this appendix.

where

$$A_1 = K_1^{20}\theta_1^{T}E^{-20}$$

$$\overline{T} = T^\circ - T_E$$

The solution to Equation 2 is obtained by rewriting 8 in the form

$$\frac{dB}{B} = (-A_1\ \theta_1^{\ Tc_3})\ dt \qquad (9)$$

and integrating both sides. The integration of the right-hand side of Equation 9 is found by approximating the function with a Maclaurin series expansion and integrating each term. The solution to Equation 2 is

$$B = B^\circ \exp\ [-\beta_o K_3 t + \sum_{n=1}^{\overline{n}} \beta_n c_3^n - \sum_{n=1}^{\overline{n}} \beta_n] \qquad (10)$$

where

$$\beta_o = A_1/K_3$$

and

$$\beta_n = \frac{\beta_o(\ell n\theta_1)^n \overline{T}^n}{n!n} \qquad n = 1, \overline{n}$$

The value of $\overline{n}$ depends on the number of members of the infinite series necessary to achieve the desired accuracy.

Note that if the temperature is assumed to be fixed at the equilibrium value then Equation 10 reduces to

$$B = B^\circ \exp\ [-A_1 t] \qquad (11)$$

which is the solution to the original Streeter and Phelps equations with temperature treated as a parameter in the system.

Given the solutions to Equations 1 and 2, the solution to Equation 3 can now be obtained. The solutions to 1 and 2 are substituted into 3 to obtain

$$\frac{dD}{dt} = A_1\theta_1^{Tc_3}\ B^\circ \exp \left[-\beta_o K_3 t + \sum_{n=1}^{\overline{n}} \beta_n c_3^n - \sum_{n=1}^{\overline{n}} \beta_n\right] - A_2\theta_2^{Tc_3}D \qquad (12)$$

where

$$A_2 = \phi F^{\mu} \theta_2^{T_E - 20}$$

If we rewrite Equation 12 in the form

$$\frac{dD}{dt} + P(t)d - Q(t) \quad (13)$$

where

$$P(t) = A_2 \theta_2^{\overline{T} c_3} \quad (14)$$

$$Q(t) = A_1 \theta_1^{Tc_3} B° \exp \left[ -\beta_o K_3 t + \sum_{n=1}^{\overline{n}} \beta_n c_3^n - \sum_{n-1}^{\overline{n}} \beta_n \right] \quad (15)$$

we see that it is a linear first-order differential equation which can be solved by multiplying both sides by u(t) where

$$u(t) = \exp\left[\int P(t)dt\right] \quad (16)$$

The left-hand side is then the exact differential of d exp $\left|\int P(t)dt\right|$ and the right-hand side is a function of only t. Both sides can now be integrated to obtain the solution to Equation 3.

The first step is to evaluate u(t). The function u(t) can be written as

$$u(t) = \exp\left[\int A_2 \theta_2^{\overline{T} c} \varepsilon dt\right] \quad (17)$$

The integral in the exponent is of the same form as the right-hand side of Equation 9. Using the solution to Equation 9 we obtain the solution to this integral:

$$\lambda_o \; (\ell n \; \overline{T} - K_3 t) + \sum_{n=1}^{\overline{n}} \lambda_n \; c_3^n$$

where

$$\lambda_o = - \frac{A_2}{K_3}$$

and

$$\lambda_n = \frac{\lambda_o (\ell n \theta_2)^n \overline{T}^n}{n!n}, \qquad n = 1, \overline{n}$$

The solution to the left-hand side of Equation 13 is

$$\exp\left[\int P(t)dt\right] = \exp\left[\lambda_o(\ell n\overline{T} - K_3 t) + \sum_{n=1}^{\overline{n}} \lambda_n c_3^n\right]D$$

The right-hand side of Equation 13 is now a function of t only and the integral with respect to t can be written as

$$\int u(t)\ Q(t)dt = \int B^\circ \alpha \exp\left[\alpha_o t + \sum_{n=1}^{\overline{n}} \alpha_n c_3^n\right] dt \qquad (18)$$

where

$$\alpha = A_1 \exp\left[\lambda_o \ell n\overline{T} - \sum_{n=1}^{\overline{n}} \beta_n\right]$$

$$\alpha_o = -K_3[\lambda_o + \beta_o]$$

$$\alpha_n = \lambda_n + \beta_n$$

Again using the Maclaurin series approximation technique, let

$$f(t) = u(t)\ Q(t)$$

The integral 18 can be written

$$\int f(t)dt = f(0)t + \sum_{m=1}^{\overline{m}} \frac{f^m(0)t^{m+1}}{(m+1)!} \qquad (19)$$

In order to evaluate 19 explicitly let

$$f(t) = B^\circ \alpha \exp[\alpha_o t + y(t)]$$

where

$$Y(t) = \sum_{n=1}^{\overline{n}} \alpha_n\ c_3^n.$$

The derivatives of f(t) are

$$f^1(t) = \rho f(t)$$
$$f^2(t) = \rho f^1(t) + y^2(t)\ f(t)$$
$$f^3(t) = \rho f^2(t) + 2y^2(t)\ f^1(t) + y^3(t)\ f(t)$$
$$f^4(t) = \rho f^3(t) + 3y^2(t)\ f^2(t) + 3y^3(t)\ f^1(t) + y^4(t)\ f(t)$$

or in general

$$f^{k}(t) = \rho f^{k-1}(t) = \sum_{i=1}^{k-1} \frac{(k-1)!}{i!(k-1-i)!} y^{i+1}(t)\ f^{k-1-i}(t),\ k > 2$$

where

$$\rho = \alpha_{o} + y^{1}(t).$$

The derivatives of y(t) are

$$y^{1}(t) = \sum_{n=1}^{\overline{n}} \alpha_{n}(-nK_{3})\ c_{3}^{n}$$

$$y^{2}(t) = \sum_{n=1}^{\overline{n}} \alpha_{n}(-nK_{3})^{2}\ c_{3}^{n}$$

or in general

$$y^{i}(t) = \sum_{n=1}^{\overline{n}} \alpha_{n}(-nK_{3})^{j}\ c_{3}^{n}$$

When t = 0, the derivatives of y(t) are

$$y^{1}(0) = \sum_{n=1}^{\overline{n}} \alpha_{n}(-nK_{3})$$

$$y^{2}(0) = \sum_{n=1}^{\overline{n}} \alpha_{n}(-nK_{3})^{2}$$

or in general

$$y^{j}(0) = \sum_{n=1}^{\overline{n}} \alpha_{n}(-nK_{3})^{j}$$

Using this information the values of $f(0)$, $f^{1}(0)$, ..., $f^{m}(0)$ can be determined.

$$f(0) = B°\alpha \exp\left[\sum_{n=1}^{\overline{n}} \alpha_{n}\right]$$

$$f^{1}(0) = \left[\alpha_{o} + \sum_{n=1}^{\overline{n}} (-nK_{3})\alpha_{n}\right] f(0)$$

$$f^2(0) = \left[\alpha_o + \sum_{n=1}^{\bar{n}} (-nK_3)\alpha_n\right] f^1(0) + \left(\sum_{n=1}^{n} (-nK_3)^2 \alpha_n\right) f(0)$$

or in general

$$f^k(0) = \left[\alpha_o + \sum_{n=1}^{\bar{n}} (-nK_3)\alpha_n\right] f^{k-1}(0)$$

$$+ \sum_{i=1}^{k-1} \frac{(k-1)!}{i!(k-1-i)!} \left(\sum_{n=1}^{\bar{n}} (-nK_3)^{i=1} \alpha_n\right) f^{k-1-i}(0)$$

The integral of the right-hand side of Equation 13 can now be expressed explicitly in terms of the expressions above.

Combining the solutions to the right- and left-hand sides of Equation 13 yields the new dissolved-oxygen equation which is

$$D \exp\left[\lambda_o(\ell nT - K_3 t) + \sum_{n=1}^{\bar{n}} \lambda_n c_3^n\right] = f(0)t + \sum_{m=1}^{\bar{m}} \frac{f^m(0)t^{m+1}}{(m+1)!} + c \quad (20)$$

The constant of integration is

$$c = D^\circ \exp\left[\lambda_o \ell nT + \sum_{n=1}^{n} \lambda_n\right] \quad (21)$$

The solution to the original Streeter-Phelps equations can be determined by adding and subtracting the constant $B^\circ\alpha/\alpha_o$ from the right-hand side and noting that the terms left, excluding the constant c, are a Maclaurin series expansion for

$$(1/\alpha_o)\ c^{\alpha_o t}.$$

Equations 7, 10 and 20 can be rewritten as

$$T = T_E + (T^\circ - T_E)\ c_3 \quad (22)$$

$$B = B^\circ\ c_4 \quad (23)$$

$$D = c_5^{-1}\left(f(0)t + \sum_{m=1}^{\bar{m}} \frac{f^m(0)t^{m+1}}{(m+1)!} + D^\circ c_6\right) \quad (24)$$

where

$$c_4 = \exp\left| -\beta_o K_3 t + \sum_{n=1}^{\bar{n}} \beta_n c_3^n - \sum_{n=1}^{\bar{n}} \beta_n \right| \quad (25)$$

$$c_5 = \exp\left| \lambda_o (\ell nT - K_3 t) + \sum_{n=1}^{\bar{n}} \lambda_n c_3^n \right| \quad (26)$$

$$c_6 = \exp\left| \lambda_o \ell nT + \sum_{n=1}^{\bar{n}} \lambda_n \right| \quad (27)$$

## APPENDIX C. COST FUNCTIONS

The cost functions in this paper represent the annual cost of the capital and operation and maintenance of the appropriate structures. The cost functions for the treatment plants were obtained from Frankel[11] and the cost functions for the piping from Linaweaver and Clark.[12] The flow augmentation dam costs were obtained from the Evansville office of the EPA for existing and proposed sites in the West Fork White Basin. For the purpose of this paper it was assumed that the expected flow from the reservoir was the upper bound and any augmentation flow less than this bound cost a proportional amount of the dam costs. The cooling tower costs were also assumed to be proportional to the flow and number of degrees cooled.

It is recognized by the authors that many of the assumptions about the cost functions are debatable. The study of cost function generating techniques appropriate for such planning models is a very important area of research since the usefulness of these models depends critically on these functions. This problem is discussed in detail in a recent paper by Marsden, Pingry and Whinston.[13]

## ACKNOWLEDGMENTS

This research was supported by the Army Research Office under contract #DA-31-124-ARO-D0477-MOD-P004.

## REFERENCES

1. Streeter, H. W. and E. B. Phelps. "A Study of the Pollution and Natural Purification of the Ohio River," *U. S. Public Health Bulletin 146* (1925).
2. O'Conner, J. "The Temporal and Special Distribution of Dissolved Oxygen in Streams," *Water Res. Res. 3(1)*:65 (1967).
3. Dobbins, W. E. "BOD and Oxygen Relationships in Streams," *J. San. Eng. Div., Proc. Amer. Soc. Civil Engr. 90(SA3)*:53 (1964).

4. Dysart, B. C. and W. W. Hines. "Control of Water Quality in a Complex Natural System," *IEEE Transactions on Systems Science and Cybernetics, SSC-6(4)* (1970).
5. Jaworski, N. A., W. J. Weber, Jr., and R. A. Deininger. "Optimal Reservoir Releases for Water Quality Control," *J. San. Eng. Div., ASCE 96(SA3)*:727 (1970).
6. Worley, J. L., F. J. Burgess, and W. W. Towne. "Identification of Low-Flow Augmentation Requirements for Water Quality Control by Computer Techniques," *J. Water Poll. Cont. Fed. 37(5)*:659 (1965).
7. Graves, G. W., G. B. Hatfield, and A. Whinston. "Mathematical Programming for Regional Water Quality Management," *Water Res. Res. 8(2)*:273 (1972).
8. Graves, G. W., G. B. Hatfield, and A. Whinston. "Water Pollution Control Using By-Pass Piping," *Water Res. Res. 5(1)*:13 (1969).
9. Graves, G. W., D. Pingry, and A. Whinston. "Application of a Large Scale Nonlinear Programming Problem to Pollution Control," *AFIPS - Conference Proceedings 39*.
10. Hass, J. E. "Optimal Taxing for the Abatement of Water Pollution," *Water Res. Res. 6(2)*:353 (1970).
11. Frankel, R. J. "Economic Evaluation of Water Quality; An Engineering-Economic Model for Water Quality Management," (Berkeley, Calif.: University of California, SERL Report No. 65-3, 1965).
12. Linaweaver, F. P. and S. C. Clark. "Cost of Water Transmission," *J. Am. Water Works Assn. 56*:1549 (1964).
13. Marsden, J. R., D. E. Pingry, and A. Whinston. "Production Function Theory and the Optimal Design of Waste Treatment Facilities," *Appl. Econ.* (in press).
14. Cootner, P. H. and G. O. G. Löf. *Water Demand for Electricity Generations: An Economic Projection Model* (Baltimore, Md.: Johns Hopkins Press, 1965).
15. Clough, D. J. and M. B. Bayer. "Optimal Waste Treatment and Pollution Abatement Benefits on a Closed River System," *Can. Oper. Res. Soc. J. 6(3)*:153 (1968).
16. Graves, G. W., D. E. Pingry, and A. Whinston. "Water Quality Control: Nonlinear Programming Algorithm," *Revue Francaise D'Informatique et de Recherche Operationnelle.* (in press).
17. Löf, G. O. G. and J. C. Ward. "Economics of Thermal Pollution Control," *J. Water Poll. Cont. Fed. 42(12)*: 2102 (1970).
18. Loucks, D. P., C. S. Revelle, and W. R. Lynn. "Linear Programming Models for Water Pollution Control," *Management Sci. 14(4)*:166 (1967).
19. Revelle, C. S., D. P. Loucks, and W. R. Lynn. "A Management Model for Water Quality Control," *J. Water Poll. Cont. Fed. 39(7)*:1164 (1967).

20. Schaumburg, G. W. *Water Pollution Control in the Delaware Estuary* (Cambridge, Mass.: Harvard University Water Program, Harvard University, 1967).

MODELS FOR ENVIRONMENTAL POLLUTION CONTROL


# CHAPTER 4

# INTRODUCTION TO SIMULATION TECHNIQUES

Gerald T. Orlob*

## I. INTRODUCTION

A basic tool of the systems approach to environmental quality management is a mathematical model of the prototype system. It provides the means for the environmental planner to simulate the behavior of the real system "before the fact." He can test the effect of operating policies, control strategies, or changes in the physical, chemical or biological conditions of the environment without actually modifying the prototype. Of greatest importance, perhaps, is the capability given the planner to compare alternatives, to assess the incremental impact between two or more courses of action. Implemented through the medium of the computer, the model expands enormously the number of alternatives that may be considered and enhances the prospect that the ultimate decision will be near optimal.

Models are of several types. *Deterministic* models seek to describe the environment in more or less real terms. Given a set of input data the model generates a response, a "simulation," that describes the behavior of the prototype in space and/or time. For example, an ecologic model of a reservoir may describe the annual cycle of algal growth and productivity at each level in the reservoir that is significant from the quality control viewpoint. *Stochastic* models are utilized to transform inputs of random character into quantifiable expressions

---

*Gerald T. Orlob is Professor of Civil Engineering at the University of California, Davis, California, and President, Water Resources Engineers, Inc., 710 So. Broadway, Walnut Creek California 94596, U.S.A.

of risk, probability, reliability, or other measures of uncertainty. The model allows the planner to make statements concerning the likelihood of an event occurring at a future date, to evaluate risk in monetary terms, or to establish the penalties or benefits derived from constraining the prototype response within preset limits. *Optimization* models find best possible solutions under explicit constraints and for given objective functions. The objective may be to minimize the cost of wastewater treatment, maximize the production of power, or minimize the wastage of water. The planner may employ the model to investigate the sensitivity of the prototype to operating policies, thus determining the restraints that are likely to be imposed, or he may use it much in the same way as deterministic models are used to explore alternatives. In practice an optimization model also performs a simulation; this becomes the deterministic solution that satisfies the objective function.

This paper is concerned primarily with deterministic models. Further, it restricts attention to those models founded on "first principles" and excludes those that are purely statistical. The goal is to minimize empiricism although empirical functions are included where knowledge of basic laws is insufficient or the simplification is consistent with planning objectives. Some basic concepts in the design of simulation models for environmental pollution control are considered first.

## II. DESIGN OF THE MODEL

There are three basic steps in model design:

Conceptual Representation

1. determining the scale of the model in time and space
2. describing the physical system as a continuum of discrete elements, a network that possesses the shape, volume, and space relationships of integral components of the system.

Functional Representation

1. defining the parameters that will characterize system response
2. describing mathematically the processes and functions of each component of the system
3. determining boundary conditions within which the simulation must be carried out.

Computational Representation
1. selecting a solution technique
2. transforming functional representations into a form suitable for solution
3. formating input and output.

A brief discussion of the first two steps will serve to illustrate the model development procedure; a discussion of the third step is outside the scope of this paper.

## III. CONCEPTUAL REPRESENTATION

### *Determination of Scale*

The detail provided in the model depends on its intended use. The general relationship between model use and time scale is illustrated in Figure 4.1, using the Texas Water Plan[1,2] as an example. If one

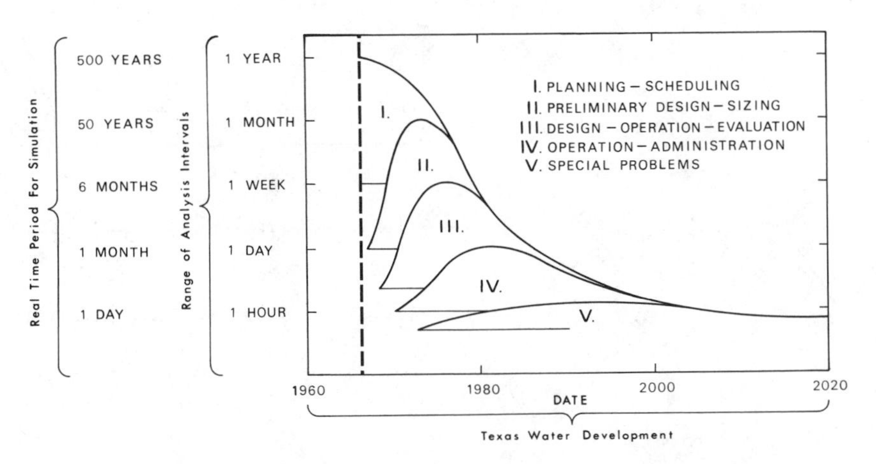


*Figure 4.1. Time interval and real time simulation periods in relation to principal activities in implementation of the Texas Water Plan.*

is concerned with large scale regional planning one may be satisfied with a model in which the smallest element is a whole reservoir, the smallest time step is a month, and the planning horizon for simulation may range from 50 to 100 years or more. On the other hand, to explore the dynamics of algal growth in a reservoir one may need to subdivide the water body into layers a foot thick and simulate algal growth at time steps of one hour over a critical summer season.

While requirements vary with the specific case it is generally true that large scale regional planning requires the least detail in both time and space. Also, since the time steps are large, transients of short duration are not of particular interest; flows are treated as steady. In design the detail is sharper, the system is subdivided, and finer and shorter time steps are taken. System operation requires capability to examine the system's response very closely; transients are important and time steps are appropriately short. Certain special problems, notably those dealing with water quality, may require an even higher degree of detail.

The spatial detail provided also varies with the use intended for the model as may be illustrated by Figure 4.2. The basin may be divided for certain planning purposes into subbasins along hydrologic

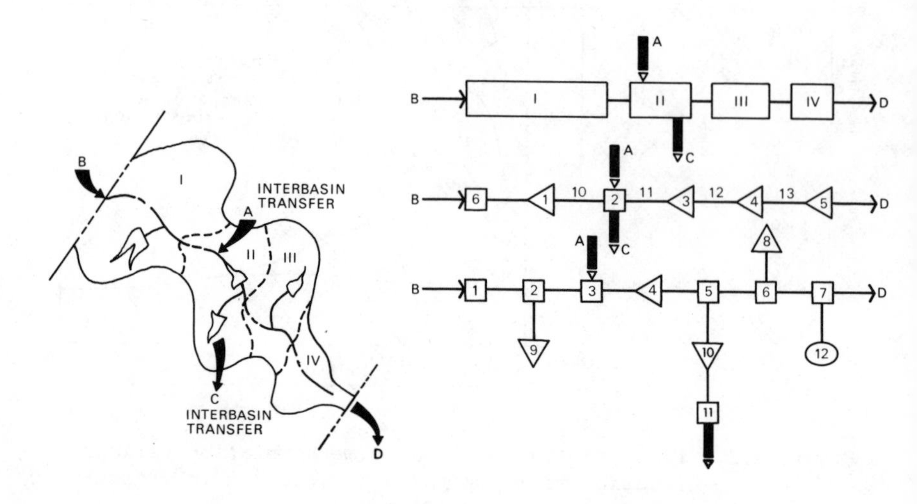


*Figure 4.2. **Model idealization of a river basin.***

divides. To simulate interbasin transfers and water balance between subbasins, the system is further refined. Finally, for purposes of an operation study each reservoir is uniquely represented as are interbasin transfer points.

*Discretizing the System*

The third sketch in Figure 4.2 identifies discrete reservoirs that for the purposes of the model are assumed to have uniform water quality properties throughout. This may be satisfactory for making a crude overall mass balance in the system but to examine environmental pollution at a scale that is truly meaningful one may need to further refine our conceptual representation of the prototype.

For illustrative purposes consider the problem of simulation of the annual cycle of temperature in a deeply stratified reservoir. Such a system experiences a definite thermal structure oriented along the vertical axis, *i.e.*, horizontal planes through the reservoir are virtually isothermal. It has proven practical in such a case to divide the reservoir into a system of slices of uniform thickness, as many as are required to describe the development of thermal stratification over the annual period. Figure 4.3 shows a conceptualization of the reservoir that has proven very satisfactory for this purpose.[3]

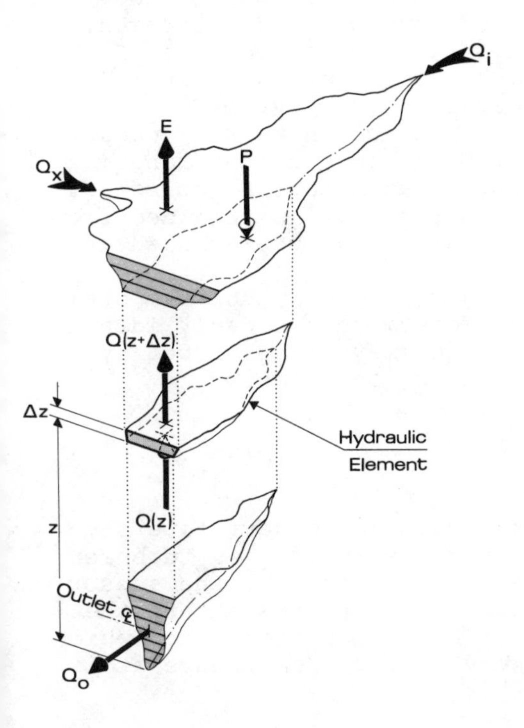


*Figure 4.3. Conceptual representation of a reservoir.*

## IV. FUNCTIONAL REPRESENTATION

### *Advection-Diffusion Equations*

Simulation of the behavior of pollutants in the environment requires a functional mathematical representation of the mechanisms of advective transport and mixing as well as the other physical, chemical, and biological processes that determine the rates of change in pollutant concentration. Once the conceptual form of the model is determined a set of equations describing these processes can be written for each element in the system.

Consider for the moment that one is dealing with the mass transport of a conservative substance, say some soluble constituent like chlorides, that may be closely identified with the fluid medium. In accord with the mass conservation principle such a substance may be neither created nor destroyed within the system under study, but it may be considered to be continually in motion. It may be redistributed in time and space in the environment by one or several phenomenological mechanisms such as molecular diffusion, turbulent diffusion, so-called eddy diffusion, and convective displacement, often termed simply "plug flow" or advection. Each of these mechanisms plays a role depending in relative importance on the scale of the system we are simulating and the detail with which we have chosen to "model" it. Further, by analogy to the random behavior of molecules in Brownian Motion (molecular diffusion) other processes that may be regarded from a practical viewpoint as random may be included in the simulation.

For a three-dimensional fluid body the temporal change of a conservative constituent closely identified with the fluid molecules may be described by the so-called advection-diffusion equation

$$\frac{\partial c}{\partial t} = - \nabla \cdot cU_i + D_m (\nabla \cdot \nabla c) + \nabla \cdot (\varepsilon_i \nabla c) \qquad (1)$$

where c is the concentration of a conservative substance, $D_m$ is molecular diffusivity, $\varepsilon_i$ is turbulent diffusivity and $U_i$ is a vector velocity; the subscript i denotes the vector direction with respect to the x, y and z axes. In most natural systems with which one must deal it is likely that $D_m \ll \varepsilon_i$; hence it is often neglected entirely.

*Effects of Scale*

The scales of the phenomena being simulated may determine the relative importance of the several terms in Equation 1 and whether additional terms are needed. If a scale is chosen for model representation, either in time or space, that is large compared to changes brought about by advection, or if turbulence is highly variable, then the analogy breaks down. On the other hand, if variations can be considered to have the same statistical properties as molecular motion, then it may be reasonable to represent the combined effects of advection and "effective diffusion" as

$$\frac{\partial c}{\partial t} = - \nabla \cdot cU_i + \nabla \cdot (E_i \nabla c) \tag{2}$$

where $E_i$ is the vector "effective diffusivity" including all mixing mechanisms, *i.e.*,

$$E_i = D_m + E_i + E_{ai} \tag{3}$$

in which $E_{ai}$ represents all other random processes analogous to molecular motion.

It is common practice to represent in $E_i$ all effects on redistribution of constituent c that cannot otherwise be accounted for by advective transport, sometimes including even numerical errors. It can be quite easily seen from inspection of Equations 2 and 3 that the relative importance of the advection or effective diffusion terms depends on how well one can describe $U_i$, *i.e.*, the velocity field within which the pollutant is being distributed. A good description of $U_i$ in a spatial and temporal sense minimizes the significance of the effective diffusion term and may justify its elimination altogether on the grounds that it may be reduced to the level of acceptable error in computation. On the other hand, a crude description of fluid flow, such as averaging across irregular cross sections or approximating transients by steady flows, may lead to a dominance of the diffusion term(s). How much reliance is placed on these terms is what most distinguishes our existing capabilities to simulate pollution of the environment, particularly with respect to the occurrence of peak concentrations and their location.

*Mass Conservation Equations*

If these principles are to be applied to an idealized reservoir such as that depicted in Figure 4.3, it is necessary to first reduce Equation 2 to one-dimensional form. Second, one may wish to provide for "sources" or "sinks" of the various quality constituents that may be carried in the system. With these modifications Equation 2 becomes

$$\frac{\partial c}{\partial t} = - \bar{u} \frac{\partial c}{\partial x} + \frac{\partial}{\partial x} [E_x \frac{\partial c}{\partial x}] \pm kc \qquad (4)$$

where kc represents the aggregate effects of sources and sinks.

In practice it has been convenient to transform Equation 4 into the mass conservation form for element i

$$\frac{dM_i}{dt} = - Q_i c - E_i A_i \frac{dc}{dx} \pm S_i \qquad (5)$$

where $M_i$ is mass, $Q_i$ is flow, c is concentration, $A_x$ is the cross section normal to flow, S represents sources and sinks and the other terms are as previously defined.

The term S can be made to represent a wide variety of processes whose reaction rates can be considered proportional to the concentration c. For the aquatic environment these include sedimentation, reaeration, decay (chemical or biochemical), chemical transformation, biological uptake, and respiration release.

Chen[4] has adopted this approach in formulating two general mass conservation equations, one for abiotic substances and one for biota.

For abiota:

$$\frac{d(\bar{V}c_1)}{dt} = \underset{\text{advection}}{\sum_{i=1}^{n} Q_i c_1} + \underset{\text{diffusion}}{\sum_{i=1}^{n} E_i A_i \frac{dc_1}{dx_i}} + \underset{\text{input}}{\Sigma Q_{in} c_{in}} - \underset{\text{output}}{\Sigma Q_{ou} c_1}$$

$$+ \underset{\text{sedimentation}}{S_1 \bar{V} c_1} \pm \underset{\text{reaeration}}{k_r A_s (c_1 - c_1^*)} - \underset{\text{decay}}{k_{d,1} \bar{V} c_1}$$

$$\pm k_{d,} \; \bar{V}c_2 \quad \pm \; \Sigma\mu_3 c_3 F_{3,1} \quad \pm \; \Sigma r_3 \bar{V} c_3 F_{3,1} \tag{6}$$

chemical transformation — biological uptake — respiration release

The terms that have not been defined previously are

- $n$ = number of adjacent elements
- $s$ = sedimentation rate
- $k_r$ = reaeration rate
- $A_s$ = surface area
- $c_1^*$ = saturation concentration of $c_1$
- $k_{d,1}$ = decay coefficient of $c_1$
- $k_{d,2}$ = decay coefficient of $c_2$
- $\mu_3$ = growth rate of biota $c_3$
- $F_{3,1}$ = conversion factor between $c_1$ and $c_3$
- $r_3$ = respiration rate of biota $c_3$.

For biota:

$$\frac{d(\bar{V}c_1)}{dt} = \sum_{i=1}^{n} Q_i c_1 + \sum_{i=1}^{n} E_i A_i \frac{dc_1}{dx_i} + \Sigma Q_{in} c_{in}$$

advection — diffusion — input

$$-\Sigma Q_{ou} c_1 + (\mu_1 - r_1 - s_1 - m_1)\,\bar{V}c_1$$

output — net growth

$$-\mu_2 \bar{V} c_2 F_{2,1} \tag{7}$$

predation

New terms are:

- $\mu_1$ = growth rate of $c_1$
- $r_1$ = respiration rate of $c_1$
- $m_1$ = mortality rate of $c_1$
- $\mu_2$ = growth rate of next higher trophic level $c_2$
- $F_{2,1}$ = conversion factor between $c_1$ and $c_2$.

Inspection of these equations reveals that they can be made to cover a wide range of processes. In the simplest form, with c = unity, they reduce to the storage equation

$$\frac{d(V)}{dt} = \sum_{i=1}^{n} Q_i + \Sigma Q_{in} - \Sigma Q_{ou} \tag{8}$$

In their most complex form they can be used to describe the complex interrelationships existing between several levels of biota in an ecosystem. Using this approach, Chen and Orlob[5] have written mass conservation equations to describe the fate of 23 quality parameters in lake and estuary systems, including ecological successions between three trophic levels (algae, zooplankton and fish).

*Motion and Continuity*

It has been stated above that a good characterization of the advective process minimizes the dependence on the empiricism of the diffusion analogy. This may be achieved in part by providing more spatial detail, such as was included in the stratified reservoir model. Or, when flows are highly variable, one may have to provide greater temporal detail, *i.e.*, take shorter time steps. Both these steps may have to be taken where the system is unusually complex as in the case of a shallow, vertically mixed tidal estuary.

Conceptually, the estuary may be idealized as a system of volume elements called "nodes" and flow elements called "links." An arrangement of nodes and links used to represent the San Francisco Bay-Delta System is illustrated in Figure 4.4. A small section is shown in Figure 4.5 for purposes of definition.

Each *link* in the system, corresponding either to a real channel or a portion of an embayment, is defined in terms of length, depth, width and roughness. Side slopes may be stipulated if the channel is trapezoidal and cross-sectional areas may be calculated from other geometric properties.

Each *node* in the system, to which several links may be connected, is identified by a depth, an elevation above a reference datum, a surface area, and certain coefficients that relate to water quality processes. Surface areas are polygons

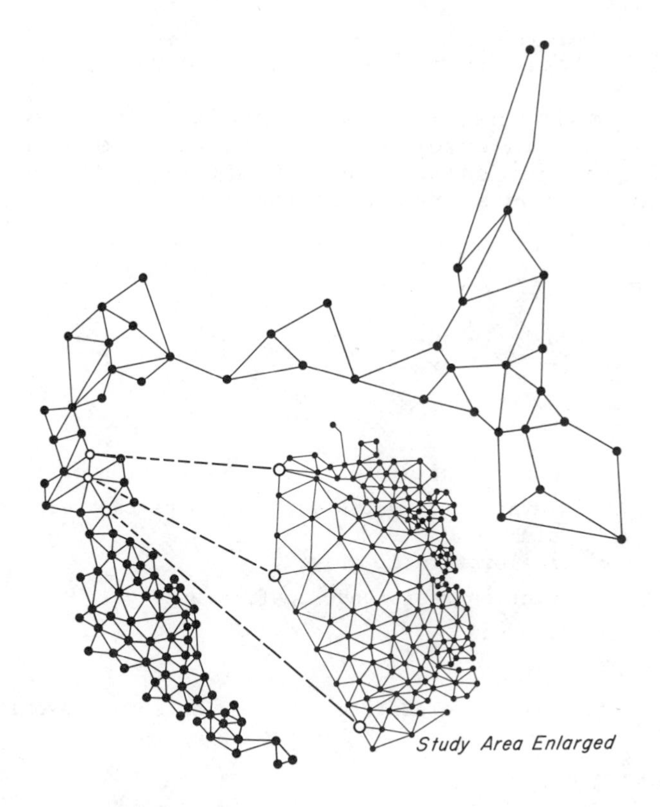


Figure 4.4. Bay-Delta estuarial mathematical model network.

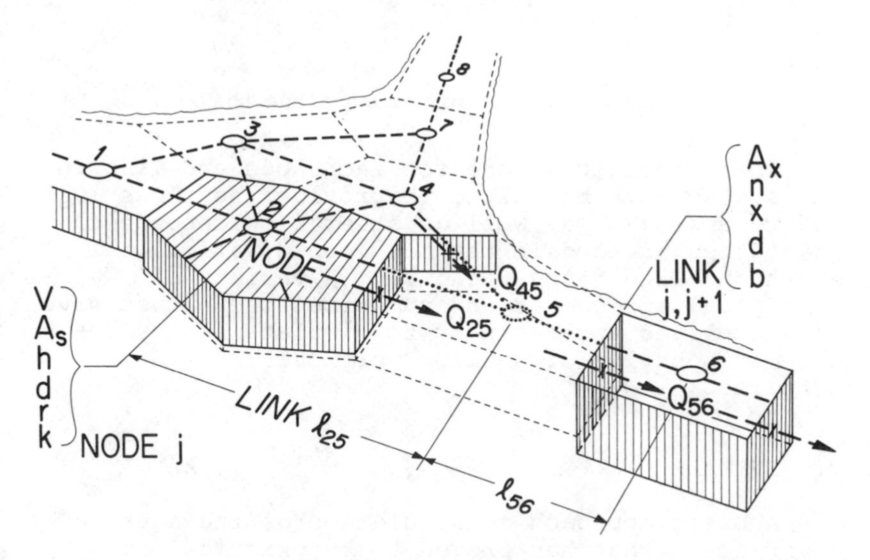


Figure 4.5. Definition sketch--WRE node-link estuarial model.

formed by the perpendicular bisectors of links in the network

To obtain the flows in channels (links) and the fluctuations in water surface (tidal) elevation at nodes, one must solve a set of motion and continuity equations. For a one-dimensional network these equations are:

Motion:

$$\frac{\partial \bar{u}}{\partial t} + \bar{u}\frac{\partial u}{\partial x} + g\frac{\partial h}{\partial x} + k|\bar{u}|\bar{u} = 0 \qquad (9)$$

where

$\bar{u}$ = mean velocity over the cross section of the link

$h$ = elevation of water surface above the reference datum

$g$ = acceleration of gravity

$k$ = friction resistance coefficient

$x$ = distance

$t$ = time.

(Note: Coriollis' acceleration and wind effects are omitted.)

Continuity:

$$\frac{\partial h}{\partial t} + \frac{\partial q}{\partial x} = 0 \qquad (10)$$

where

$q_x$ = tidal flow per unit width along the x-axis.

These equations, one for each node and link in the system, can be solved to provide the flows ($Q_S$) and volumes ($\bar{V}$), as well as other flow dependent quantities, needed in the mass conservation equations (Equations 6 and 7). A variety of conceptual approaches and numerical techniques have been employed successfully to simulate the hydrodynamic behavior of shallow vertically mixed estuaries.[6]

## V. SUMMARY

A basic approach to simulation of the aquatic environment that has proven to be practical in planning for pollution control has been outlined.

Following this approach useful models have been developed for many classes of problems and are now being applied in planning studies in the United States. Among those that may be cited are models for:

1. *Stream Water Quality*. An outstanding example is QUAL I, developed by F. D. Masch and Associates for the Texas Water Development Board.[7] This model is being adapted to about a dozen river basins in the United States under the auspices of the Environmental Protection Agency. Capability is being added to simulate some biological parameters.

2. *Reservoir Temperature*. The WRE Reservoir Temperature Model is in wide use, notably by the Corps of Engineers in design of reservoir outlet works. It has been applied to 8 reservoirs in the Pacific Northwest and to 18 in the Ohio River Division.

3. *Estuarial Hydrodynamics and Quality*. The Bay-Delta Models cited earlier have been applied extensively to many Pacific Coast estuaries in the United States, to Bay Biscayne in Florida, and to two major embayments in Australia.[8] The Delaware Estuary Model (DECS) has been applied successfully in studies of that estuary and several shallow bays along the eastern seaboard.[9] A unique set of two-dimensional models has been developed and applied by F. D. Masch and his associates to Gulf Coast estuaries.[10] Estuarial modeling is currently being advanced to include density effects and ecologic responses.

4. *Ground Water Quality*. WRE developed a model for simulation of ground water quality in gravity aquifers.[11] The model has been applied successfully to the Santa Ana River Basin in Southern California. As far as is known it is the only model yet developed that has been used in planning for pollution control in ground water basins.

5. *Aquatic Ecology*. The models developed by Chen and Orlob, cited above, are currently being applied to a number of reservoirs in the United States. Among these are two pumped storage projects and several proposed impoundments in developing areas. The Lake Ecologic Model has been demonstrated on Lake Washington in the Pacific Northwest. The Estuary Ecologic Model has been demonstrated on the San Francisco Bay Delta System and is currently being adapted to the simulation of Puget Sound.

Models of environmental systems have gained wide acceptance, are being usefully applied and, in the process of application, are being improved. Much improvement is still needed, however. In particular, the work on modeling ecosystems needs extension. There is a need to improve the capability of simulating three-dimensional systems, particularly stratified flows. More work is needed in the area of ground water modeling and serious attention needs to be given to the coastal environment. It appears that there is no lack of challenging problems in the field of environmental pollution control. Mathematical models for simulation of alternative strategies will no doubt be of great use in finding meaningful and acceptable solutions.

REFERENCES

1. Water Resources Engineers, Inc., "Simulation and Technical Management Concepts for the Texas Water Plan," report to the Texas Water Development Board (November, 1967).
2. Water Resources Engineers, Inc., "System Simulation for Management of a Total Water Resource," completion report to the Texas Water Development Board (August, 1969).
3. Orlob, G. T. and L. G. Selna. "Temperature Variations in Deep Reservoirs," *J. Hydr. Div., ASCE, 96,* HY2, 391 (1970).
4. Chen, C. W. "Concepts and Utilities of Ecologic Model," *J. San. Eng. Div., ASCE, 96,* SA5, 1085 (1970).
5. Chen, C. W. and G. T. Orlob. "Simulation of Aquatic Ecosystems," 1st Annual Report, OWRR Proj. C-2044, Office of Water Resources Research, U.S. Dept. of the Interior (August, 1971).
6. Orlob, G. T. "Mathematical Modeling of Estuarial Systems," Proc. Intnl. Symp. on Mathematical Modeling Techniques, Ottawa, Canada (May, 1972).
7. Masch, F. D. and Associates. "Simulation of Water Quality in Streams and Canals," Rep. No. 128, Texas Water Development Board (May, 1971).
8. Orlob, G. T. and R. P. Shubinski. "Water Quality Modeling of Estuaries," Proc. Intnl. Assoc. on Scientific Hydrology, Conf. on Hydrology of Deltas, Bucharest (May, 1969).
9. Thomann, R. V. "Mathematical Model for Dissolved Oxygen," *J. San. Eng. Div., ASCE, 89,* SA5, 1 (1963).
10. Masch, F. D., *et al.* "Tidal Hydrodynamic and Salinity Models for San Antonio and Matagordo Bay, Texas," report to Texas Water Development Board, Austin, Texas (June, 1971).

11. Water Resources Engineers, Inc., "An Investigation of Salt Balance in the Upper Santa Ana River Basin," Final Report to State Water Resources Control Board and Santa Ana River Basin Regional Water Quality Control Board (March, 1969).

# CHAPTER 5

# MATHEMATICAL MODEL FOR OXYGEN BALANCE IN RIVERS

G. Chevereau*

## I. INTRODUCTION

The mathematical model described in this study is based on the Streeter-Phelps equation which has been extended to allow also for diffusion. First a computer program was established which simulates unsteady flow in rivers. The parameters of the unsteady flow are then introduced into the pollution model in which the Streeter-Phelps equation is approximated in a finite difference form and integrated by the so-called "two steps explicit method."

The model was tested on the French river Vienne in a reach downstream of a large paper mill, in connection with a specialist team of chemists and biologists. A significant discrepancy was observed between the rates of biodegradation of organic matter measured in laboratory tests and those obtained by adjustment of the model. Auxiliary phenomena make the problem more complex so that rational determination of the rate of biodegradation of organic matter is hazardous.

Thus, at the present time, it seems best to evaluate "bulk rates" of biodegradation by adjustment of a model on observed situations in the river. This method necessitates numerous field measurements and requires great care in extrapolating the results of the model.

---

*Mr. G. Cheverau is an engineer with the Scientific Application Department of SOGREAH, Avenue Léon Blum, 38042 Grenoble, France.

In Section II of the paper a general review of modeling of pollution in a river is given. Section III deals with the mathematical aspect and problems of the model. In Section IV the actual application of the model to the river Vienne is discussed, and Section V contains the conclusions.

## II. THE PARAMETERS OF ORGANIC POLLUTION

### *Defining Pollution*

Choosing a criterion to characterize the degree of pollution of a river is not easy because several quite different kinds of pollution (organic matter, radioactivity, heat, poisonous chemicals, pathogenic germs) are found in a river. The effects also are quite different, for example, turbidity, death of the fauna, odors, or eutrophication. In fact, pollution of rivers by organic matter (most municipal and industrial effluents) is certainly the most significant in terms of quantity and by effect on the river.

In river water, organic matter is naturally eliminated by a biochemical degradation performed by bacteria. As long as dissolved oxygen is available, the biochemical breakdown can be considered equivalent to an oxidization reaction which will lower the level of dissolved oxygen and therefore deteriorate the ecological balance in the river. When there is no oxygen left in the water, the breakdown of organic matter becomes anaerobic. Fortunately atmospheric oxygen enters the water through the surface, and thus a river is capable of eliminating a definite amount of organic matter by itself; this ability of the river to regenerate is called selfpurification.

Since the ecology of the river depends largely on the quantity of dissolved oxygen in its water, this dissolved oxygen (DO) seems to be a convenient criterion for measuring the degree of pollution of a river as far as organic pollution is concerned. However, even the term organic pollution embodies a great number of different materials and the question of assessing this load of pollutants is raised. Considering that the effect of all kinds of organic matter will be the consumption of dissolved oxygen, it is usual to measure the load of organic pollution by the quantity of oxygen necessary to completely oxidize this load by bacteriological breakdown, *i.e.*, by its biological oxygen demand (BOD).

The purpose of this study is to develop a mathematical model that simulates the variations of these two parameters over time at each point of a river (or reach of a river).

### *Streeter-Phelps Equations*

Consider a particle of polluted water in a river with a concentration of organic pollution (BOD) denoted by L. This particle undergoes several processes which will change its concentration L. Among these, the main ones are advection, diffusion, biodegradation of organic matter, and other inputs of organic matter. Let C be the concentration of dissolved oxygen of the particle. The processes which will change C are mainly advection, diffusion, oxidization of organic matter, reaeration through the surface, and other inputs or outputs of organic matter.

A mathematical model describing the variations of L and C over time was first developed by Phelps and Streeter. Only the diffusion term has been added in the following equations:

$$\frac{\partial L}{\partial t} = D_L \frac{\partial^2 L}{\partial x^2} - U \frac{\partial L}{\partial x} - K_1 L + L_A \qquad (1)$$

$$\frac{\partial C}{\partial t} = D_L \frac{\partial^2 C}{\partial x^2} - U \frac{\partial C}{\partial x} - K_1 L + K_2 (C_s - C) - C_A \qquad (2)$$

In these equations

- $D_L$ is the longitudinal diffusion coefficient
- $U$ is the mean velocity
- $K_1$ is the biodegradation coefficient
- $K_2$ is the surface reaeration coefficient
- $L_A$ is the rate of input/output of BOD
- $C_A$ is the rate of input/output of DO
- $C_S$ is the saturation concentration of oxygen
- $t$ is time
- $x$ is the abscissa along the river.

Note that:

-- the above equations are one-dimensional, with the basic hypothesis being that C and L are uniform in the cross section

-- U is constant over time and the flow is assumed steady
-- when the river is depleted of oxygen, Equation 1 is no longer valid and must be replaced by Equation 3:

$$K_1 L = K\ (C_S - C) - C_A \qquad (3)$$

which states that the quantity of oxygen consumed is equal to the quantity of oxygen introduced during the same time.

The different terms of the above equations will now be examined in more detail.

## The Advection Term

This type of term is obvious, at least as long as the velocity is uniform as was assumed above.

## The Diffusion Term

Since the equations are one-dimensional, the diffusion term can only be longitudinal diffusion; however, in this case, longitudinal diffusion covers a complex reality because the nonuniformity of velocities in the section cannot be ignored.

In natural streams, molecular diffusion is very weak compared with turbulent diffusion and therefore only turbulent diffusion will be considered. The study of diffusion goes back to Prandtl's first semiempirical turbulence studies. The bases of Prandtl's turbulence theory are clearly shown in Bakhmeteff's book.[1]

It should be noted that most complete theoretical and experimental studies are devoted to the problem of turbulence in pressurized circular pipes and that few studies consider turbulence in open channel flow; however, the turbulent diffusion in a very wide rectangular channel assuming two-dimensional flow was investigated and reported by Elder.[2] Based on this study, the coefficient of diffusion D can be expressed in the following form:

$$D = Ahu^* = Ah\sqrt{ghS} \qquad (4)$$

in which:

A is an empirical coefficient

h is the depth

u* $\left(= \frac{\sqrt{\tau}}{\rho} = \sqrt{ghS}\right)$ is the wall velocity

g is the weight acceleration

S is the slope of the energy line.

In the case of a very wide open channel, Elder has proposed the following semiempirical values:

-- for the mean vertical diffusion coefficient over the depth: A = 0.067
-- for the mean transverse diffusion coefficient: A = 0.23
-- for the mean longitudinal diffusion coefficient: A = 10.

If the vertical and transverse diffusions depend essentially on the turbulent fluctuations of the vertical and transverse components of the velocity, the longitudinal diffusion depends very little on the fluctuations of the longitudinal component of the velocity since this coefficient comes mainly from the differential convection due to the nonuniformity of velocities in the cross section. Therefore it is easy to conceive that the value A = 10 given by Elder for a wide channel where the velocities are assumed to be two-dimensional will not be convenient for natural rivers. The value of A will then increase with the nonuniformity of the velocity in a cross section. Some authors have found values of up to 400 for A. A mean value of 20 in a river with a regular and almost rectangular cross section and a mean value of 60 in a river with some islands was found in this study.

One may also wonder whether Elder's transverse coefficient is not too weak for natural rivers. Indeed, in some cross sections of a river, transverse velocities can appear (in bends for instance) and modify the transverse diffusion. It does seem, though, that for nearly straight and regular parts of a river Elder's value of 0.23 for A is fairly accurate.

Streeter and Phelps, and many authors after them, neglected the diffusion term in the pollution equations. One might ask, then, why there is a need to deal with it here. In most river pollution studies, the discharges of pollutants are nearly continuous or vary slowly with time. In this case, the effect of diffusion is perfectly negligible.

However, in the case of an accidental, instantaneous, concentrated point injection of pollutants, diffusion is the principal cause of the dilution of the pollutants. Therefore the computer program developed allows inclusion or exclusion of the diffusion term. Second, knowledge of how transverse diffusion works in a river is necessary for choosing the points where field concentration measurements must be taken. The minimum distance an effluent travels downstream before the concentration in the cross section can be considered uniform is often several miles.

## Surface Reaeration

In the Streeter-Phelps equation, this term is written as:

$$K_2 \ (C_S - C)$$

where $K_2$ is a coefficient which depends on local hydraulic conditions.

The molecular diffusion of dissolved oxygen in water is very small ($2.10^{-5} cm^2/s$) but it is generally acknowledged that, except close to the surface and to the river bed, turbulent diffusion is sufficient to mix the dissolved oxygen in a uniform manner over a vertical section. Field measurements support this statement. Thus the reaeration phenomenon is mainly dependent on the flow structure of the boundary layer near the immediate free surface. The difficulty is that little is known about the distribution of turbulence in this layer.

Numerous formulas have been proposed by different authors for $K_2$, most of them being empirical and based on laboratory or field experiments. O'Connor and Dobbins[3] developed a formula based on a theoretical analysis which yields good results when compared to experiments. However, most formulas have the following similar form:

$$K_2 = C \frac{U^m}{H^n} \tag{5}$$

in which m would be approximately 1 and n approximately 1.5. Most authors consider C as a constant coefficient a coefficient depending on the diffusion coefficient, or on Froude's number. With m = 1, n = 1.5 and C being a constant coefficient, a "mean formula" is obtained which is satisfactory in practical cases.

It is generally admitted that the coefficient $K_2$ is dependent on water temperature according to the following equation

$$K_2\ (T) = K_2\ (20°C)\ \theta^{(T-20)} \tag{6}$$

in which θ is approximately 1.025. Some research has been done to determine the most accurate value for θ or to determine whether θ was dependent or not on turbulence or on the temperature, but as will be shown in the next paragraph, pollution models do not need such accuracy.

## Biochemical Breakdown

Streeter and Phelps' Assumption. In the Streeter-Phelps equation, the biochemical breakdown term is written:

$$\frac{\partial L}{\partial t} = -K_1 L \tag{7}$$

The solution of this differential equation is:

$$L = L_o\ e^{-K_1 t} \tag{8}$$

This assumes that the pollution load, measured by its total BOD (BOD∞) is degraded according to an exponential law (Figure 5.1) at a constant rate, $K_1$, called biodegradation coefficient.

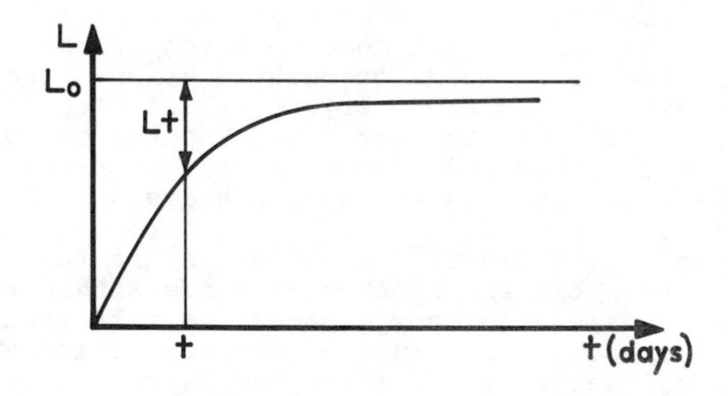


*Figure 5.1. Exponential law of biodegradation; Streeter and Phelps assumption.*

Equation 8 is equivalent to:

$$L = L_o \ 10^{-k_1 t} \tag{9}$$

provided that $K_1 = 2,3 \ k_1$. Streeter and Phelps estimated, based on laboratory experiments, that $k_1$ was approximately equal to 0.1 $day^{-1}$. Under these conditions, there is a relation between the initial pollution load $L_o$ measured by BOD∞ and the load $L_t$ after a time t of incubation. For instance, if $t = 5$ days, then:

$$BOD_\infty = 1.45 \ BOD_5 \tag{10}$$

Therefore, the determination of BOD∞, which theoretically needs an infinite time (which is impossible), can be obtained by determining the $BOD_5$ by way of laboratory $BOD_5$ tests and applying Equation 10. Unfortunately this easy method is contested and subject to criticism.

Laboratory BOD Tests. As more and more laboratory BOD tests were carried out with increasing accuracy and duration, it became apparent that the data could not be described by Equation 9 alone. Figure 5.2 illustrates two examples of discrepancies between real curves and the Streeter-Phelps exponential curve. In example 1, the curve shows at time $t_1$ a sudden acceleration of oxygen consumption, and sometimes the same phenomenon occurs again at time $t_2$. An explanation has been found for this. The first part of the curve (from 0 to $t_1$) results from the oxidation of carbonaceous matter (organic matter other than $NH_3$) by the normal bacterial fauna in the water. This part of the decomposition process is known as the first stage of deoxygenation. The latter part of the curve (beyond $t_1$) is associated with the growth of the nitrifiers and the oxidation of $NH_3$ to nitrites and, subsequently, to nitrates; this stage is known as the second stage or nitrification stage.

Example 2 of Figure 5.2 shows that if the results of a BOD test fit an exponential curve within the first five days, the deoxygenation is more rapid than would be expected from the latter part of the exponential curve that fits the first few days. It is also shown that the value of BOD∞ given by Equation 10 is too small and is, in fact, reached very early. Moreover, it should be pointed out that $k_1$ is no longer

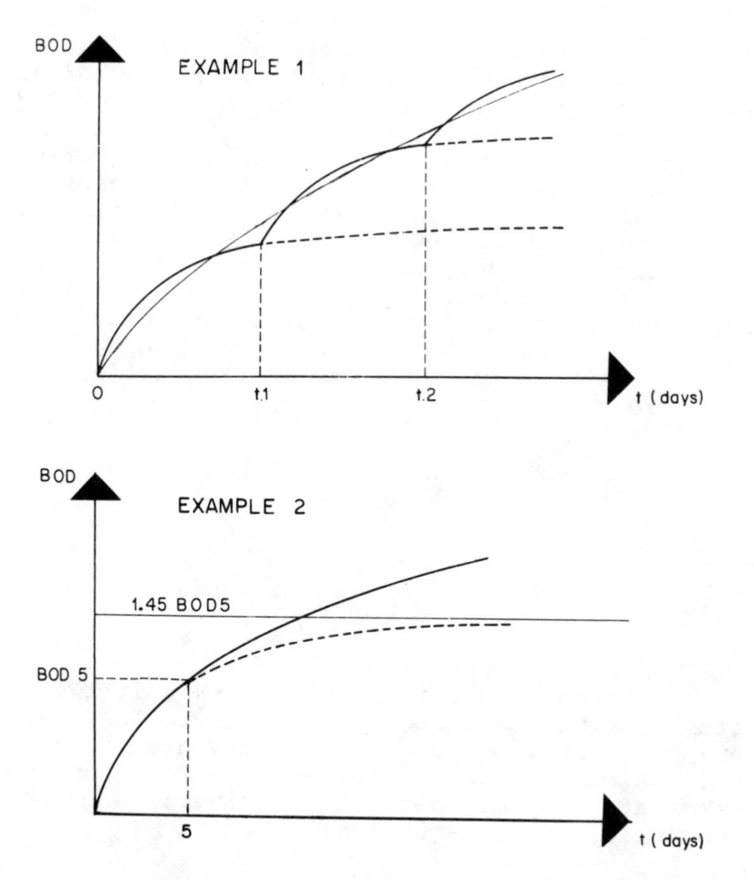


*Figure 5.2. Examples of observed laboratory BOD test curves.*

regarded as a universal constant; Camp,[4] Gannon,[5] and many others have found $k_1$ coefficients ranging from 0.01 to 1.0 $day^{-1}$.

Some authors have proposed a description of the curves shown in example 1 by a sum of exponential functions with a lag time:

$$L_t = L_o \left[ p' e^{-k_1' t} + p'' e^{-k_1''(t-\alpha_1)} + p''' e^{-k_1'''(t-\alpha_2)} \right] \quad (11)$$

However, such a formulation requires the knowledge of seven more coefficients which all depend on the kinetics of the bacterial process.

In a batch, the microbial population can change with time since the nutrients themselves, being oxidized, change. There is no evidence that in a natural river where conditions are steady or almost steady the oxidation of carbonaceous matter and nitrification cannot proceed simultaneously.

No great error is incurred if a mean curve is considered when compared to the other uncertainties of the model. It seems more useful to simulate more accurately the mean BOD test curve and this not only up to the fifth day, as the Phelps and Streeter assumption would suggest, but also after this fifth day because the consumption of oxygen continues to be far from negligible. Therefore, biodegradation will be represented by a sum of exponential functions as follows:

$$L_t = L_o \sum_i (p_i \, e^{-k_{1,i} t}) \qquad (12)$$

with

$$\sum_i p_i = 1$$

It was found that this formula with only two terms fits observed laboratory curves up to 20 days. For example, the laboratory curve presented in Figure 5.3, which is a mean obtained from four samples taken at the same point of the river, is close to the curve with an equation:

$$L = L_o \; (0.3 \times 10^{-0.06t} + 0.7 \times 10^{-0.006t}) \qquad (13)$$

This equation states that pollution behaves as if it were composed of two parts, 30:70, with biodegradation coefficients equal to 0.06 and 0.006 days$^{-1}$, respectively.

Three curves (each obtained from four samples) were determined at three different points downstream of a paper mill effluent along a river as shown in Figure 5.4. From upstream to downstream, the curves become progressively lower and the curvature less pronounced as shown in Figure 5.5, a fact consistent with and confirming Equation 12.

Equation 12 also has a horizontal asymptote, but this is not well-defined by the beginning of the curve and it is preferable to assess $L_O$ or the BOD$\infty$ by other means. The value of BOD$\infty$ has only a slight influence on the beginning of the curve, and an accurate definition of BOD$\infty$ is not required.

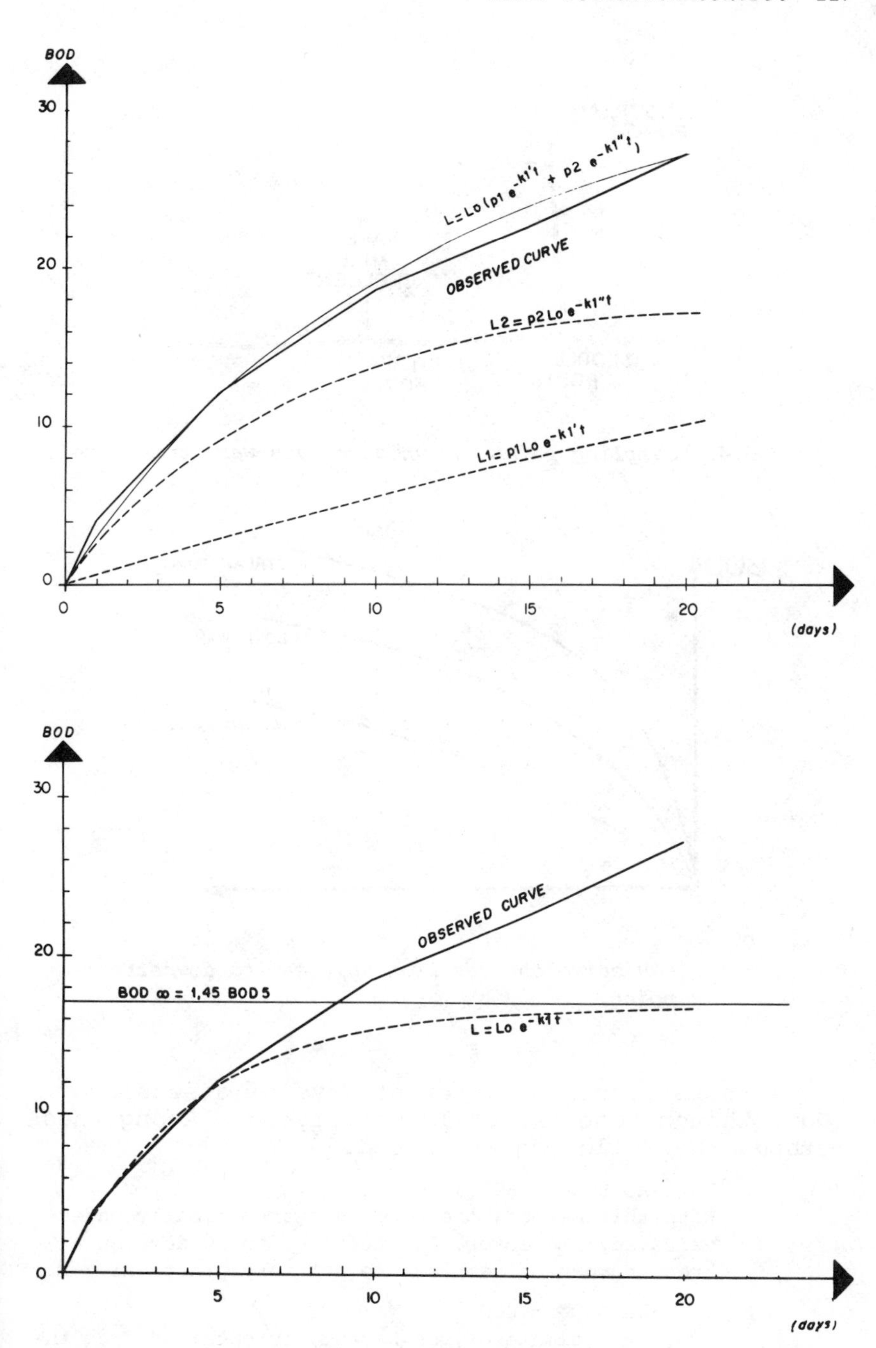


*Figure 5.3. Laboratory BOD test curves.*

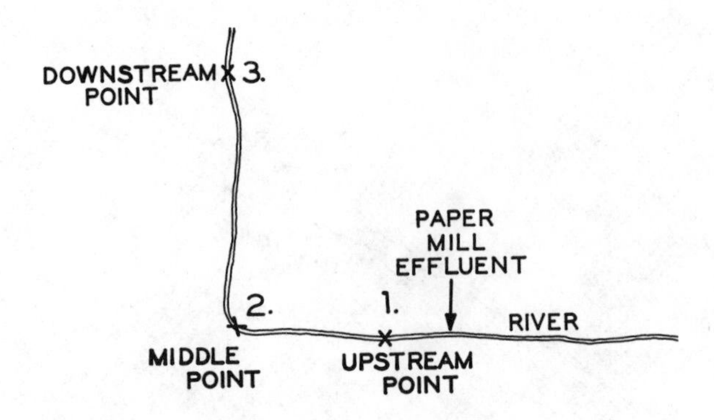


*Figure 5.4. Sampling points at which curves were determined.*

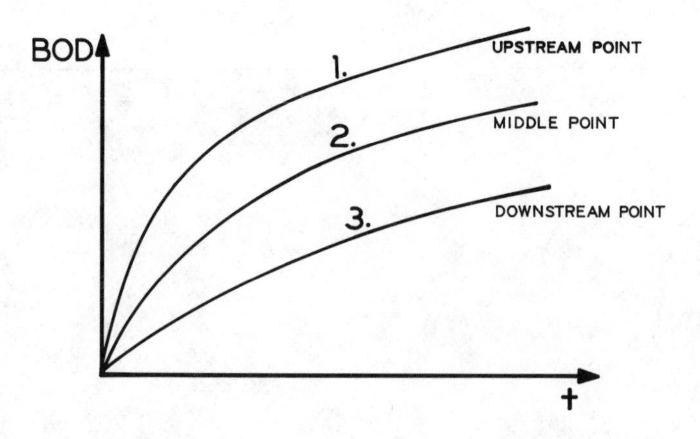


*Figure 5.5. BOD curve changes from upstream to downstream point.*

Authors advocate different ways of assessing $BOD_\infty$, though none is really satisfying. Among these methods the following are cited:

1. To take $BOD_\infty = BOD_{20}$
   With this method, the $BOD_\infty$ is systematically undervaluated. Moreover, BOD tests up to 20 days on a great number of samples are not very practical.

2. To take $BOD_\infty = UOD$
   UOD, or Ultimate Oxygen Demand, is obtained from the chemical formula of pollutants by assuming that all the matter is completely oxidized. This method assumes knowledge of the chemical composition of the

pollutants, and with its use BOD∞ would be over-evaluated. It has been developed by the Water Pollution Research Laboratory of Stevenage in England.[6]

3. To take $BOD_\infty$ = COD
   COD, or Chemical Oxygen Demand, is obtained by oxidization of organic matter with a chemical oxidizer. The advantage of this method is its rapidity allowing it to be applied to a large number of samples. There is no evidence, however, that all organic matter oxidized biologically would be oxidized by chemical reagents and vice versa. Camp[7] stated that for most organic compounds oxidation is 95-100% complete with the dichromate as oxidizer.

This last method seems to be the best approach to the assessment of BOD∞ and was used in these studies.

Field Evaluation of Biodegradation Coefficient. All the researchers who worked on natural streams agree that "apparent" biodegradation coefficients evaluated in rivers can be much greater (3 to 10 times) than those evaluated by laboratory tests, but no satisfactory explanation has been given for this discrepancy. Secondary phenomena have been invoked, the influence of which there is no doubt, but the way these secondary phenomena proceed is not known. One obvious difference between biodegradation in a laboratory batch and biodegradation in a natural stream is agitation. For oxidation to proceed, three elements must be present: oxygen, nutrients, and bacteria. The agitation in a natural river favors the meeting of these three elements. Conversely in a batch the absence of agitation favors instead sedimentation of particles which would not have settled in a river.

Agitated laboratory BOD tests carried out at the Water Pollution Research Laboratory, Stevenage, have shown that agitation could partly explain the observed discrepancy. There are additional secondary phenomena to be considered also.

## Secondary Phenomena

As they are little known, secondary phenomena do not appear as separate terms in Equations 1 and 2. They are considered as inputs or outputs of oxygen or pollution and are embodied in the last terms $L_A$ and $C_A$, respectively. In fact these terms cover

1. the increase of BOD by sources along the river
2. the decrease of BOD due to sedimentation (sometimes due to flocculation). Dobbins advocated that the loss of BOD by sedimentation be expressed by the following law:

$$\frac{\partial L}{\partial t} = -K_3 L \tag{14}$$

   It would seem that this equation neglects the influence of velocity and is inadequate to describe the sedimentation.
3. the increase of BOD by scouring
4. the decrease of BOD by biological extraction
5. the consumption of BOD and of oxygen by mud deposits
6. production and consumption of oxygen by algae (photosynthesis and respiration)
7. point reaeration at weirs.

Although called secondary, these phenomena can have a great influence on the process of self-purification of rivers. Except for item 1, for which a mathematical formulation is evident, and for items 6 and 7 for which acceptable ones are available, the mathematical formulation of secondary phenomena is hazardous. It is suggested that they be embodied in the biodegradation term which would become a self-purification term. Thus the assumption is made that the variations of BOD due to the purification phenomena are proportional to present BOD.

## III. MATHEMATICAL ASPECT OF POLLUTION MODELING

### *Basic Equations*

Based on Equations 1 and 2, a computer program called PROSPER has been developed for steady or unsteady flow. It was necessary to allow for the variations of hydraulic conditions to adjust the model correctly. Indeed, retention times depend heavily on the rate of flow, and the reaeration formula, for instance, is very sensitive to variations of discharge and above all the presence of weirs in the river. Since SOGREAH already has a program, called SIMOUN, capable of computing unsteady flows in a river, it was decided to take advantage of the possibility of coupling the two programs. Therefore the complete procedure is as follows.

1. The program SIMOUN is used to simulate the flow in the river during the period of field measurements. This provides a magnetic tape on which the hydraulic conditions at each point of the model for the duration of the simulation are stored.
2. The pollution equations are processed with the exact hydraulic conditions.

This procedure in adjusting the pollution model circumvents the uncertainties due to approximate hydraulic conditions. For unsteady flow, Equation 1 is written

$$\frac{\partial(SL)}{\partial t} = \frac{\partial(SD_L \frac{\partial L}{\partial x})}{\partial x} - \frac{\partial(USL)}{\partial x} - SK_1L + SL_A \qquad (15)$$

Developing derivatives yields

$$L\frac{\partial S}{\partial t} + S\frac{\partial L}{\partial t} = \frac{\partial(SD_L \frac{\partial L}{\partial x})}{\partial x} - US\frac{\partial L}{\partial x} - L\frac{\partial(US)}{\partial x} - SK_1L + SL_A \qquad (16)$$

If U and S satisfy the mass equation

$$\frac{\partial S}{\partial t} = -\frac{\partial(US)}{\partial x} \qquad (17)$$

as it is even for unsteady flow in the procedure proposed above, then Equation 16 becomes

$$S\frac{\partial L}{\partial t} = \frac{\partial(SD_L \frac{\partial L}{\partial x})}{\partial x} - US\frac{\partial L}{\partial x} - SK_1L + SL_A \qquad (18)$$

Equation 2, describing the oxygen concentration, has an identical structure, and the same reasoning applies to it. Therefore only Equation 1 is considered here. It should be pointed out that Equation 1 is independent of the oxygen concentration C and that the two equations can be solved separately. Analytical solutions of the equation can be found in some simple cases, but for practical cases an approximate solution can be found by numerical analysis methods, methods which can be used thanks to the help of computers.

Expressed in terms of finite differences Equation 18 becomes

$$S\frac{\Delta L}{\Delta t} = \frac{\Delta(SD_L \frac{\Delta L}{\Delta x})}{\Delta x} - US\frac{\Delta L}{\Delta x} - SK_1L + SL_A \qquad (19)$$

in which $\Delta x$ and $\Delta t$ have finite values and are the space increment and the time increment, respectively. This equation was integrated by the so-called "two steps explicit method" advocated by Dobbins.[8]

*Integration of the Equations*

Only the main features of the method are highlighted:

- -- The method is called the "two steps explicit method" because the convection and the diffusion behave as if they were not simultaneous but successive phenomena, so that for each increment of time the computation is performed in two steps: convection (+ processing terms other than diffusion) and diffusion.
- -- The pseudodiffusion problem.

The only difficulty of this method is the appearance of a parasitic diffusion called pseudodiffusion which is due to the computation of convection only. Let us consider that at time n the concentration C is known and equal to 0 at each point of the river except at point I where $C = C_o$ as represented in the (x,t) plane on Figure 5.6 and on Figure 5.7 by curve 1. If $\Delta x = U\Delta t$, a particle of water will be

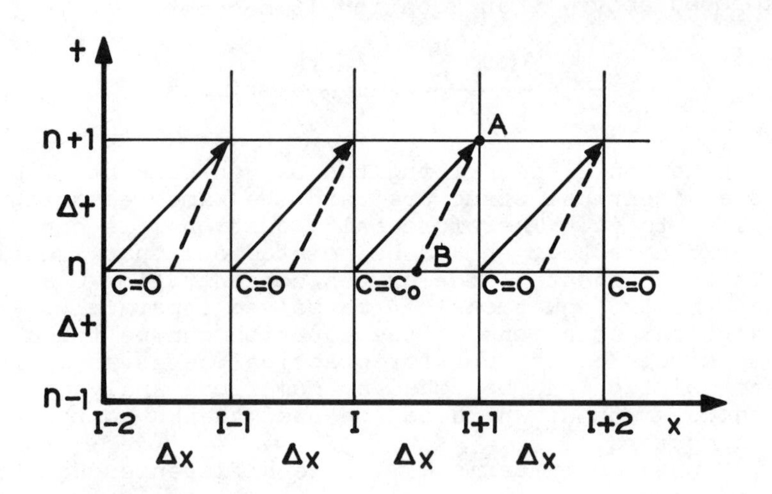


*Figure 5.6. Scheme used for numerical computation of convection.*

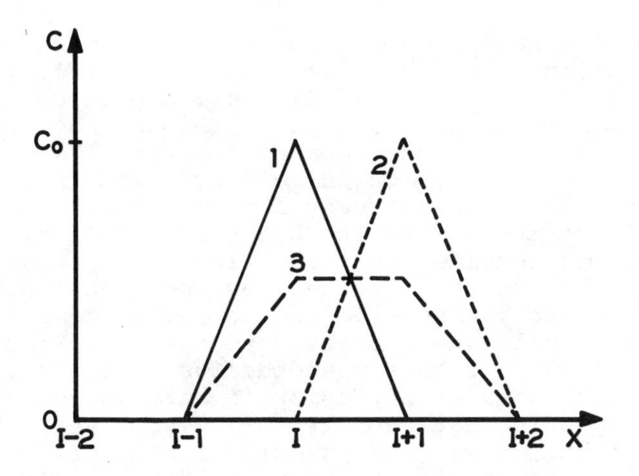


*Figure 5.7. The pseudodiffusion phenomenon.*

shifted through a distance $\Delta x$ during time $\Delta t$. As it undergoes a pure convection, its concentration will not change, giving

$$C_{I+2}^{n+1} = C_{I+1}^{n} = 0$$

$$C_{I+1}^{n+1} = C_{I}^{n} = C_o$$

$$C_{I}^{n+1} = C_{I-1}^{n} = 0$$

$$C_{I-1}^{n+1} = C_{I-2}^{n} = 0$$

and at time n+1 the distribution of concentration along the river will be represented by curve 2 on Figure 5.7. It can be seen that the particles have effectively undergone a pure convection.

But what about if $\Delta x \neq U\Delta t$?

To illustrate this case, assume $\Delta x = 2U\Delta t$. The trajectories of particles in the (x,t) plane are plotted in dotted lines; for example, the particle arriving at point A was at point B at time n. It is simplest in this case to admit that the concentration of this particle can be interpolated between the two immediate lower and upper points:

$$C_A = C_B = \frac{C_I^n + C_{I+1}^n}{2} = \frac{C_o}{2}$$

If this is done for each point, the distribution of concentration along the river is represented by curve 3 in Figure 5.7. If $C_A = C_B$, a pure convection should have occurred, but it is obvious that curve 3 has also diffused when compared to curve 1.

This pseudodiffusion is met each time $\Delta x \neq U\Delta t$. When considering only steady flow one can manage to choose $\Delta x$ so that $\Delta x = U\Delta t$, but since it is also desirable to consider unsteady flow it is impossible to avoid the pseudodiffusion. By choosing adequate $\Delta x$ and $\Delta t$, one can only lower its influence so that it becomes small when compared to natural diffusion. It should be noted that pseudodiffusion can be much greater than natural diffusion if care is not taken.

By expanding Equation 18 in a Taylor series up to the second order, an approximation $D_p$ of the coefficient of pseudodiffusion can be obtained. The value of $D_p$ depends on the relative position of point B between points I and I + 1 (see Figure 5.6). If B is in the middle, $D_p$ is maximal and equal to

$$D_{p\ max} = \frac{\Delta x^2}{8\Delta t} \tag{20}$$

By choosing $\Delta x$ small and $\Delta t$ great $D_{p\ max}$ can be minimized, but unfortunately this is limited by the condition of computation stability of the diffusion term which is expressed

$$\frac{\Delta x^2}{\Delta t} \geq 2\ D \tag{21}$$

So it appears that pseudodiffusion cannot be indefinitely minimized and that $D_{p\ max}$ cannot be inferior to D/4.

As long as diffusion is not essential, this limitation can be accepted. If not, the two-steps-explicit method must be replaced by a more convenient one. Such methods have been investigated and are available. They are more sophisticated and the computation times would be increased.

### *Analytical Solutions - Sag Curves*

Although analytical solutions can rarely be applied to practical cases, study of them can help in understanding the mechanism of self-purification in rivers and the influence of the main parameters.

Consider the case in which the variations of concentration in BOD and oxygen are due only to biodegradation and reaeration through the surface. Assume that the flow is steady and that diffusion is ignored. If $K_1$ and $K_2$ are constants, the concentration in oxygen C of a particle moving at velocity U is given by

$$\frac{dC}{dt} = -K_1 L + K_2 (C_s - C) \tag{22}$$

Replacing the concentration C by the oxygen deficit D (= $C_S$ - C), Equation 22 becomes

$$\frac{dD}{dt} = K_1 L - K_2 D \tag{23}$$

An integral of Equation 23 is

$$D = \frac{K_1 L_o}{K_2 - K_1} (e^{-K_1 t} - e^{-K_2 t}) + D_o e^{-K_2 t} \tag{24}$$

$L_o$ and $D_o$ being the values of L and D at time t = 0. Using $k_1$ and $k_2$ instead of $K_1$ and $K_2$:

$$D = \frac{k_1 L_o}{k_2 - k_1} (10^{-k_1 t} - 10^{-k_2 t}) + D_o 10^{-k_2 t} \tag{25}$$

This equation is that of a sag curve as represented in Figure 5.8.

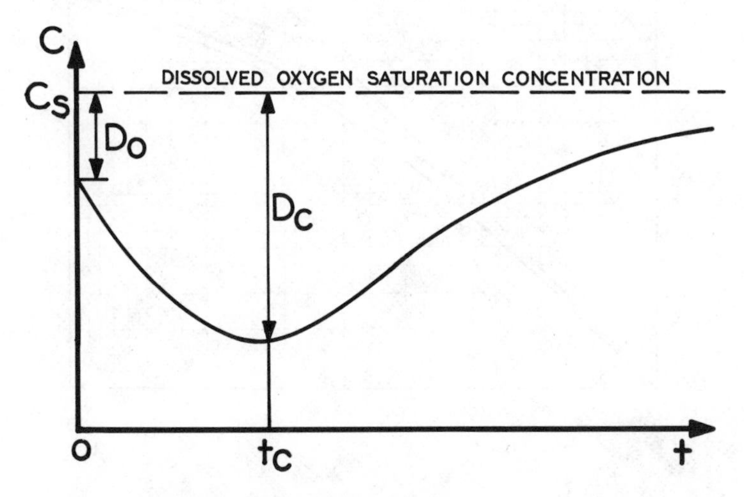


*Figure 5.8. Dissolved oxygen sag curve.*

Worst conditions in a river are observed at a point where the dissolved oxygen concentration is a minimum. Therefore this point of the curve, when D is equal to a maximum $D_c$, called the critical deficit, appears at the critical time $t_c$. Writing for this point that dD/dt = 0, the values of $t_c$ of $D_c$ can be calculated:

$$t_c = \frac{1}{k_2 - k_1} \log\left\{\frac{k_2}{k_1}\left[1 - \frac{D_o(k_2 - k_1)}{L_o k_1}\right]\right\} \tag{26}$$

$$D_c = \frac{k_1}{k_2} L_o \cdot 10^{-k_1 t_c} \tag{27}$$

From these equations, diagrams (Figures 5.9 and 5.10) can be plotted which easily give approximate values of $D_c$ and $t_c$ against $D_o$, $L_o$, $k_1$ and $k_2$.

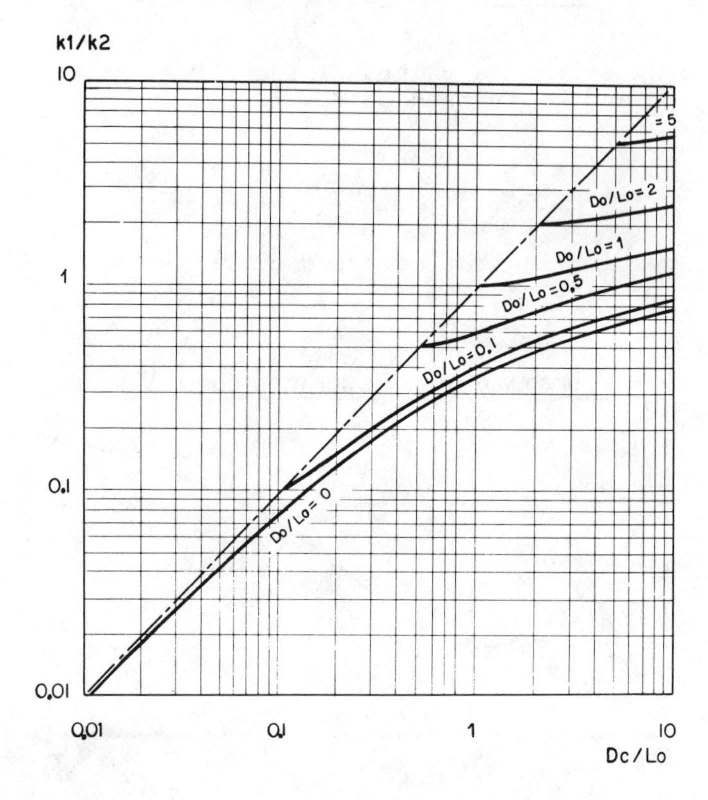


*Figure 5.9. Graph showing the relation between $k_1/k_2$ and $D_c/L_o$.*

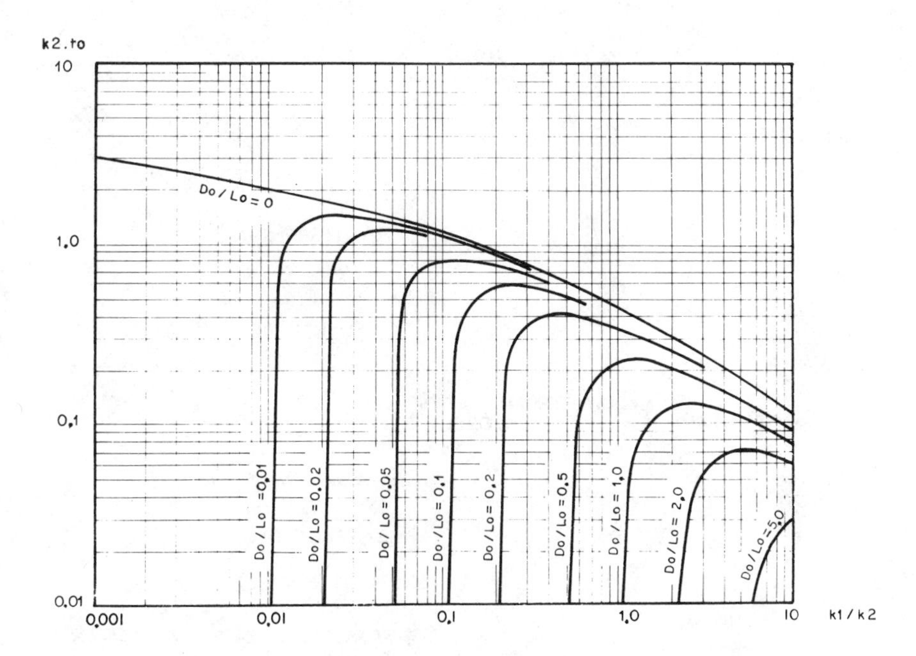


*Figure 5.10. Graph showing $k_2 t_c$ as a function of $k_1/k_2$.*

## IV. APPLICATION OF THE MODEL TO A NATURAL STREAM

### *Description of Field Measurements*

During the same time that the model was theoretically developed, three field measurement surveys were carried out on a reach of about 27 km along the river Vienne, 3 km downstream of a large paper mill. Ten weirs divide this reach in a succession of pools and riffles (Figure 5.11).

Among the three surveys, the second was chosen for model simulation because of its particularly good measurement conditions. Indeed during this second eight-day survey, the discharge was very low (an average of about 13 $m^3/s$), so that oxygen sag curves with a minimum of dissolved oxygen concentration close to 0 could be observed and significant BOD concentration could be measured.

During the survey more than 2000 chemical and biochemical measurements were carried out, dissolved oxygen and temperature at three points were measured continuously, discharge was recorded by level

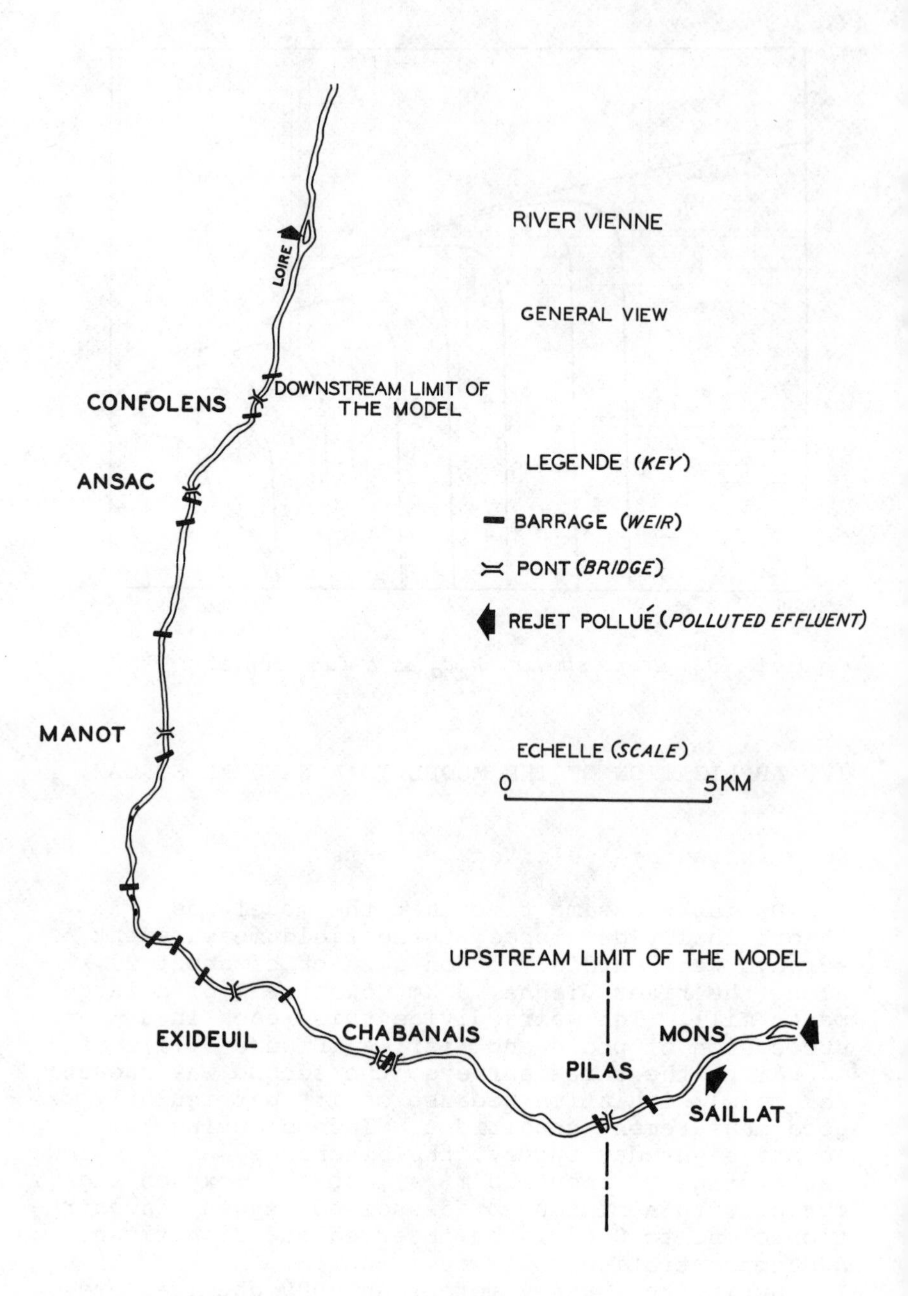


Figure 5.11. Model of the river Vienne.

recorders, and topographical data of the reach were collected to build a hydraulic mathematical model of the reach. In addition, rhodamine tests were performed to evaluate the longitudinal diffusion coefficients.

Every two hours, mean samples were taken by an automatic sampling apparatus, night and day, for seven days at several points of the model. The samples were automatically refrigerated and processed in a 20°C conditioned laboratory of the paper mill as soon as possible. The following parameters were measured for each sample: $BOD_1$, $BOD_2$, $BOD_5$, COD, the suspended solids, the pH, the dissolved oxygen and the temperature.

### *Adjustment of the Model*

A hydraulic model was first established from the data collected during the survey. The river reach was defined by 26 cross sections and was divided into 61 computation points. Then the model was adjusted so that the computed levels and retention times fit those observed during the survey. When the adjustment was satisfactory, the hydraulic regime of the river during the survey was simulated and the water surface level, the section, the mean velocity and the mean height for each computation point at every 15 minutes were stored on a magnetic tape.

The pollution model was adjusted within each of its terms, a discussion of which follows.

#### The Diffusion Term

The river entering the model at Pilas was polluted partly by well-mixed effluents from a big town and from small industries dispersed along the river far enough upstream of the studied reach and partly by the effluents from a big paper mill situated only a few kilometers upstream of Pilas leading to a non-uniform distribution of pollution in the Pilas section. Lateral diffusion of the paper mill effluent and longitudinal diffusion were studied by injecting a tracer (rhodamine) into the paper mill effluent. Then the passage of the plume was recorded at six points along the width at Pilas bridge and at three points along the width at Chabanais bridge.

It appeared effectively that the concentration was not uniform in the Pilas section and some care should be taken in order to obtain a mean value in this section. On the contrary, at Chabanais the concentrations appeared to be fairly uniform in the section.

A mean longitudinal diffusion coefficient between Pilas and Chabanais was calculated from the rhodamine measurements and a value of 2.5 $m^2/s$ for $D_L$ was adopted. By replacement in expression 4 a value of approximately 20 is given for A. It also appeared by comparison of two computation runs, the first one allowing for the above value of $D_L$ and the second one neglecting diffusion, that diffusion could be ignored in simulating the evolution of pollution during the survey.

## The Reaeration Term

Formula 5 was used with the coefficient that the authors calculated from field measurements which were carried out by Churchill, Elmore, and Buckingham[9] in several natural rivers. Some trial and error computation tests showed that $K_2$ could not diverge by more than 50% of this value without giving obvious erroneous results.

## The Biodegradation Term

The formulation adopted was the sum of two exponential functions (13). The values of $k_1'$ and $k_1''$ were taken as 0.4 and 0.015 respectively instead of 0.06 and 0.006 as was suggested by the laboratory tests. These values have been adjusted by trial and error computation tests to fit the numerous observed measurements.

The question arises that in this active disappearance of the BOD which is the part due to biodegradation and which is the part due to other phenomena? Note that if BOD is really biodegraded an equivalent quantity of dissolved oxygen disappears, but if BOD is simply settled or adsorbed no oxygen is needed.

Although it is quite certain that some small quantities of BOD settled in the upper pools of the reach, it was assumed in the computation that the total consumption of BOD required an equal quantity of oxygen. The balance of dissolved oxygen in the upper part of the reach shows that even in allowing for possible errors on $K_2$ and a possible action of algae, the real biodegradation coefficient must be greater than the laboratory one.

### The Secondary Phenomena

In the case of river Vienne, only sedimentation was supposed to have played a nonnegligible role in the self-purification process. An evaluation of the action of algae was attempted by the dark and light bottle method and appreciable photosynthesis was found in laboratory batches. However, in view of the turbidity of the water, this action is possible only in a superficial layer of some inches and the input of oxygen by photosynthesis in the natural river was in fact very low. Usually characteristic cyclic variations in measured dissolved oxygen betray an active photosynthesis.

Numerous weirs participated in the reaeration process and to allow for them the Gameson formula[10] was used:

$$\frac{Db}{Da} = 1 + 0.11K\ (1 + 0.046T)\ h \tag{28}$$

in which

- Da and Db are the oxygen deficits above and below the weir
- K is a coefficient depending on the shape of the weir and on the quality of the water
- T is the water temperature in °C
- h is the height of the fall.

The coefficient K was evaluated by field measurements.

### *Prospective Use of the Model*

After having been adjusted satisfactorily, the model can be used for prospective computation. Two examples will be discussed.

### Influence of Pollution Abatement

In this case it was assumed that the constant discharge in the river was a flow of 10 $m^3/s$. The hydraulic model gave the hydraulic parameters corresponding to this situation along the river. Then a series of sag curves for an organic pollution of 80, 60, 40, 30 and 20 ppm COD entering the reach was computed. Such abatements of pollution can be obtained by treatment of the pollution upstream of the reach. It was assumed that the initial value of the dissolved oxygen concentration was in every case 4 ppm. Resulting computed sag curves are shown in Figure 5.12.

### Influence of Weirs

One might think that because oxygen enters the water at weirs, they can improve the oxygen content. However this supposition ignores the fact that upstream of weirs retention times are increased and reaeration through the surface is decreased.

A computation was run to learn what the sag curve would have been if, during the field measurement survey, the Chabanais weir had not existed. Figure 5.13 shows what would have been the surface of the river in this case: upstream of Chabanais the mean height which appears in the denominator of the reaeration formula is much less and the velocity which appears in the numerator is much greater.

Figure 5.14 shows a considerable change in the sag curve. The minimum dissolved oxygen content is increased from 0 to 3 ppm and is shifted downstream. The oxygen demand, however, instead of being satisfied in Chabanais pool is also shifted downstream, bringing about a slight decrease in oxygen content in the downstream part of the river.

## V. CONCLUSION

The validity of the prospective use of the model depends heavily on the validity of the equations which have been used. Thanks to the knowledge of accurate hydraulic parameters, advection, diffusion, and reaeration are fairly well-known by a theoretical approach when compared to biodegradation and other phenomena.

At the present time, we think that a detailed comprehensive field measurement survey is necessary to appreciate the influence of auxiliary phenomena and to determine empirically the "bulk biodegradation coefficients" to be introduced into the model. Under these conditions a complete understanding of

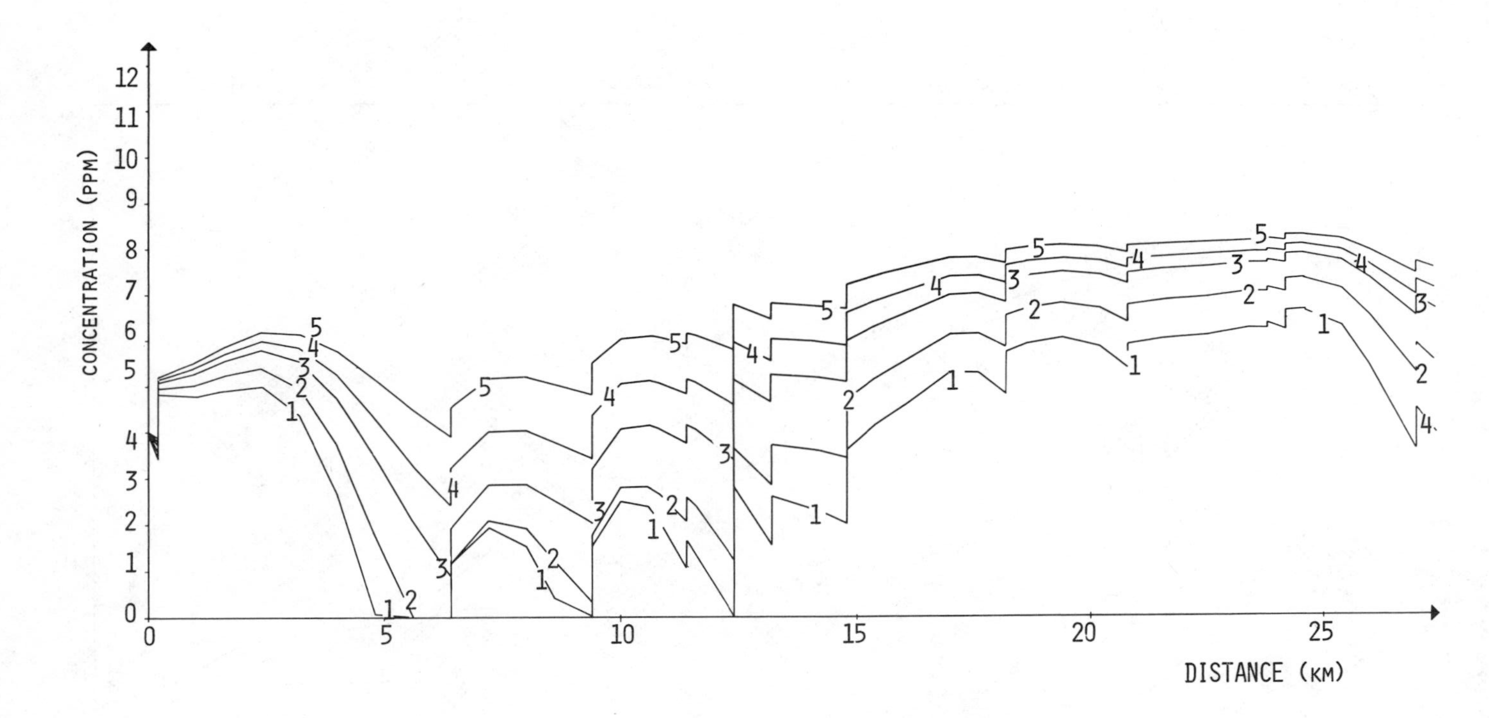


*Figure 5.12.* Sag curves for various hypotheses of effluent treatment.

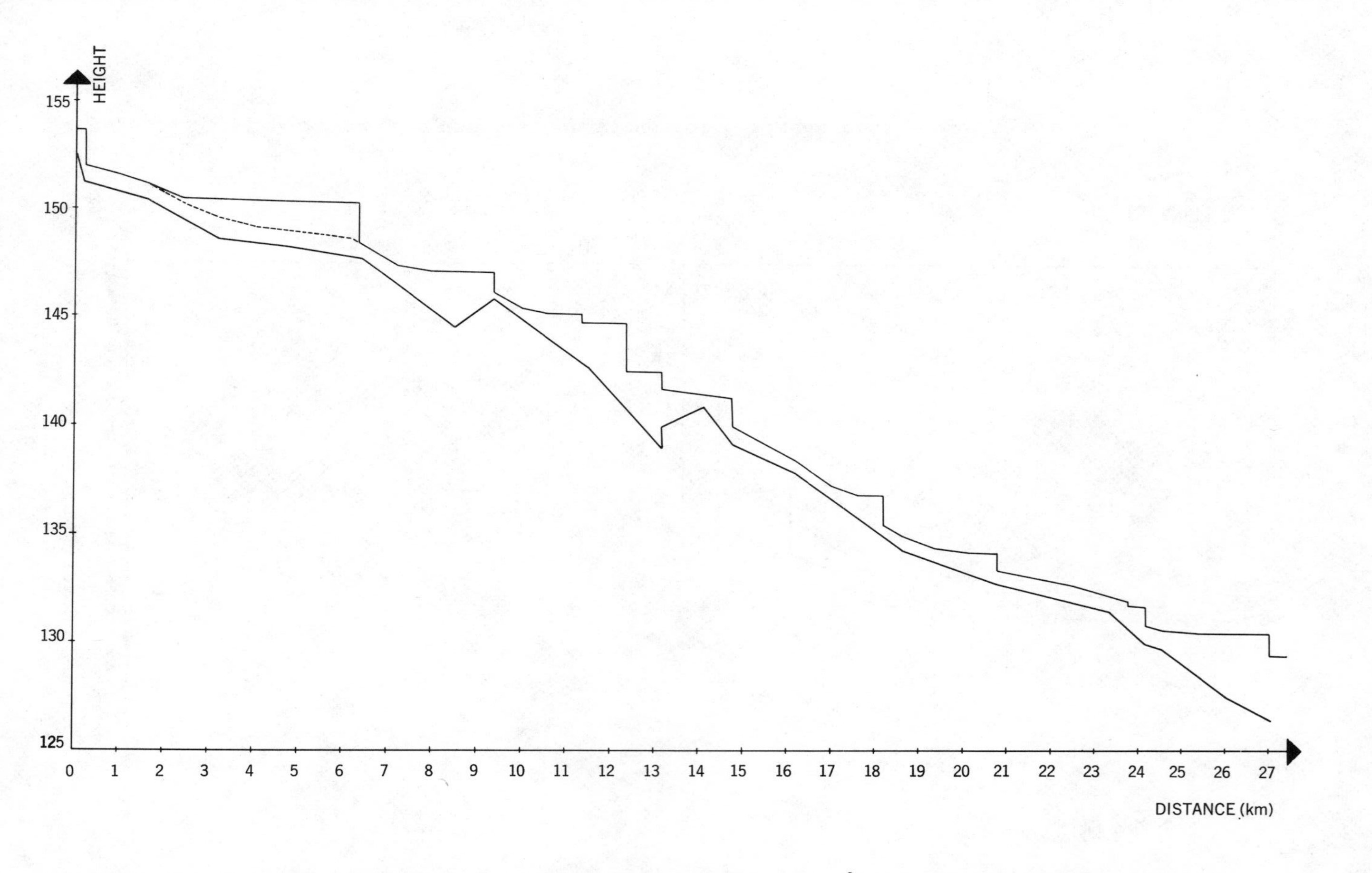


*Figure 5.13. Water surface elevations for a discharge of 12 $m^3/s$ with and without the weir at Chabanais.*

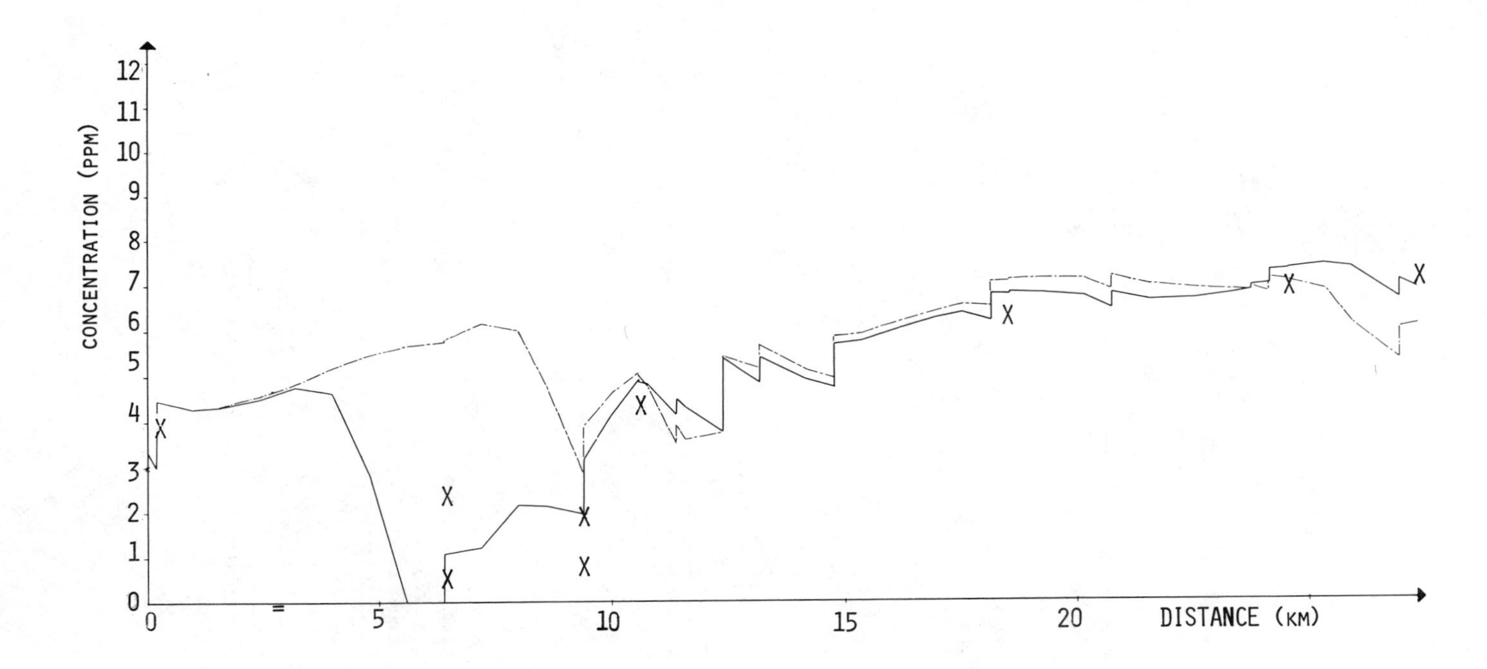


*Figure 5.14. Sag curves in the river Vienne with and without the Chabanais weir.*

the mechanism of self-purification can be obtained. The validity of extrapolations depends upon this understanding.

## ACKNOWLEDGMENTS

This study has been carried out by a team of SOGREAH specialists (Société Grenobloise d'Etudes et d'Applications Hydrauliques) for hydraulics and data processing and IRCHA specialists (Institut de Recherche de Chimie Appliquée) for chemistry and biochemistry. It was financed in part by a grant from DGRST (Délégation Générale à la Recherche Scientifique et Technique, contract n° 690 199 500 212 7501 (1971)).

## REFERENCES

1. Bakhmeteff, B. A. *Mécanique de l'écoulement turbulent des fluides* (Paris: Dunod, 1940).
2. Elder, S. W. "The Dispersion of Marked Fluid in Turbulent Shear Flow," *J. Fluid Mechanics 5(4)*:544 (1959).
3. O'Connor, D. J. and W. E. Dobbins. "Mechanism of Reaeration in Natural Streams," *Transactions of the ASCE 123*:641 (1958).
4. Camp, T. R. "Field Estimates of Oxygen Balance Parameters," *J. San. Eng. Div., Proc. ASCE 92(SA1)*:1 (1966).
5. Gannon, J. J. "River and Laboratory BOD Rate Considerations," *J. San. Eng. Div., Proc. ASCE 92(SA1)*: 135 (1966).
6. Water Pollution Research Laboratory of Stevenage (England. "Effects of Polluting Discharges on the Thames Estuary," Water Pollution Research Technical Paper 11 (1964).
7. Camp, T. R. *Water and Its Impurities* (New York: Reinhold Publishing, 1963).
8. Dobbins, W. E. and D. A. Bella. "Difference Modelling of Stream Pollution," *J. San. Eng. Div., Proc. ASCE 94(SA5)*: 995 (1968).
9. Churchill, M. A., H. L. Elmore, and R. A. Buckingham. "The Prediction of Stream Reaeration Rates," *J. San. Eng. Div., Proc. ASCE 88(SA2)*:1 (1962).
10. Gameson, A. L. H. "Weirs and the Aeration of Rivers," *J. Inst. Water Eng. 11*:477 (1957).
11. Brebion, G., G. Chevereau, B. Lebrun, and A. Preissmann. "Modèles Mathématiques de Pollution," Report SOGREAH n° 10 406 (1971).

## CHAPTER 6

# WATER QUALITY MODELS AND AQUATIC ECOSYSTEMS STATUS, PROBLEMS AND PROSPECTIVES

R. Diaz, M. Bender, R. Jordan and D. Boesch*

## I. INTRODUCTION

Although many elaborate biological models have been developed during the last few years,[1] most are too restricted in scope to be directly useful in evaluating pollutant-related stresses in aquatic environments. Models exist that describe nutrient levels and related response parameters such as algal biomass in specific bodies of water.[2] Although such models have proven useful in waste management they are limited in two respects: (1) they usually describe only one problem area (trophic level) within the system, and (2) they often must be ramified, at considerable expense, before applying them to other similar environments. There is every reason to believe, however, that in specific types of aquatic environments certain stresses that are considered natural in nature, *e.g.*, those created by oxygen-consuming wastes, are understood sufficiently to allow for the construction of total ecosystem models. Such models can account for the major functional changes such as respiration and nutrient regeneration as functions of variables such as time and concentration. However, nonfunctional changes, such as alterations in community composition, are much less easily modeled although often equally important. Also, it will be some time before we have enough knowledge to construct models

---

*The authors are with the Virginia Institute of Marine Science, Gloucester Point, Virginia 23062, U.S.A.

capable of describing the effects of pollutants of unnatural origin or of being applied directly to widely varied ecosystems.

To illustrate the present status of practical water quality models that consider biotic responses, two examples of key concern were chosen: biotic indices and eutrophication. Although the examples presented can at best be considered as subsystem models, they are representative of the types of models being utilized to solve water pollution problems.

## II. BIOTIC INDICES AND WATER QUALITY

Mathematical techniques have been used to describe community structure or to express the relationships among samples or species. Community structure means the way in which the biological resources (numbers, biomass, energy flow, etc.) are distributed among the constituent species in a community. Some of the more popular techniques will be described and then a few examples of their use in describing water quality changes will be given. First, however, it should be pointed out that these measures are often needed to explain the responses of the communities to stresses when the specific causative factor or factors are unknown.

### *Community Structural Indices*

#### Measures of Dominance

The relative abundance (per cent abundance) of a species is usually used to express its structural (not functional) dominance in a community. If a number of samples are taken in a community, the species can be ranked within each sample and assigned a rank score (or bio-index). These scores are then summed over the series of samples to express its overall dominance. Usually the species are scored on a 5 or 10 point basis. The most abundant species is scored 5 (in the 5 point system), the next most abundant 4, and so on. For example, a 5 point bio-index for one sample with 7 species is indicated below.

| *Species* | *Number* | *Bio-index* |
|---|---|---|
| A | 19 | 5 |
| K | 15 | 4 |
| Y | 10 | 2.5 |
| N | 10 | 2.5 |
| Q | 3 | 1 |
| P | 2 | 0 |
| Z | 1 | 0 |

## Species Diversity Measures

Species diversity has often been related to water quality in biological surveys. Measurement of species diversity is popular because of its utility and attractive because of the theoretical relationships between diversity and community stability. The most widely accepted concept of species diversity is that it is a function of the number of species present (species richness or abundance) and the evenness with which individuals are distributed among the species (evenness or equitability).

The most commonly used diversity index is that of Shannon,[3] which expresses the amount of information content per individual. In other terms it is the amount of uncertainty in predicting the specific identity of a randomly chosen individual from an assemblage. The more species there are and the more evenly they are represented, the higher this uncertainty. The index generally indicated by H' is given by:

$$H' = -\sum_{i=1}^{S} P_i \log P_i$$

where S = number of species in the sample and $P_i$ = the proportion of the $i^{th}$ species in the sample. Base 2 logarithms are often used and the index has a dimension of bits/individual. Many additional diversity indices have been used and many have been reviewed by Hurlbert.[4] Comparisons of species diversity indices should be done with caution. In addition to depending on sample size, diversity measures are also sensitive to sampling technique.

The simplest measure of the species richness component of species diversity is the number of species in a collection. But because this depends on sample size this number is usually standardized by relating it to the number of individuals in the sample. Margalef's richness index,[5] which is widely used, defines species richness as

$$SR = S-1/\ln N$$

where S is the number of species and N the number of individuals in a sample.

Alternately, Sanders[6] has developed a technique by which the number of species in a smaller or rarefied sample is predicted. This rarefication technique generates a curve that predicts the number of species taken in smaller and smaller hypothetical samples (Figure 6.1). In this manner the species

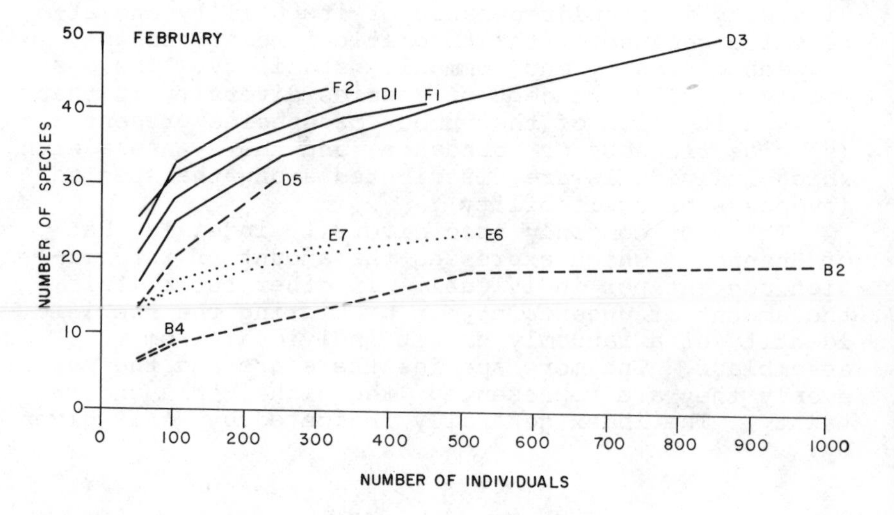


*Figure 6.1. Rarefication curves demonstrating the species richness of macrobenthos at 15 stations in the Hampton Roads area, Virginia.*

richness samples of disparate sizes can be directly compared. It should be noted that Sanders' method consistently overestimates the number of species in rarefied samples, but this can be overcome by using Hurlbert's[4] modification of the rarefication method.

The evenness component can be measured by examining the relationship of the measured species diversity H' and the maximum diversity possible given the number of species in the collection, *i.e.*, the diversity if all species are equally represented. Evenness (J) can be expressed

$$J = H'/H\ max = H'/\log S$$

### *Classification and Ordination Techniques*

The relationships between samples or between species can be investigated by computing similarity indices based on the species composition of these samples or the distribution patterns of the species over a series of samples, respectively. When the relationships between samples are considered it is called a "normal" analysis, and when interspecies relationships are considered it is called an "inverse" analysis. A similarity index is computed between all pairs of samples (species in the inverse analysis), producing a symmetrical (sample by sample) matrix of similarity values.

A great many similarity indices have been used that reflect either the qualitative composition of samples (cooccurrence of species in the reverse analysis) or, to varying degrees, the quantitative composition of samples. The simplest qualitative measure is Jaccard's coefficient

$$S_J = c/(a + b - c)$$

where a = number of species at station A
b = number of species at station B
c = number of species occurring at both A and B.

Another often used qualitative measure is Sorenson's coefficient

$$S_S = 2\ c/a + b.$$

The most often used quantitative measure is "percentage similarity" in which the percentage of the total sample attributable to each species is computed and the lesser of the percentage values for each species common to the pair of samples being considered is summed. We must reiterate that there are many more similarity indices, and their values may change depending on data transformations and standardizations. The above measures are simple and widely used and will serve our purposes of illustration.

The traditional practice has been to reorder the computed similarity matrix so that samples most similar are adjacent, *i.e.*, the highest similarity values are near the principal diagonal of the matrix.

The matrix is then coded by shading or coloring and presented as a "trellis diagram" (Figure 6.2). Techniques have been developed recently to classify samples (species in the inverse analysis) into groups (classification) or to order them in an imaginary multidimensional spatial representation (ordination) based on the computed similarities.

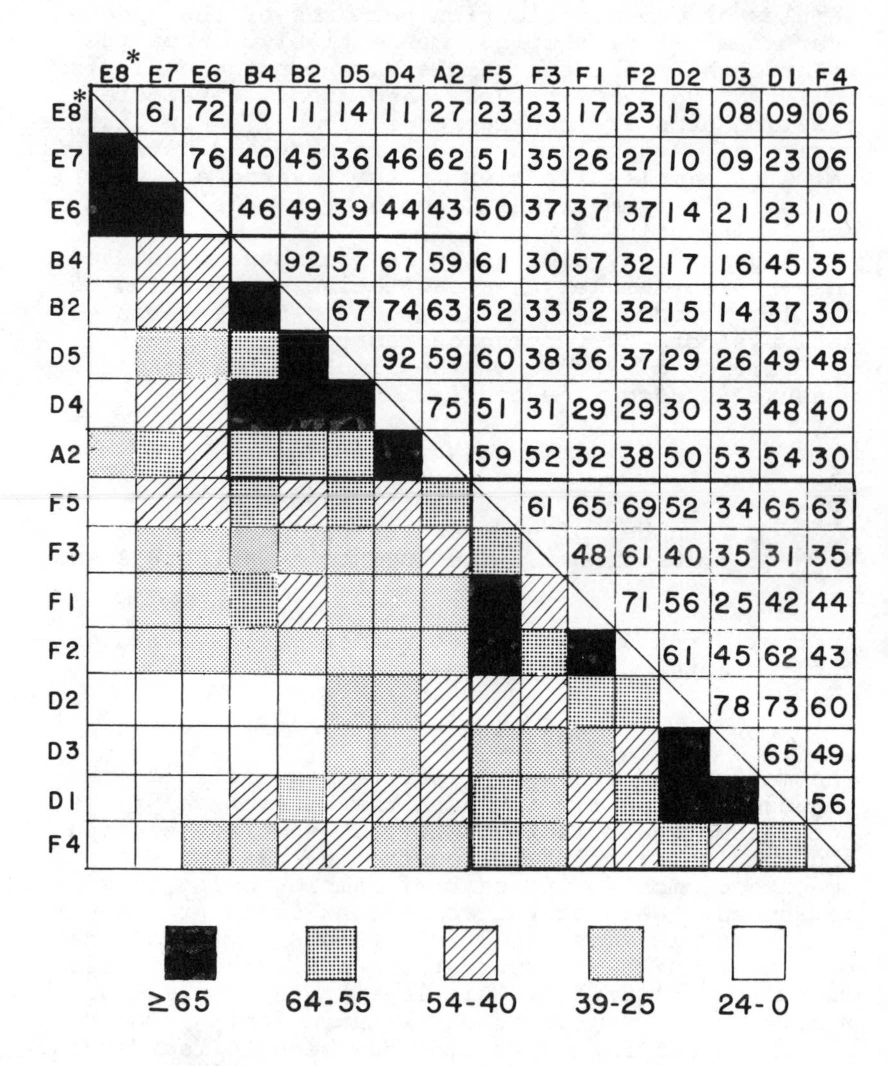

| | E8* | E7 | E6 | B4 | B2 | D5 | D4 | A2 | F5 | F3 | F1 | F2 | D2 | D3 | D1 | F4 |
|---|---|---|---|---|---|---|---|---|---|---|---|---|---|---|---|---|
| E8* | | 61 | 72 | 10 | 11 | 14 | 11 | 27 | 23 | 23 | 17 | 23 | 15 | 08 | 09 | 06 |
| E7 | | | 76 | 40 | 45 | 36 | 46 | 62 | 51 | 35 | 26 | 27 | 10 | 09 | 23 | 06 |
| E6 | | | | 46 | 49 | 39 | 44 | 43 | 50 | 37 | 37 | 37 | 14 | 21 | 23 | 10 |
| B4 | | | | | 92 | 57 | 67 | 59 | 61 | 30 | 57 | 32 | 17 | 16 | 45 | 35 |
| B2 | | | | | | 67 | 74 | 63 | 52 | 33 | 52 | 32 | 15 | 14 | 37 | 30 |
| D5 | | | | | | | 92 | 59 | 60 | 38 | 36 | 37 | 29 | 26 | 49 | 48 |
| D4 | | | | | | | | 75 | 51 | 31 | 29 | 29 | 30 | 33 | 48 | 40 |
| A2 | | | | | | | | | 59 | 52 | 32 | 38 | 50 | 53 | 54 | 30 |
| F5 | | | | | | | | | | 61 | 65 | 69 | 52 | 34 | 65 | 63 |
| F3 | | | | | | | | | | | 48 | 61 | 40 | 35 | 31 | 35 |
| F1 | | | | | | | | | | | | 71 | 56 | 25 | 42 | 44 |
| F2 | | | | | | | | | | | | | 61 | 45 | 62 | 43 |
| D2 | | | | | | | | | | | | | | 78 | 73 | 60 |
| D3 | | | | | | | | | | | | | | | 65 | 49 |
| D1 | | | | | | | | | | | | | | | | 56 |
| F4 | | | | | | | | | | | | | | | | |



*Figure 6.2. Trellis diagram showing similarities of the benthic fauna between all pairs of 16 stations in the Hampton Roads area, Virginia.*

Classifications may be discreet and nonhierarchical (*e.g.* Fager's[7] recurrent group analysis) or more often hierarchical. Hierarchical classifications present the intersample relationships in the form of a branching dendrogram (Figure 6.3). Again a number of clustering strategies exist by which dendrograms are formed from a similarity matrix.[8,9] Because of the

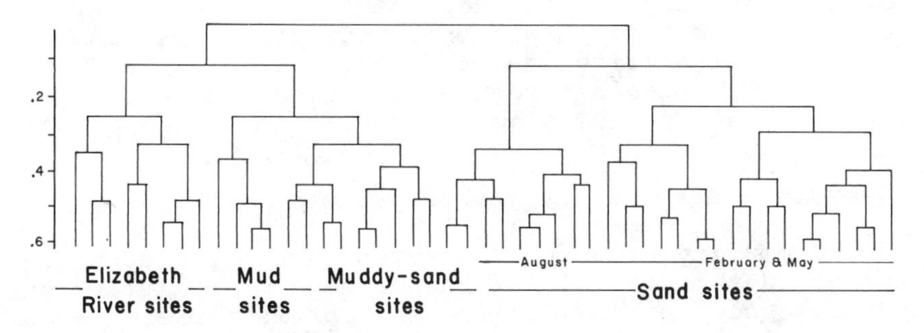


*Figure 6.3. Dendrogram depicting a hierarchical classification of 47 samples of macrobenthos from the Hampton Roads area, Virginia.*

great number of possible combinations to be considered, a computer is needed to perform classifications of more than a few samples. Depending on the similarity measure (including data transformations and standardizations) and the clustering strategy, it is possible to achieve vastly different classifications. Therefore, we caution against the acceptance of the results of a classification strategy, often selected on the sole basis of availability, as a completely objective or statistically rigorous analysis. Instead one should be aware of optional techniques and their biological implications.

Ordination involves even more complicated mathematics, and again most methods require automatic computation. Most techniques involve the ordering of the individuals or samples along redefined imaginary axes (eigenvectors). The first few axes usually account for most of the information in the collection of individuals. The results are often presented as scatter diagrams of points in two- or three-dimensional space as indicated by their coordinates (loadings) on the first two or three axes. The relative proximity of these points in the ordinated spaces gives a conceptual picture of the similarity of the individuals they represent. Some ordination techniques are principal components analysis, and the mathematically less rigorous Bray-Curtis ordination.[10,11] Often the defined axes have been correlated with abiotic factors, presumably of primary importance to the distribution of the organisms being investigated. Although this ability to extract factors from assorted collections is attractive, some serious problems of nonlinearity of the data occur even when one factor predominates overwhelmingly.[12] Experience indicates that, in general, classification produces more interpretable results with the heterogeneous collections usually encountered in biological surveys than does ordination

## Uses in Water Quality Investigations

Community structure measures have received wide usage in the analysis of the effects of pollution on aquatic communities. Wilhm and Dorris[13] presented a rationale for the use of species diversity measures as indices of water quality. Wilhm[14] reviewed ranges of diversity values (H') of freshwater benthic invertebrate communities and found that H' ($=\bar{d}$) usually varies between three and four bits/individual in clean water stream areas and is usually less than one in polluted stream areas. The decrease in richness of fish and benthic populations below outfalls from both a dye casting plant and a sewage discharge is shown in Figure 6.4.

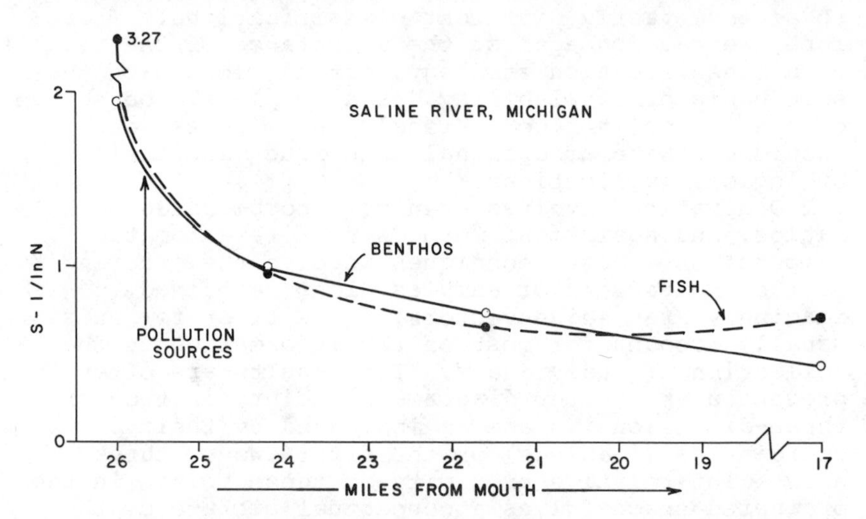


*Figure 6.4. Effect of multiple pollution source on species richness of fish and benthos in the Saline River, Michigan.*

Boesch[15] reported H' values for a range of marine and estuarine benthic communities in Virginia and reviewed values for marine benthos from the literature. Species diversity is highly variable but is generally higher on sand bottoms, at higher (poly- and euhaline) salinities, and on the outer continental shelf. Diversity was lower in the Elizabeth River, an active port area receiving heavy loadings of pollutants, than on adjacent bottoms of similar substrate and salinity conditions. The between-habitat differences in diversity were largely differences in the species richness component. It seems from this and similar findings by Wilhm[16] that in polluted streams specific attention should be paid to the richness component of diversity in water quality investigations.

Armstrong, *et al.*[17] investigated benthic species diversity and water quality relationships in San Francisco Bay and found relationships between diversity and salinity and sediment type similar to those found by Boesch. They developed a regression equation to relate diversity as a dependent variable with chlorosity and percentage sand as independent variables. Diversity at locations subject to pollution stress departed from that predicted by the regression equation. Similarly, Bechtel and Copeland[18] used regression to relate fish species diversity and wastewater concentration (as predicted by mathematical model) in Galveston Bay, Texas. It should be pointed out that because statistical distributions of these community structural indices are not well known, use of probabilistic parametric statistical tests to test significance of observed trends must be viewed with caution. However, some statistical methods, such as multiple regression, can be satisfactorily used as hypothesis generating (versus hypothesis testing) tools or as predictive models.

Methods of classification and ordination have not received much use in water quality research. These techniques are rapidly developing and evolving in community ecology. Boesch[19] used a classification strategy to classify samples and species of macrobenthos from Hampton Roads, Virginia. The normal analysis clearly separated those samples in the Elizabeth River from those elsewhere (Figure 6.3). In addition to the reduction in species diversity from that typical of similar muddy-sand bottoms nearby, the fauna was clearly of different species composition. Similarly, the inverse classification pinpointed both those species found nearly exclusively in the Elizabeth River (indicators) and those largely excluded from the Elizabeth River.

Relationships between diversity or richness and certain water quality variables have been demonstrated As an example, Figure 6.5 graphically depicts the decrease in species richness of benthic animals as a function of fluoride concentration in certain freshwater streams. However, all too often more than one factor or pollutant influences community composition, and frequently synergistic or antagonistic influences may also enter into the picture. Hence, such a straightforward relationship as shown in Figure 6.5 is more often the exception than the rule.

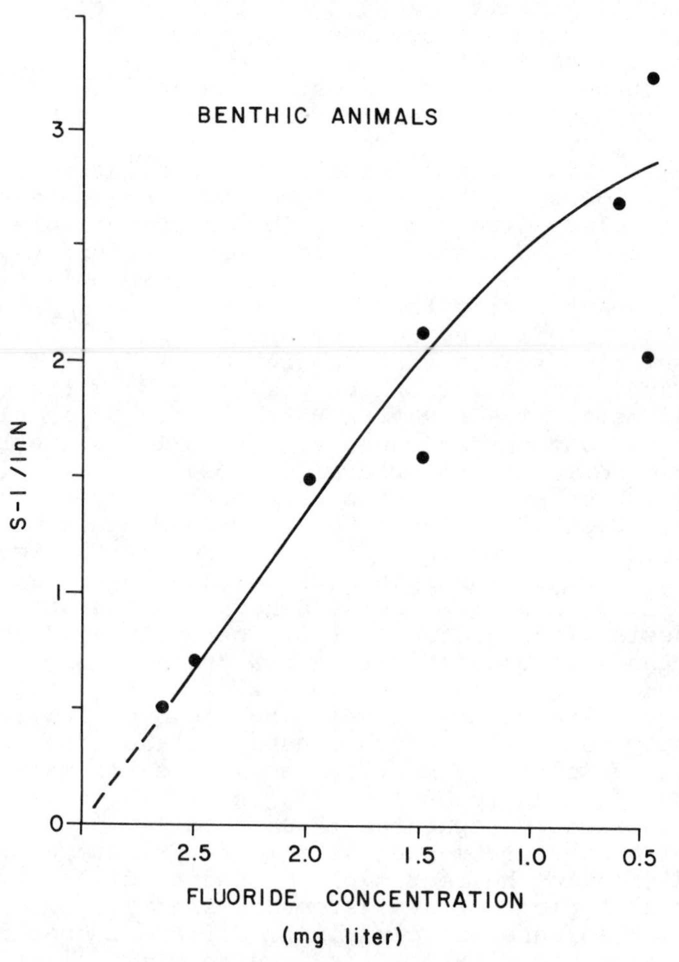


**Figure 6.5.** *Effect of fluoride concentration on species richness of benthos in certain freshwater streams.*

## III. EUTROPHICATION

### *The Enrichment Problem*

The biological productivity of a body of water is influenced by many factors.[20] Among these are morphometry, the climate in which it is found, and the content of dissolved and suspended matter, loosely referred to as water quality. Morphometry and climate tend to change relatively slowly, but water quality can change rapidly in response to changes in the surrounding drainage basin. Man is an increasingly widespread and potent effector of changes in drainage basins, but he has failed to control his activities sufficiently to avoid unplanned consequences in neighboring waters. Figure 6.6 depicts the relationships between nutrients and algal growth. Although the nutrient values and growth patterns are hypothetical, the figure serves as an example of biological response to nutrient enrichment. As a result numerous pollution problems have developed that reduce the suitability of natural waters for indigenous organisms and for use by man.

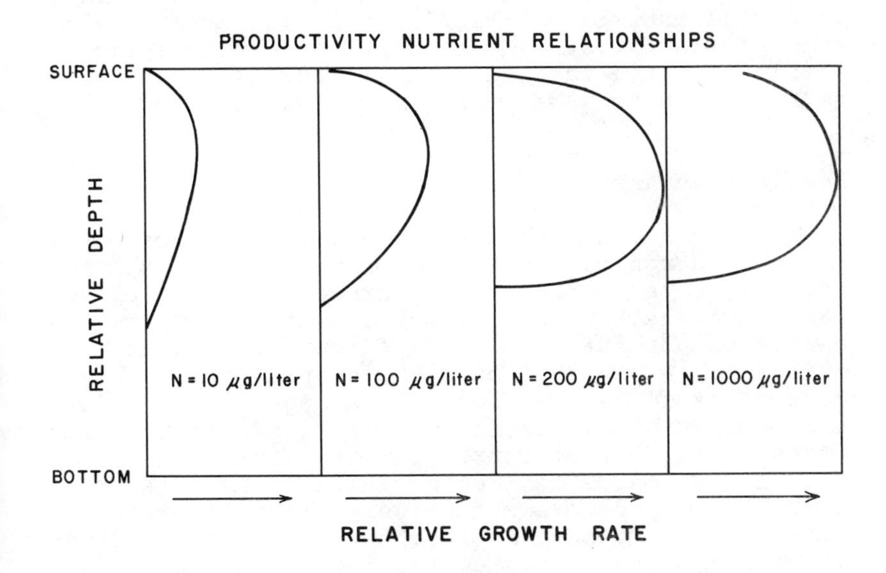


*Figure 6.6. Generalized relationship between nutrient enrichment and phytoplankton algal growth in lakes.*

One consequence of uncontrolled human activities is the enrichment of a water body with plant nutrients a process referred to as eutrophication.[21] The relationship between human activities and the release of nutrients into natural waters is most easily established where sewage effluents enter lakes and streams, less readily confirmed where fertilizers applied to agricultural soils are washed in,[23] and even more obscure where natural terrestrial nutrient cycles are destroyed, permitting soil nutrients to escape into drainage waters.[24]

Eutrophication is not inherently bad and in fact has been done deliberately in many instances in attempts to increase fish production. However, enough cases of inadvertent fertilization resulting in undesirable changes in natural waters have occurred so that nutrient enrichment is now recognized as one of our most serious water quality problems.

Most of the unfavorable consequences of eutrophication are not direct effects of the nutrients themselves but side effects of resultant excessive growths of aquatic plants.[25] Excessive algae in drinking water supplies can increase the clogging rate of sand filters, increase the chlorine demand of the water, and contribute color, tastes, and odors to the finishe water. In recreational lakes algal accumulations can impair aesthetic qualities as well as cause deterioration of fisheries. In the long run eutrophication accelerates the rate at which lakes fill up with sediment, thereby hastening their extinction.

### *Nutrient Management*

The pressure for corrective action has become intense enough to involve government agencies,[26,27] which have been encouraged by recent reports of the rapid improvement of Lake Washington following elimination of the inflow of sewage treatment plant effluents.[28] In this case the complete effluents were diverted to a different drainage basin, solving the problem by exporting it rather than by instituting improved methods of nutrient management.

In most cases effluents cannot be exported but must be managed more effectively within a drainage basin. The ultimate approach will most probably have to be recycling of wastewater into drinking water without releasing wastes into natural waters, but this is in the future. The nutrient control method that is the most compatible with contemporary wastewater management practice, the foundation of which is the production and discharge of an effluent, is removal of nutrients prior to effluent release. This can be done by reducing the levels of nutrients entering wastewaters, the approach that is the basis of the current pressure for removal of phosphates from detergents, and by improving wastewater treatment to remove the nutrients that cannot be initially excluded.

It is generally believed that not all of the nutrients that enter natural waters as a result of human activities are responsible for the resultant problems because not all nutrients are present in natural waters in the proportions required by plants. Natural supplies of some nutrients are in great excess relative to supplies of other nutrients, and their abundance is therefore unlikely to be limiting to plant growth. Wastewater inflows are more likely to enhance aquatic plant growth by supplying nutrients that are naturally scarce than nutrients that are naturally abundant, and it is logical to conclude that removal of the naturally scarce nutrients from wastewater should improve conditions in receiving waters.

A further point that must be considered, however, is that there are sources of nutrient enrichment other than wastewaters, some of which, such as rainfall, are impossible to control. Consequently it is not sufficient to identify the scarcest nutrient in a given water body and remove it from wastewater inflows; it is also necessary to compare the extent to which the total inputs of different nutrients can be reduced by removing them from wastewater.

An example of the integration of these two aspects of enrichment, nutrient limitation and nutrient controlability, is provided by a study of eutrophication in the Potomac estuary.[2] Nitrogen was found to be more influential than phosphorus in controlling phytoplankton growth, but phosphorus input could be reduced more effectively than nitrogen input by improved waste treatment. Therefore in the proposal for future nutrient management, phosphorus removal from effluents was emphasized more strongly than nitrogen removal.

As long as the approach to nutrient management is to be selective, emphasizing control of only the key substances, there is a need for methods that can be used to evaluate each body of water for which nutrient control is desired. In order to alleviate existing problems, techniques are needed for

(1) identifying limiting nutrients, (2) identifying nutrient sources, (3) identifying the substances in nutrient sources that are stimulating growth, and (4) evaluating the effectiveness of nutrient management practices. In order to protect aquatic communities from future problems, methods are needed for (1) evaluating the sensitivity of an aquatic system to additional enrichment and (2) screening potential future environmental contaminants for possible effects.

*Research Methods*

Limiting nutrients can be identified with some degree of confidence by investigating nutrient changes in the system of interest. For example, phosphorus (see Figure 6.7) rather than nitrogen, was identified as the key nutrient in the eutrophication of Lake Washington[28] because its decline subsequent to nutrient diversion was more closely correlated with the decline in phytoplankton biomass, whereas nitrogen was found to be more influential than phosphorus in the coastal waters off New York City, where nitrogen levels decline more rapidly than phosphorus levels with distance from the

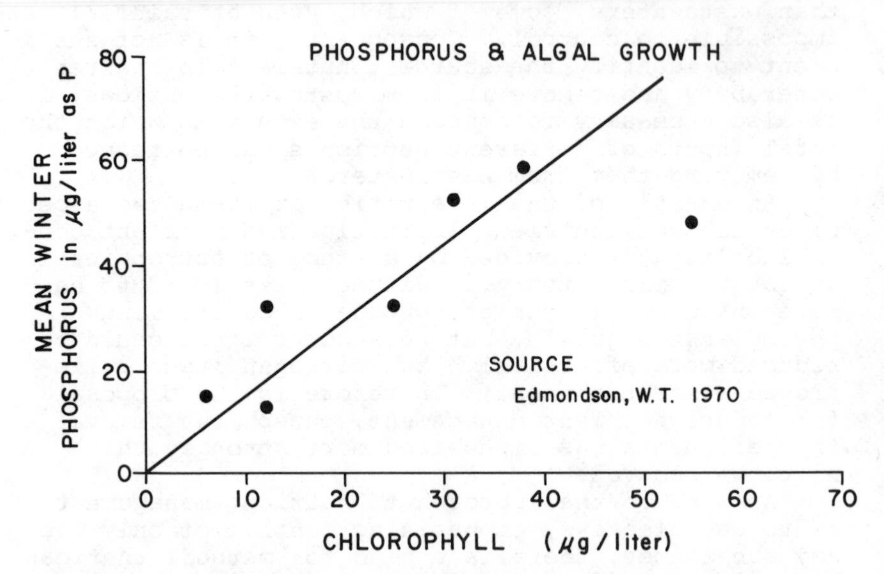


*Figure 6.7. Relationship between mean winter phosphorus concentration and summer phytoplankton standing crop in Lake Washington.*[28]

harbor.[29] In the former case, however, the limiting nutrient was identified after its input had been curtailed.

Several techniques applicable to eutrophication studies employ algae as analytical tools. One group of these can be classed as biochemical assays because they exploit various biochemical traits of algae to assess their nutritional state. For example, limitation of growth by phosphate can be inferred from (1) high activity of alkaline phosphatase enzymes that are produced by phosphate deficient cells and function by releasing inorganic phosphate from organic compounds or (2) essential absence of the intracellular stores of surplus inorganic phosphates that are characteristic of cells growing in an environment rich in phosphate.[30] In a similar way nitrogen limitation can be inferred from (1) high rates of ammonia utilization in the dark[31] or (2) low activity of the nitrate reductase enzyme.[32] However, these techniques are difficult to apply to natural diverse algal communities because of the physiological variations that occur among species. Thus a high enzyme level for one species may be low for another species. Also, different species within a community may be limited by different nutrients at the same time,[33] precluding any conclusion about what is limiting to the community. Nonetheless they have been used successfully to complement other types of measurements, such as in the Potomac study[2] where high rates of ammonia utilization in the dark in conjunction with low alkaline phosphatase activity and the presence of stored phosphate in algal samples supported the conclusion that nitrogen was limiting in the reaches where ambient nitrogen levels were low.

A second group of approaches employing algae, which can be applied to most of the questions that arise in eutrophication studies, consists of the productivity bioassays, or enrichment experiments. In these techniques algal growth in response to a treatment is measured, and the methods vary according to the algae used (single species cultures or samples of natural communities) and the growth conditions (batch or continuous culture, laboratory or *in situ* incubation). The Joint Industry-Government Task Force on Eutrophication[27] has been developing a series of standardized algal assay methods, concentrating so far on a bottle test employing separate cultures of four standard species grown under laboratory conditions.[34] The procedural and statistical aspects of the method have been studied intensively, but the problems of using the results to evaluate or predict events in natural waters have not yet been resolved.

The second method under study is similar to the first except that it employs continuous rather than batch culture conditions. The third method, however, differs fundamentally from the other two in employing samples of natural algal communities as the test organisms.

## REFERENCES


1. Patten, B. G. *Systems Analysis and Simulation in Ecology.* (New York: Academic Press, 1971).
2. Jaworski, N. A., D. W. Lear, Jr., and O. Villa, Jr. "Nutrient Management in the Potomac Estuary," In G. E. Likens (ed.) *Nutrients and Eutrophication; The Limiting Nutrient Controversy.* Amer. Soc. of Limnol. and Oceanogr. Special Symposia Vol. I. (1971).
3. Pielou, E. C. *J. Theoret. Biol. 13:*131 (1966).
4. Hurlbert, S. H. *Ecology 52:*577 (1971).
5. Margalef, D. R. *Gen. Syst. 3:*36 (1958).
6. Sanders, H. L. *Amer. Nature. 102:*243 (1968).
7. Fager, E. W. *Ecology 38:*586 (1957).
8. Sokal, R. R. and P. H. A. Sneath. *Principles of Numerical Taxonomy.* (San Francisco: W. H. Freeman, 1963).
9. Williams, W. T. *Ann. Rev. Ecol. System 2:*303 (1971).
10. Whittaker, R. H. *Biol. Rev. 42:*207 (1967).
11. Rohlf, F. J. *Syst. Zool. 21:*271 (1972).
12. Austin, M. P. and I. Noy-Mier. *J. Ecol. 59:*763 (1971).
13. Wilhm, J. L. and T. C. Dorris. *Bio. Sci. 18:*477 (1968).
14. Wilhm, J. L. *J. Water Pollut. Contr. Fed. 42:*R221 (1970).
15. Boesch, D. F. *Ches. Sci. 13:*206 (1972).
16. Wilhm, J. L. *J. Water Pollut. Contr. Fed. 39:*1673 (1967).
17. Armstrong, N. E., P. N. Storrs, and E. A. Pearson. *Development of a Grass Toxicity Criterion in San Francisco Bay.* Proc. 5th Int. Conf. Water Pollut. Res. III. (1971).
18. Bechtel, T. J. and B. J. Copeland. *Contr. Mar. Sci., U Univ. Texas 15:*103 (1970).
19. Boesch, D. F. *Classification and Community Structure of Macrobenthos in the Hampton Roads area, Virginia.*
20. Rawson, D. S. *A.A.A.S Bull. 10:*9 (1939).
21. Stewart, K. M. and G. A. Rohlich. "Eutrophication--A Review," California State Water Quality Control Board, Pub. 34 (1967).
22. Hasler, A. D. *Ecology 28:*383 (1947).
23. Stanford, G., C. B. England, and A. W. Taylor. "Fertilizer Use and Water Quality," U.S. Dept. of Agriculture, Agricultural Research Service Pub. ARS 41-168 (1970).

24. Bormann, F. H. and G. E. Likens. *Sci. Amer. 223:*92 (1970).
25. Lee, G. F. "Eutrophication," U. of Wis. Water Resources Center, Eutrophication Information Program, Occasional Paper No. 2 (1970).
26. U.S. Congress. House. Committee of Government Operations. Conservation and Natural Resources Subcommittee. *Phosphates in Detergents and the Eutrophication of America's Waters.* (Washington, D.C.: U.S. Govt. Printing Office, 1970).
27. Bueltman, C. G., *et al. Provisional Algal Assay Procedure.* (New York: Joint Industry-Government Task Force on Eutrophication, 1969).
28. Edmondson, W. T. *Science 169:*690 (1970).
29. Ryther, J. H. and W. M. Dunstan. *Science 171:*1008 (1971).
30. Fitzgerald, G. P. and T. C. Nelson. *J. Phycol. 2:*32 (1966).
31. Fitzgerald, G. P. *J. Phycol. 4:*121 (1968).
32. Eppley, R. W., J. L. Coatsworth and L. Solorzano. *Limnol. Oceanogr. 14:*194 (1969).
33. Fitzgerald, G. P. *Limnol. Oceanogr. 14:*206 (1969).
34. Environmental Protection Agency. National Eutrophication Research Program. *Algal Assay Procedure Bottle Test.* (Washington, D.C.: U.S. Govt. Printing Office, 1971).

# PART II

# WATER SUPPLY AND WATER RESOURCES DEVELOPMENT

MODELS FOR ENVIRONMENTAL POLLUTION CONTROL


# CHAPTER 7

# MODELS OF WATER SUPPLY SYSTEMS

Ralf G. Cembrowicz*

## I. INTRODUCTION

This paper deals with the optimal design of water supply systems with several specific parts of the overall water supply system treated in detail.

In Section II the problem of a cost-benefit analysis for water supply projects is discussed briefly. Section III considers the problem of the selection of different sources and the choice of the most economical combination of alternate sources.

In Section IV the focus is on the optimization of unit processes in a water treatment plant. Section V investigates the economically optimal design of the water distribution system.

And finally, Section VI deals with the problem of the capacity expansion of a water supply system.

## II. COST-BENEFIT CRITERIA FOR WATER SUPPLY SYSTEMS

Inherent difficulties exist in comparing costs of water supply systems to the benefits of ample and safe supply. In recent decades standards of living in Western nations have resulted in the customers' "willingness to pay" for an adequate supply of water to all customers at any time. Monetary cost-benefit analysis has been easily accomplished for industrial and commercial usage. But only now have competing demands for water in recreation, irrigation, waste

*Dr. Ralf G. Cembrowicz is a research associate at the Institut für Siedlungswasserwirtschaft, Universität Karlsruhe, 75 Karlsruhe 1, Germany.

disposal, hydro-power, industrial and domestic use and exploitation of the available resources in densely populated areas initiated the need for comprehensive analysis of the entire scale of sources and demand.

With regard to developing countries the problem definition seems different. Overcrowded cities receive totally inadequate supply, fluctuating in quality and quantity. There, the construction of a system covering, for example, the demand mean already provides obvious health and socioeconomic benefits. However, scarcity of monetary resources, high opportunity costs from alternative national investments, and repayment conditions of foreign creditors often require the planning of water supply systems based on a strict balance of costs and revenue income from the system.[1]

Another practice is to equate the benefits of a water supply system to the cost of the least-expensive alternative supply. This may underestimate the criterion of "willingness to pay," which comprises monetary, socioeconomic, amenity, and health benefits. Needless to say, it is rather difficult to obtain information regarding these benefits. Commonly, in planning water supply projects, water demand is set at a predicated level to be met without consideration of water needs for other purposes. This concept of minimum requirements and standards can be treated in a systems analysis approach using constraints. Hence, if the objectives are minimum costs, the resulting optimal costs are associated with specified design parameter levels of water quantity, quality, peak load, etc. Subsequent sensitivity tests of the optimization model determine the marginal cost influence of these parameters as a possible measure of their appropriate level.

Similarly, marginal imputed benefits can be computed if the water supply source is used for alternative purposes. If competing water demands of the environment necessitate multipurpose projects, the predication of a set level of water supply is tantamount to an imputed benefit for the water supply system.[2] Provided other water uses generate monetary benefits, the imputed marginal benefit of the water supply system is equal to the marginal monetary increase of these uses if they receive one unit of water that was originally diverted to the water supply system.

The result can show that a justifiable level of water is used for water supply by comparing health, amenity and monetary benefits of the water supply to the imputed value. But the imputed benefits may also exceed a desirable level, *e.g.*, in developing countries with strong needs for a judicious distribution of investments. The imputed benefits also provide an additional monetary value judgment to the previous concept of equating benefits to the least expensive alternative supply.

Assume the equation $x = (x,..,x_i,..,x_n)$ denotes n possible draft uses from an existing reservoir of volume V. Benefit functions $B_i(x_i)$ are available. In addition, y units of water supply are required. Hence,

$$\sum_{n=1}^{n} x_i + y = V \tag{1}$$

Using the Lagrange method,[3] we find the optimality of the function

$$L(x,\lambda) = \sum_i B_i(x_i) + \lambda \left(\sum_i x_i + y - V\right) \tag{2}$$

is obtained from the necessary conditions:

$$\frac{\partial(\Sigma B_i(x_i^*))}{\partial x_i} + \lambda = 0 \tag{3}$$

Also, at the optimum x*:

$$\frac{\partial(\Sigma B_i(x_i^*))}{\partial y} = \lambda \tag{4}$$

The Lagrangian multiplier $\lambda$ represents a measure of the imputed marginal benefits associated with changes of the original water supply constraint y.

## III. SELECTION OF WATER SUPPLY SOURCES

If fresh water is scarce or existing and projected water demands exceed accessible resources, an investigation of all possible supply alternatives in an entire region becomes desirable. Alternatives may include ground-water fields, advanced waste treatment, or desalination of sea water. To find the most economic combination of resource and supply points, mathematical programming techniques can be efficiently applied.

The objective of economic optimality was chosen to be cost minimization. Cost functions are commonly composed of capital and OMR terms (operation, maintenance, repair), including factors due to unit costs, amortization rates, interest rates, choice of design periods. In addition, the minimization is subject to constraints denoting physical, technological, economic relationships and requirements, like water demand, hydraulic head losses, and maximum water supply.

Objective function and constraints set may be either linear or nonlinear. Efficient algorithms exist for solving both linear programming and nonlinear programming models.[3] A probabilistic approach will not be considered in this presentation.

Given points of water demand, $i \in I^d$, and possible locations of water supply sources, $i \in I^s$, a schematic network of pipe connections j between source and demand points can be established as a directed graph.[4] Hence, once the location alternatives of the supply points are included in the network, the variables of interest remain the amounts of water $x_j$ being transported from source to demand points. The set of links j originating at source i shall be denoted by $J_i^s$, and connections j ending at demand points i shall be contained in $J_i^d$. Accordingly, total supply from point i:

$$X_i = \sum_{j \in J_i^s} x_j, \forall i \in I^s \tag{5}$$

and total demand at point i,

$$X_i = \sum_{j \in J_i^d} x_j, \forall i \in I^d \tag{6}$$

It is convenient to compound capital and OMR costs of the raw water development and treatment in one term:

$$C_i(X_i) = \alpha_i X_i^{\beta_i}, \quad \forall i \in I^s \tag{7}$$

Second, piping costs arise including first costs, amortized capital and OMR costs of pipes and pumps, pipe installment and energy costs:

$$C_j(x_j) = \alpha_j x_j^{\beta_j}, \forall j \tag{8}$$

Note that amortization as well as the time stream of OMR and energy costs require knowledge of interest and depreciation rates. On one hand depreciation is considered a realistic measure to secure necessary replacement after the economic lifetime of the equipment. But it is also held that technological advances cannot be clearly foreseen, and that rather OMR costs and the existing final "scrap value" of the equipment over a specified time horizon are sufficiently close approximations of future costs. Also, the cost functions imply economic and technological assumptions concerning economic lifetimes, design periods, time horizons. This will be discussed in more detail in Section VI.

A further breakdown of the cost in separate capital and OMR and/or energy cost terms may be justified depending on the quality of additional cost parameter estimates.

The parameters $\alpha_i$, $\alpha_j$ compound factors due to first costs, amortized capital, OMR and energy costs. The exponents $\beta_i$, $\beta_j$ may denote economies of scale. The magnitude of these exponents determines whether the response surface is convex or nonconvex, which is tantamount to expecting global or local cost optima. The additional constraints express the upper limits on the maximum amount of water $B_i^s$ available from source i

$$X_i \leq B_i^s, \forall i \in I^s \tag{9}$$

and the lower limits on the water demand $B_i^d$,

$$X_i \geq B_i^d, \forall i \in I^d \tag{10}$$

Hence, a NLP follows: minimize

$$\sum_{i \in I^s} \alpha_i X_i^{\beta_i} + \sum_j \alpha_j x_j^{\beta_j} \tag{11}$$

subject to

$$X_i = \sum_{j \in J_i^s} x_j, \forall i \in I^s \tag{12}$$

$$X_i = \sum_{j \in J_i^d} x_j, \forall i \in I^d$$

$$X_i \leq B_i^s, \quad \forall i \in I^s$$

$$X_i \geq B_i^d, \quad \forall i \in I^d$$

Note that the constraints are linear, which greatly facilitates the solution. If the $\beta$ exponents are larger or equal to unity, a global minimum cost solution follows. Otherwise, different initial solutions are selected to iterate towards local minima. At this point external judgment is helpful to locate initial solutions close to potential optima. The large number of local minima commonly prohibits their total evaluation. However, a sample of local optima may contain valid information about the global.[5]

Examples of the previous model were recently developed for the James River region, Virginia,[6] and the regional North Atlantic plan of the U.S.[7]

## IV. WATER TREATMENT PLANT DESIGN

The formulation of the technological functions describing the hydraulic, chemical, physical performances and reactions of the unit processes involved is an essential part of the systems analysis approach. A general problem arises when trying to avoid inscrutable numerical intricacy but not at the expense of an adequate mathematical model. Aiming at optimization, it is most useful to devise models amenable to existing solution algorithms. The consideration of available solution techniques during the modeling process may substantially abbreviate the optimization.

As an example, Geometric Programming is introduced with an application in the optimization of the hydraulic design of sedimentation tanks. This problem occurs both in water and wastewater treatment. Geometric Programming, a recent optimization method,[8] is most suited for solving typical unit processes design problems.

### *Geometric Programming*

Consider the minimization of a cost function consisting of i = 1...n positive terms: minimize

$$f(x) = \sum_{i=1}^{n} u_i = \sum_{i=1}^{n} c_i \prod_{j=1}^{m} x_j^{a_{ij}} \tag{13}$$

where $x_j$, j = 1...m, are *primal* design variables and $c_i$ cost parameters.

Introducing the *dual* variables $w_i$ so that

$$\sum_{i=1}^{n} w_i = 1$$

an inequality relationship between the arithmetic and geometric mean can be formed:

$$\sum_{i=1}^{n} \left(\frac{u_i}{w_i}\right) w_i \geq \prod_{i=1}^{n} \left(\frac{u_i}{w_i}\right)^{w_i} \tag{14}$$

$$\sum_{i} u_i \geq \prod_{i} \left(\frac{u_i}{w_i}\right)^{w_i}$$

$$\sum_{i=1}^{n} c_i \prod_{j=1}^{m} x_j^{a_{ij}} \geq \prod_{i=1}^{n} \left(\frac{c_i}{w_i}\right)^{w_i} \prod_{j=1}^{m} x_j^{D_j}$$

where

$$D_j = \sum_{i=1}^{n} w_i a_{ij} \tag{15}$$

Proposing $D_j = 0, \forall\, j$, it follows that

$$\sum_{i=1}^{n} u_i \geq \prod_{i=1}^{n} \left(\frac{c_i}{w_i}\right)^{w_i} \tag{16}$$

The left side of the inequality is called *primal* and the right side the *dual* function. A mathematical proof shows that the minimum of the primal and the maximum of the dual are identical. Hence, maximization of the dual function with respect to the dual variables $w_i$ can be used to obtain the optimal solution to the primal cost minimization problem. In addition it is necessary to maximize

$$d(w) = \prod_{i=1}^{n} \left(\frac{c_i}{w_i}\right)^{w_i} \tag{17}$$

subject to the m + 1 linear constraints:

$$\sum_{i=1}^{n} w_i = 1, \text{ normality} \tag{18}$$

$$\sum_{i=1}^{n} w_i a_{ij} = 0, \forall\ j, \text{ orthogonality} \tag{19}$$

Note that if n = (m + 1), the linear constraints set is necessary and sufficient to solve for the n dual variables $w_i$.

Once the optimum $d^*(w^*)$ of the dual function has been obtained, it follows from

$$d^* = \sum_{i=1}^{n} u_i = \sum_{i=1}^{n} \left(\frac{u_i}{w_i^*}\right) w_i^* \tag{20}$$

that

$$d^* = \frac{u_i}{w_i^*} \tag{21}$$

This relationship is used to evaluate the primal variables. Also, it shows that the optimal dual variables $w_i^*$ equal the proportion of the optimum due to the *i*th term in the primal objective function.

If the primal cost function f(x) is subject to primal constraints, the constraints are first rewritten in the form:

$$u_k(x) \leq 1, \ k = 1 \ldots t \tag{22}$$

Then, again, it has been shown that the primal problem can be transformed to maximizing the following dual function:

$$d(w) = \prod_{i=1}^{n} \left(\frac{c_i}{w_i}\right)^{w_i} \prod_{k=1}^{t} c_K^{\,w_K} \tag{23}$$

subject to the constraints

$$\sum_i w_i = 1 \tag{24}$$

$$\sum_i w_i a_{ij} + \sum_k w_k a_{kj} = 0, \ \forall j$$

*An Example*

This example was developed by Harrington.[9]

A rectangular sedimentation tank of length $x_1$, width $x_2$ and depth $x_3$ is to be designed at minimum cost for an influent Q. The cost function is given as

$$f(x) = x_1x_2 + 4x_1x_3 + 4x_2x_3 \tag{25}$$

In addition the following design rules are to be observed:

1. the theoretical detention time must be greater than or equal to 2 hours
2. the ratio of length to width must not be less than 4
3. the weir loading rate, in gallons per day per foot of weir length, must not exceed 30,000 gallons per day per foot.

Adjusting the dimensional units and substituting Q = 4mgd, the design rules yield the constraints set:

$$\frac{44560}{x_1x_2x_3} \leq 1 \tag{26}$$

$$\frac{4x_2}{x_1} \leq 1 \tag{27}$$

$$\frac{33.33}{x_2} \leq 1 \tag{28}$$

Hence, the associated dual problem follows: maximize

$$d(w) = \left(\frac{1}{w_1}\right)^{w_1}\left(\frac{4}{w_2}\right)^{w_2}\left(\frac{4}{w_3}\right)^{w_3}(44560)^{w_4}(4)^{w_5}(33.33)^{w_6} \qquad (29)$$

Subject to

$$\begin{aligned} w_1 + w_2 + w_3 &= 1 \qquad (30) \\ w_1 + w_2 - w_4 - w_5 &= 0 \\ w_1 + w_3 - w_4 + w_5 - w_6 &= 0 \\ w_2 + w_3 - w_4 &= 0 \end{aligned}$$

The constraints set can be employed to eliminate four variables in d(w) leaving the dual function with two remaining variables and without constraints. Gradient techniques[3] will deliver the global maximum. Solution vector:

$$w^* = (0.40, 0.48, 0.12, 0.60, 0.28, 0.20) \qquad (31)$$

$$d^*(w^*) = 11127.$$

$$x^* = (133.33, 33.33, 10.33) \text{ ft.}$$

## V. WATER DISTRIBUTION SYSTEMS

Water distribution networks usually contribute the largest share of the total water supply systems cost. Therefore, optimal designs are highly desirable.

Traditionally, trial and error selection of diameters and subsequent verification of the hydraulic network balance have been used for optimization. This method is still valid if a small number of conduits is involved but proves infeasible for large scale designs.

Historically and analytically network optimization developed in three steps:

1. hydraulic balance of a predesigned network (graph) for given diameters

2. cost optimization for a given graph, but unknown diameters, observing the hydraulic laws
3. development of optimal graphs that lead to optimal costs, observing the hydraulic laws.

Sufficient experience is available concerning point 1.[10] To the author's knowledge research has not yet produced applicable results concerning point 3.

Over several decades problem 2 was investigated (*e.g.* by Camp,[11] Lischer,[12] and Neigut[13]), and research increased with the advent of modern computers and the application of mathematical programming methods. Pitchai[14] applies a random sampling model to solve the general problem of finding optimal diameters at minimum energy and capital cost for multiple load patterns. Jacoby[15] formulates the problem as an unconstrained sequential optimization, and Smith[16] compares different solution techniques: linear programming, steepest descent, random sampling.

The common difficulty encountered by previous researchers arises from the nonconvexity of the network cost minimization problem. The solution methods that have been suggested yield local minima and involve considerable numerical intricacy.

The following approach uses graph theory to decompose the original nonconvex version into convex subproblems amenable to standard solution algorithms.[17] As in previous models, diameters are treated as continuous variables rather than integers.

*The Tree Principle*

The incidence matrix A is introduced to describe the topology of N nodes, B branches and L loops of a distribution network. An element $a_{ij}$ of A is equal to +1 if the flow in branch j points to node i, -1 if the flow points away from i, 0 if branch j is not incident upon node i. The matrix A is nonsingular if a reference is omitted, and subsequently A will have the dimension (N-1)xB.

Let the flows in the pipes be given by vector $f = (\ldots, f_j, \ldots)$, the head loss by $h = (\ldots, h_j, \ldots)$ and the piezometric pressures at the nodes by $p = (\ldots, p_i, \ldots)$.

Kirchhoff's laws propose flow balance at the nodes and pressure balance along any loop:

$$Af = q \tag{32}$$

$$A^T p = h \tag{33}$$

with $A^T$ denoting the transpose of A.

By definition, a network *tree* contains a subset of all branches that connect all the nodes without generating a loop. Hence, partitioning A into an (N-1) square matrix E and into an (N-1) x (B-N+1) rest matrix G,

$$A = [E|G] \tag{34}$$

it follows that E may represent a tree. If it does, it can be shown for the determinant $|E|$ that

$$|E| = \pm 1.$$

The total number of the trees T of a network is given by Weinberg:[18]

$$T = |AA^T| = \sum_i |E_i E_i^T| \tag{35}$$

A listing of the branches belonging to a tree may be derived from Trent:[19]

$$M = |ADA^T| \tag{36}$$

D is an (N-1) square matrix carrying the branch designations on the main diagonal. The value of the determinant M is the sum of products of branch designations, any product being the sequence of branch names of a tree.

Branches that are not contained in a tree are called *chords*. An alternative way to record trees by digital computers is suggested by assigning zero values to (B-N+1) potential chord elements in A. Then, any of the $\binom{B}{N-1}$ possible column combinations in A forming an (N-1) square matrix E with no zero rows must represent a tree.

Let the flows be partitioned according to Equation 34 into tree flows $f_E$ and chord flows $f_G$. From Equation 32 it follows that

$$[E|G] \begin{pmatrix} f_E \\ f_G \end{pmatrix} = q \tag{37}$$

$$Ef_E + Gf_G = q$$

$$f_E = E^{-1}q - E^{-1}Gf_G. \tag{38}$$

The set of linear equations (38) determines the flow in any tree given the flows in the chords.

Any loop has one degree of freedom, *i.e.*, any loop contains one chord, if a tree is formed. This is in accordance with Euler's law:[4]

$$L = B - N + 1 \tag{39}$$

*Convex Capital Cost Minimization*

Assuming the pipe diameter, d, to be continuous, and introducing $\ell$, the pipe length, c, the price per unit length including installment, the costs of a pipe line are given by Camp:[11]

$$C = c\ell d^{w} \tag{40}$$

Scaling factors w typically range between

$$1 \leq w \leq 1.5,$$

where the lower bound may indicate smaller pipe sizes and water distribution systems,[15] and the upper bound larger diameter of overland pipe lines.[20,21]

To relate head loss h (ft), friction factor r, flow f (cfs), and diameter d (ft) of a pressure conduit, Darcy-Weisbach's formulation

$$h = 0.0252\ r\ell f^2/d^5 \tag{41}$$

is employed, but the following considerations hold for any equivalent hydraulic expression. The exponent of f, for example, varies approximately between 1.7 and 2.0 in different versions of Equation 41, like the Hazen-Williams formula or the Colebrook-Moody diagram.[22]

Substituting the diameter in Equation 40 by Equation 41:

$$C = \gamma \frac{f^{0.4\ w}}{h^{0.2\ w}} \tag{42}$$

$$\gamma = c\ell\ (0.0252\ r\ell)^{0.2\ w}$$

The cost of a distribution network, with branches j = 1,...,B, follows accordingly:

$$C\ (h,f) = \sum_{j=1}^{B} \gamma_j \frac{f_j^{\,0.4\ w}}{h_j^{\,0.2\ w}} \tag{43}$$

The nonconvex function C (h,f), Equation 43, is concave with respect to the flows, f, and convex with regard to the head losses, h, within the indicated range of w and including different versions of the hydraulic relationship, Equation 41.

The objective is to minimize costs, *i.e.* Equation 43. Convexity of the function C (h,f) and of any associated constraints space is, therefore, desirable. Given f, Equation 43 will be convex:

$$C\ (h|f) = \sum_{j=1}^{B} \gamma_j \frac{f_j^{\,0.4\ w}}{h_j^{\,0.2\ w}} \tag{44}$$

Concavity of expression 43 with respect to f can be interpreted as economies of scale. Hence, the most economical solution must show flows concentrated in pipes with high economies of scale compared to minimum flows in pipes with small economies of scale.

In a distribution system minimum flows are limited to quantities required by minimum supply. Also, the economies of scale still depend upon the elements of vector h that are quantitatively unknown at this point for each individual pipe. Hence, the problem is to find all possible combinations of distributing minimum flows in a network.

According to the Tree Principle, the flows in a network are determined by the flows in the chords. Assuming that demand and supply are being exerted at the nodes, each loop contains one chord whose flow may be chosen. Assigning minimum flows to the chords offers the solution to finding all possible combinations of minimum flows that may now be evaluated by expressions 36 and 38.

It may appear desirable to combine minimum flows with maximum admissible head losses to obtain minimum diameter constraints according to expression 41. If the flows are determined by the Tree Principle, Equation 38, using minimum flows for the chords, the original nonconvex objective function, expression 43, can be replaced by the convex equivalent, Equation 44: minimize

$$C(h|f) = \sum_{j=1}^{B} \gamma_j \frac{f_j^{0.4\,w}}{h_j^{0.2\,w}}$$

In addition, the following set of linear constraints must be observed:

*Type a,* Kirchhoff's loop equations:

$$\sum_j h_{jk} = 0, \; j \in \text{loop } k, \; \forall\, k \tag{45}$$

*Type b,* maximum admissible head loss:

$$\sum_j h_{j\pi} \leq H_\pi, \; j \in \text{path } \pi, \; \forall\, \pi, \tag{46}$$

$\pi$ leading from a reference node to specified nodes of the network.

The resulting set of nonlinear minimization problems are convex with linear constraints. Standard mathematical programming algorithms are available for solutions.[23] The global minimum of the original nonconvex cost function, expression 43, follows by comparing the solutions of the decomposed convex versions.

The number of the convex problems are derived by evaluating the trees of the network and may be relatively large. Numerical examples[17] showed that the number is substantially reduced by controlling the flow incidences to avoid uneconomical flow recirculations. For example, as a simple case, flow vectors must point away from supply input nodes.

Since diameters are considered continuous in the model a final conversion of the resulting optimal diameters to integer values will be necessary. Cost increases due to rounding off diameters to the closest upper standard pipe size were experienced to remain within ten per cent. Adjusting the continuous optimal diameters to upper and lower standard sizes in accordance with the hydraulic constraints will further reduce cost differences.

### *Conclusions*

The nonconvex capital cost function of a hydraulic network is transformed to subsets of nonlinear convex functions by a decomposition principle from graph theory. The variables are the flows and head losses in each pipe and the constraints are linear expressions of the head losses. The global minimum cost solution follows using standard nonlinear programming algorithms. Pumps, standpipes, pressure controls, etc. can be incorporated since flow inputs at each node enter the formulation explicitly and pressure potentials are controlled by the constraints set.

Also, the model allows the minimization of pumping energy to be included by augmenting the original objective function, Equation 44, by the linear terms

$$\sum_{j=1}^{B} e_j f_j h_j$$

$e_j$ denoting conversion and cost factors of the power consumption of a pressure conduit j.

## VI. CAPACITY EXPANSIONS

The following model implies the summary of the yearly capital and OMR-energy costs of a complete water supply system of capacity Q placed into service at time $T_{i-1}$ as $\alpha_i Q(T_i)^{\beta i}$, and costs accruing subsequently from expansions and extensions of the system introduced at time $T_i$ as $\alpha_{i+1} Q(T_{i+1})^{\beta i+1}$ (Figure 7.1). This formulation is closely related to

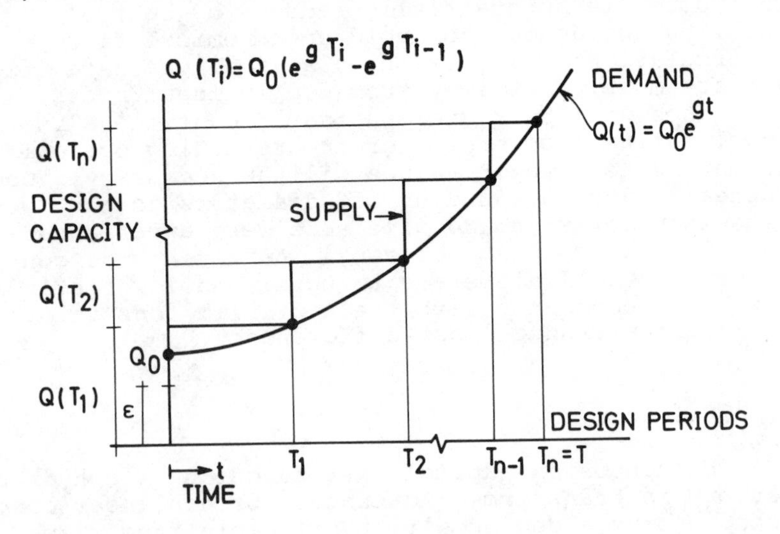


*Figure 7.1. Capacity expansion model.*

capacity expansion models presented in the literature.[24,25] The parameters $\alpha$ and $\beta$ may be derived following an analysis as presented in previous chapters, which implies that they will include characteristics of the design as well as their response to the water demand selected for the design.

In general, it may also be desirable to investigate alternative solutions for further development and expansion of a system of future design periods, which may include, for example, the introduction of different raw water sources at more than one location, various layouts and interactions of storage and pumping units, and options to reinforce and extend the distribution grid. It follows that selecting one particular set and arrangement of components at stage i may not only constrain the design at stage (i+1) but also influence $\alpha_{i+1}$ and $\beta_{i+1}$.

The physical lifetime, $t_j$, also called economic life in the literature, is used so that any system component j is assumed to be perpetually replaced after $t_j$ years at the same cost. This approach is meant to take into account the fact that once a major component is built its function in the system will be sustained for a period that appears to be, for all practical purposes, infinite. In the meantime, alterations and extensions of the same component j may take place at different design periods i, and the model assumes that such activities incur a constant cost rate as far as the existing capacity is concerned. However, expansions and extensions introduced at $T_i$ that increase the system's capacity are assumed to start a new timestream of costs. Although alterations or major repair works may also be categorized and accounted for as maintenance, the introduction of physical lifetimes in expression 47 is considered a realistic concept and flexible enough to adjust the model to any desirable range. For example, $t_j = \infty$ is a common assumption. If one wishes to include a salvage value at the end of $t_j$, he must adjust the cost, $c_j$, of component j in order to maintain the constant timestream of costs in Equation 47. However, one may also consider the retention of a salvage value to account for costs arising from scrapping the old unit or for inflationary price increases.

The present value of a set of components j constituting a supply system and placed into service at time, $T_{i-1}$, may now be stated as

$$\alpha_i \int_{T_i}^{\infty} Q(T_i)^{\beta_i} e^{-zt} dt = \int_{T_i}^{\infty} \left\{ \sum_j \frac{c_j}{\int_o^{t_i} e^{-zt} dt} \right\} e^{-zt} dt \tag{47}$$

where z is the instantaneous discount rate.

The formulation of this model is presented primarily as an alternative means to further facilitate decisions related to the selection of design periods in a preliminary screening process. It is doubtful whether refinements would clearly improve the model, considering the imponderables involved in projecting estimates of parameters for developing countries in particular and the relative insensitivity of capacity expansion models in general. However, the detailed analysis of the previous chapters may be used to distill a wide range of information into the lump-sum cost formulation $\alpha Q^{\beta}$.

Assuming a planning horizon of T years with n construction stages or design periods and disallowing supply backlog at future stages, the problem may be stated: minimize

$$\sum_{i=1}^{n} \left\{ \int_{T_i}^{\infty} \alpha_i Q(T_i)^{\beta_i} e^{-zt} dt \right\} \tag{48}$$

The model also implies the assumption, common to other capacity expansion models in the literature, that $T_i$ denotes the intersection of supply and demand as well as the point where the total cost of stage (i+1) occurs. It is recognized that the construction period requires a finite period of time, and that costs and increases in demand may occur over that period, but this is not explicitly considered.

Using a geometric demand growth (see Figure 7.1) so that $Q(T_i) = Q_O(e^{gT_i} - e^{gT_{i-1}})$ and recognizing a partial initial capacity, $\varepsilon$, so that $0 \le \varepsilon \le Q_O$, expression 48 may be restated as the following mathematical programming problem: minimize

$$F(\ldots,T_i,\ldots) = \frac{\alpha_1}{z} \left( Q_O e^{gT_1} - \varepsilon \right)^{\beta_1}$$

$$+ \sum_{i=2}^{n} \left\{ \frac{\alpha_i Q_O^{\beta_i}}{z} \left( e^{gT_i - (z/\beta_i)T_{i-1}} - e^{T_{i-1}[g-(z/\beta_i)]} \right)^{\beta_i} \right\} \tag{49}$$

subject to

$$T_{i+1} \geq T_i, \quad i=1,\ldots,n$$

$$T_n = T. \tag{50}$$

To facilitate the optimization, available information about the curvature of F would be helpful and, in general, be provided by the second derivative. However, unless evaluated for a specific numerical example, the resulting expressions proved analytically inscrutable in general terms.

An expensive approach for solving 49, subject to 50, would consist, for example, of "Random Methods for Nonconvex Programming."[5] But the character of the constraints and the extreme smoothness of the response surface encountered suggested the use of a simple scanning algorithm for exhaustive sampling. Its efficiency was tested in a numerical example for T=50 years: evaluating the objective function for a series of discrete points $T_1,\ldots,T_n = T$, and various n so that

$$\left\{\ldots\; T_1 = \Delta t + i\Delta t, \quad \left[i = \Delta t,\ldots,\Delta t\left(\frac{T_n}{\Delta t}\right)\right]\right. \tag{51}$$

$$\left. T_2 = T_1 + j\Delta t, \quad \left[j = \Delta t,\ldots,\Delta t\left(\frac{T_n - T_1}{t}\right)\right],\ldots,T_n\right\}$$

will automatically observe the set 50.

Local minima were further refined, or "sweetened," by applying a gradient method to 49. The results listed in Table 7.1 and 7.2 were obtained for the following data input: T=50 yrs., $\alpha_1=\alpha_2=\ldots=\alpha_n=9.6\text{x}10^4$, $\beta_1=\beta_2=\ldots=\beta_n=0.7$, z=0.07, g=(0.055,0.065, 0.075,0.085), ε=(0,2.0) mgd., $Q_0$=4.0 mgd., n=3, ..., 7.

The structure of the optimal solution vector (Table 7.1) indicates some inherent limitations of a capacity expansion model. With initial backlog and constrained time horizon, the optimal solution does not consist of equidistant time intervals as in the balanced case and for an infinite time horizon[26] but of a series of decreasing elements. The numerical results often show shorter time periods in the distant future. But, relatively short design periods

*Table 7.1*

*Cost for Total Time Horizon, T (50 yrs.), and First Design Period, $T_1$ in \$$10^6$*

$\varepsilon = 0$

| *n = no. of design periods in T* | *g = 0.055* | *0.065* | *0.075* | *0.085* |
|---|---|---|---|---|
| 7 | 10.92<br>5.745 | 13.33<br>6.248 | 16.37<br>6.448 | 20.24<br>6.964 |
| 6 Optimum | 10.38<br>5.745 | 13.280<br>6.248 | 16.31<br>6.448 | 20.17<br>6.964 |
| 5 | 10.88<br>5.970 | 13.30<br>6.248 | 16.35<br>6.795 | 20.27<br>7.391 |
| 4 | 10.98<br>6.204 | 13.48<br>6.843 | 16.65<br>7.162 | 20.73<br>7.844 |
| 3 | 11.39<br>6.701 | 14.12<br>7.495 | 17.62<br>8.835 | 22.16<br>9.952 |

$\varepsilon = 2$

| *n = no. of design periods in T* | *g = 0.055* | *0.065* | *0.075* | *0.085* |
|---|---|---|---|---|
| 7 | 9.83<br>4.422 | 12.27<br>4.616 | 15.33<br>5.067 | 19.23<br>5.545 |
| 6 Optimum | 9.79<br>4.422 | 12.22<br>4.906 | 15.27<br>5.067 | 19.16<br>5.545 |
| 5 | 9.801<br>4.660 | 12.25<br>4.906 | 15.34<br>5.423 | 19.29<br>5.976 |
| 4 | 9.930<br>4.906 | 12.47<br>5.521 | 15.68<br>6.187 | 19.80<br>6.910 |
| 3 | 10.39<br>5.695 | 13.16<br>6.541 | 16.72<br>7.478 | 21.32<br>8.517 |

*Table 7.2*

*Optimal Vector $T_i$*

| $i$ | $g = 0.055$ | $0.065$ | $0.075$ | $0.085$ |
|---|---|---|---|---|
| 1 | 12.0 | 11.6 | 10.8 | 10.0 |
| 2 | 19.7 | 18.8 | 18.5 | 18.0 |
| 3 | 27.7 | 26.8 | 26.5 | 26.0 |
| 4 | 35.7 | 34.8 | 34.5 | 34.0 |
| 5 | 43.7 | 42.8 | 42.5 | 42.0 |
| 6 | 50.0 | 50.0 | 50.0 | 50.0 |
| Cost in $\$10^6$ T | 10.88 | 13.28 | 16.31 | 20.18 |
| $T_1$ | 5.745 | 6.143 | 6.381 | 6.56 |

$\varepsilon = 0$, no. of design periods in T: $n = 6$.

can hardly be expected to correspond to a realistic planning policy at that future point, even for centralized economies with long-term planning goals. Yet, the period of immediate concern, the first design period, will be influenced by applying such an optimization scheme. One may, therefore, tend to advocate longer time horizons, only to arrive at a diminishing reliability of parameter estimates and decreased influence of a constrained time horizon.[27]

REFERENCES

1. Shipman, H. R. "Water Supply Problems in Developing Countries," *J. Am. Wat. Works Assoc.*, *59(7)*:767 (1967).
2. Thomas, H. A. "Operations Research in Water Quality Management," Harvard Water Resources Group, Report to Public Health Service, Department of Health, Education and Welfare (1963).
3. Hadley, G. *Nonlinear and Dynamic Programming*. (Reading, Mass.: Addison-Wesley Publishing Co., 1964).
4. Berge, C. and A. Chouila-Houri. *Programming Games and Transportation Networks*. (London: Methuen and Co. Ltd., 1965).

5. Rogers, P. "Random Methods for Nonconvex Programming," Harvard Water Resources Group (Cambridge, Mass.: Harvard University, 1966).
6. Young, G. K. and M. A. Pisano. "Nonlinear Programming Applied to Regional Water Resource Planning," *Water Resour. Res. 6(7):*32 (1970).
7. Delucia, R. J. and P. Rogers. "North Atlantic Regional Supply Model," *Wat. Resour. Res. 8(3):*760 (1972).
8. Duffin, R. J., E. L. Peterson, and C. Zener. *Geometric Programming* (New York: Wiley, 1967).
9. Harrington, J. J. and M. B Fiering. "Systems Analysis in River Basin Planning," Pan Health Organization (1968).
10. Cembrowicz, R. G. and J. J. Harrington. "Discussion. Comparative Analysis of Water Distribution Systems," *J. Hydr. Div., Proc. ASCE, 98:*1890 (1972).
11. Camp, T. R. "Economic Pipe Sizes for Water Distribution Systems," *Trans. Am. Soc. Civ. Eng., 104:*190 (1939).
12. Lischer, V. C. "Determination of Economical Pipe Diameters in Distribution Systems," *J. Am. Wat. Works Assoc. 40(8):*848 (1948).
13. Neigut, E. G. "Distribution Designed by a New Method," *Wasteworks Wastes Eng. 3:*46 (1964).
14. Pitchai, R. "A Model for Designing Water Distribution Pipe Networks," Harvard Water Resources Group (Cambridge, Mass.: Harvard University, 1966).
15. Jacoby, L. S. "Design of Optimal Hydraulic Networks," *J. Hydr. Div., Am. Soc. Civ. Eng. 94(3):*647 (1968).
16. Smith, D. V. *Minimum Cost Design of Linearly Restrained Water Distribution Networks.* (Cambridge, Mass.: Massachusetts Institute of Technology, 1966).
17. Cembrowicz, R. G. "Mathematical Model of a Water Supply System Under Fluctuating Demand," Environmental Systems Program (Cambridge, Mass.: Harvard University, 1971).
18. Weinberg, L. *Network Analysis and Synthesis* (New York: McGraw-Hill, 1962).
19. Trent, H. M. "A Note on the Enumeration and Listing of all Possible Trees in a Connected Linear Graph," *Proc. Nat. Acad. Sci. 40:*10 (1954).
20. Thomas, Harold A. Personal communication. (1969).
21. Linaweaver, F. P. and C. S. Clark. "Cost of Water Transmission," *J. Am. Wat. Works Assoc. 56:*12 (1964).
22. Fair, G. M., J. C. Geyer, and D. A. Okun. *Water and Wastewater Engineering*, Vol. 1 (New York: John Wiley & Sons, 1966).
23. Zoutendijk, G. *Methods of Feasible Directions* (Amsterdam: Elsevier Publishing Co., 1960).
24. Manne, A. S., ed. *Investment for Capacity Expansion-Size, Location and Time-Phasing.* (Cambridge, Mass.: M.I.T. Press, 1967).

25. Muhich, A. J. "Capacity Expansion of Water Treatment Facilities," Harvard Water Resources Group (Cambridge, Mass.: Harvard University, 1966).
26. Srinivasan, T. N. "Geometric Rate of Growth of Demand," in *Investments for Capacity Expansion*, Manne, A. S., ed. (Cambridge, Mass.: M.I.T. Press, 1967).
27. Ortolano, L. "Artificial Aeration and the Capacity Expansion of Wastewater Treatment Facilities," Harvard Water Resources Group (Cambridge, Mass: Harvard University, 1969).
28. Birkhoff, G. "A Variational Principle for Nonlinear Networks," *Quart. Appl. Math. 21(2)*:160 (1963).
29. Rogers, P. "Rational Design of a Water Distribution System," (Cambridge, Mass.: Harvard University, 1963).
30. Wilde, D. J. *Foundations of Optimization* (Englewood Cliffs, N.J.: Prentice-Hall, 1967).

MODELS FOR ENVIRONMENTAL POLLUTION CONTROL


# CHAPTER 8

# AN ECONOMIC APPRAISAL OF CHANGES IN WATER USE THROUGH INVESTMENTS IN NAVIGABLE RIVERS AND CANALS

W. Schmid, G. Kuhlmann, J. Mühlbauer*

## I. INTRODUCTION

This paper describes a research study of the economic appraisal of changes in water use caused by construction of weirs and canals. For the purpose of this study, the term "water use" covers urban and industrial water supply as well as agricultural irrigation and surface run-off; it does not cover transportation, hydroelectric power generation, flood control, recreation and ecology. The objective of the study was to determine the net benefits accruing to urban, industrial and agricultural water users from the construction of weirs and canals for navigation purposes.

## II. APPROACH

Canals and weirs influence the water supply and receiving water capacity of a channel-reservoir system. (Receiving water capacity is the quantitative capability of a channel-reservoir system to receive waste waters.) An obvious example of this is the regional redistribution of water and the creation of receiving water capacity by the construction of canals. But apart from quantitative changes in the regional availability of surface water and receiving water capacity, there is another impact,

---

*Messrs. Schmid, Kuhlmann, and Mühlbauer are with F. H. Kocks KG, Consulting Engineers, Augusta Strasse 30, Düsseldorf, Germany.

namely the alteration of water supply in qualitative terms. Hence, for the availability of water and receiving water capacity within a well-defined area there are three major consequences of weir and canal construction:

- an alteration of water supply in quantitative terms
- an alteration of water supply in qualitative terms
- an alteration of the receiving water capacity.

These alterations may or may not have an influence on water usage. If they have no effect on water use, because they take place within the range of excess supply, then they are economically irrelevant. If, however, these alterations do influence water utilization, then the effect is economically relevant; it can be negative or positive from each user's point of view, meaning benefits or costs to him. The individual benefits and costs add up to total benefits and costs and form the components of the net benefit accruing to urban, industrial and agricultural users. Since the purpose of this study is to determine this net benefit, one can exclude immediately those alterations that do not simultaneously influence water utilization.

Benefits accrue to water users from the constructio of weirs and canals if

- these investments cause an increase of water supply in quantitative terms and/or an improvement of water supply in qualitative terms and/or an increase of the receiving water capacity

and if

- this additional water supply and receiving water capacity is utilized.

Weirs and canals that make such a contribution to water supply and wastewater disposal save economic resources that otherwise would have been used to produce the same results by other means.

On the other hand, costs are imposed by weirs and canals to water users if--as a consequence of their construction--the quantity of raw water is reduced, its quality deteriorated and the receiving water capacity reduced so that economic resources, which otherwise could have been saved, must be sacrificed in order to restore the same level of water supply and wastewater removal. Only if alterations in water supply and receiving water capacity happen to occur within the range of excessive quantity or quality are they economically irrelevant.

In order to determine the monetary net benefit accruing to urban, industrial and agricultural water users from the construction of weirs and canals the following approach was used.

In the first step the interdependence between the construction of weirs and canals and alterations of water supply and receiving water capacity was investigated; economically relevant changes of water supply and receiving water capacity were separated from economically irrelevant changes. In the second step a simulation model was developed to determine costs and benefits in physical terms (*i.e.*, in terms of quantity, quality and receiving water capacity). In the third step a pricing concept was elaborated for the sake of transforming benefits in physical terms into costs and benefits in monetary terms.

Steps one and two are briefly outlined in this paper, while step three, the transformation of the physical benefit into monetary benefits, will be skipped.

In order to derive the physical benefits accruing to urban, industrial and agricultural water users, classical tools of cost-benefit-analysis were applied and a simulation model was developed for a WITH-WITHOUT comparison of

- supply and demand of water (quantity, quality) and receiving water capacity in the situation with the canal(s) and the weir(s) (WITH-case)
- supply and demand of water (quantity, quality) and receiving water capacity in the hypothetical situation which would prevail if the canal(s) or the weir(s) to be evaluated were nonexistent (WITHOUT-case).

For purposes of a workable WITH-WITHOUT comparison a basic assumption was made which *prima vista* may appear to be unrealistic, but which is in fact only a minor simplification of the problem. This assumption concerns the demand for water (quantity, quality) and receiving water capacity which was assumed to be the same for each user in both situations. This simply means that demand for water is not related to supply of water and implies, therefore, that the effect of excessive resources on the location of industry and population be ignored. It must, however, be kept in mind that, apart from the availability of water and receiving water capacity, there are other factors influencing locational choice. These are the availability of labor, raw materials and energy, nearby markets and immediate access to transportation networks such as domestic waterways. If one ignores the interdependence between excessive supply and demand of water resources by assuming the demand for water and receiving water capacity to be the same in both cases, one only assumes that all the other factors of locational choice are more important than the availability of water resources.

Thus, the error introduced by our assumption is confined to the small group of users who would be nonexistent (or only existent) in the WITHOUT situation for exclusively one reason, namely the alteration of the regional availability of water resources. Since there is little scope left for water supply oriented locational choice in most of the highly industrialized countries because of mounting water supply and pollution problems in all regions, the error made by ignoring interrelations between supply and demand for water is not important and perhaps even marginal.

Consider now the simulation model that was developed in order to determine the net benefit in physical terms accruing to water users from the construction of weirs and canals.

## III. THE SIMULATION MODEL

The simulation model needs specific qualifications in order to be able to transform given input data into desired output data. The input consists of the channel-reservoir-system and of the use of water and receiving water capacity for each year of the period under investigation. The desired output data are the components of the net-benefits in physical terms.

As may be seen from the diagram in Figure 8.1 there are five main elements forming the simulation model. The first element consists of the input data. The second element is devoted to the computation of water availability in both situations, WITH and WITHOUT the canal(s) or weir(s). The third element is directed at the determination of the region affected, *i.e.*, the area embracing all water users affected. The boundaries of the channel-reservoir-system are defined in subsequent steps. Step one is a preliminary definition of the area affected; obvious preliminary border points, which could be final border points, are for example points at which

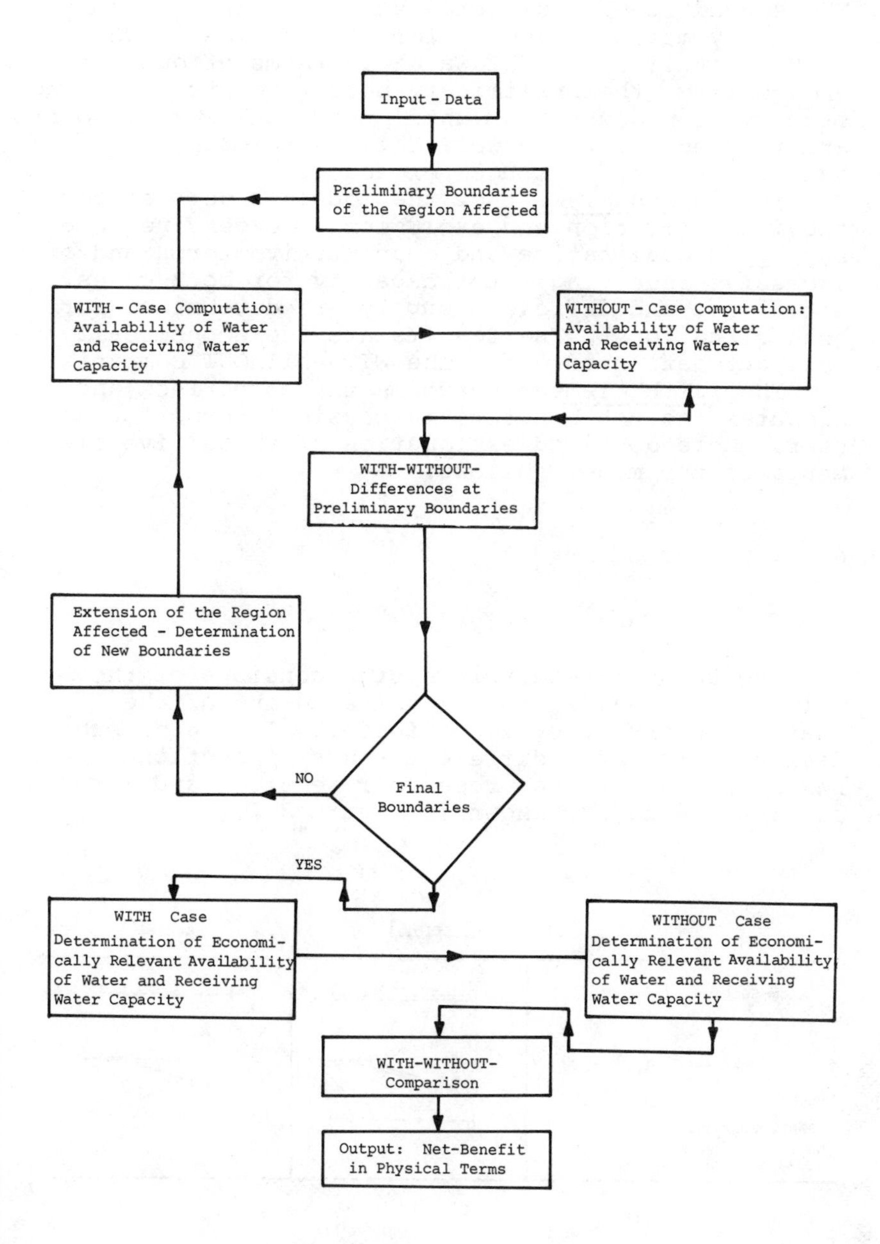


*Figure 8.1. Simulation model diagram.*

rivers meet or rivers are tapped or fed by canals. The second step is directed at deriving final border points by means of the differences between WITH and WITHOUT availability of water in terms of quantity and quality; the preliminary border points are moved step by step downstream until the final border points are reached, at which well-defined tolerances of sensitivity are no longer exceeded.

The fourth element of the model is devoted to the identification and exclusion of excessive water supply in qualitative and quantitative terms and of excessive receiving water capacity for both cases. Actual use and deficient supply are determined for both situations. The results are the economically relevant data needed for the WITH-WITHOUT comparison.

The fifth element performs the comparison and computes the net benefits in physical terms for each user. More detailed explanation of these five elements of the model follows.

## A. Input Data

### *The Channel-Reservoir-System*

The channel-reservoir-system consists of the weir(s) or canal(s) to be evaluated and of the channel-reservoir-system affected. It can be subdivided into three different kinds of sections, namely flow sections, reservoir sections and zero sections which are shown in Figure 8.2.

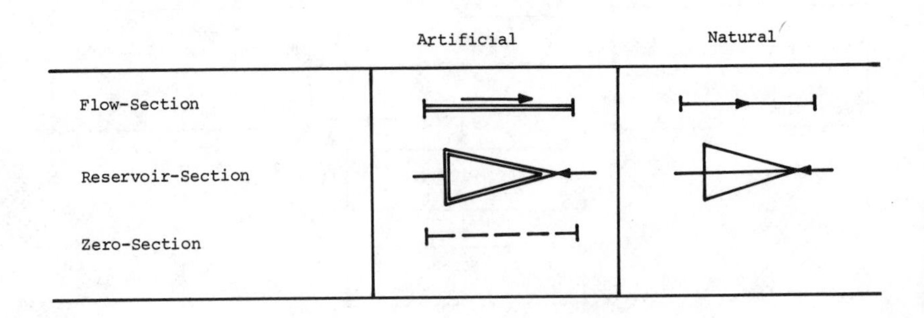


*Figure 8.2. Elements of the channel-reservoir-system.*

- flow sections or flow links are natural or artificial channels without reservoirs
- reservoir sections or reservoir links are natural or artificial reservoirs
- zero sections or zero links are hypothetical canals or canal sections. Zero links are needed to define the channel-reservoir-system in the WITHOUT situation; they are identical with the axis of the canal existing in the WITH case but assumed to be non-existent in the WITHOUT case.

The channel-reservoir system of the WITHOUT situation is different from the one existing in the WITH situation: Reservoir links of rivers describing reservoirs for navigation purposes existing in the WITH situation are transformed into flow links in the WITHOUT situation by assuming the weir(s) to be nonexistent. Canals described by either reservoir links or flow links in the WITH situation are transformed into zero links of the WITHOUT case.

### *Demand for Water and Receiving Capacity*

Demand for water and receiving water capacity is equal in both cases and assumed to correspond with actual water use. The demand for water is described for each user in terms of the quantity and quality needed by each user in the WITH situation while the demand for receiving water capacity is described for each user by the amount and quality of wastewater fed back into the channel-reservoir-system in the WITH situation. Users or homogeneous groups of users are depicted in Figure 8.3.

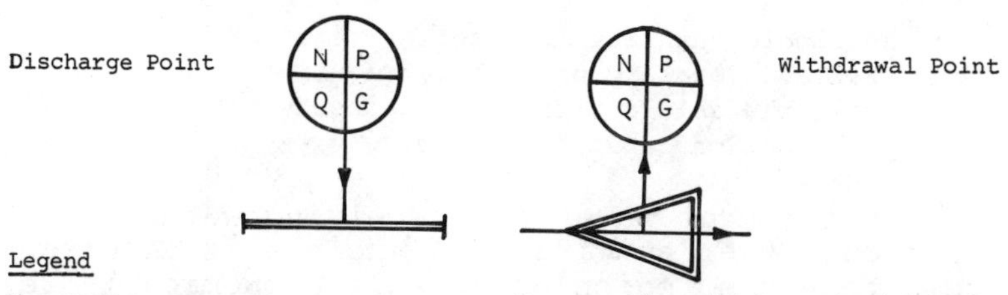


Legend

N = user
P = withdrawal-point or discharge point (location at the channel-reservoir-system)
Q = quantity withdrawn or fed back
G = water quality required at withdrawal points and quality of wastewater at discharge points

*Figure 8.3. Water and receiving water utilization of user.*

Figure 8.4 describes a hypothetical channel-reservoir-system affected by the construction of a canal and the users linked to it. The left side depicts the situation WITH the canal; the right side shows the situation WITHOUT the canal.

## B. Computation of WITH and WITHOUT Availability of Water and Receiving Water Capacity

The region affected was depicted by a linear system of channels and reservoirs with users or homogeneous user groups linked to it at specific points of withdrawal and wastewater discharge. In order to determine the availability of water and receiving water capacity there is no need to know the exact location of each user within the region. What is needed is simply the exact location of each user's withdrawal point for water and each user's discharge point for wastewater. Each user was therefore transferred to its withdrawal point and to its discharge point at the channel-reservoir-system so that for the purpose of this study each user may have two locations if his withdrawal and discharge points do not coincide geographically.

Since the availability of water and receiving water capacity is altered by the construction of weirs and canals, there are differences in supply between the two situations while demand is the same in both cases. Actual use of water and receiving water capacity may coincide with either supply or demand, depending on which of the following two factors is the limiting factor and, therefore, economically relevant:

a. available quantity or quantity required
b. available quality or quality required
c. available self-purification capacity or self-purification capacity required.

Availability of water and receiving water capacity is recorded either at withdrawal points or at discharg points within the region affected. Economically relevant differences in quantity and quality of water are computed only for points of withdrawal; differences of receiving water capacity are computed only for discharge points. Differences in quality of the wastewater returned to the channel-reservoir-system are not recorded at discharge points but at withdrawal points located farther downstream.

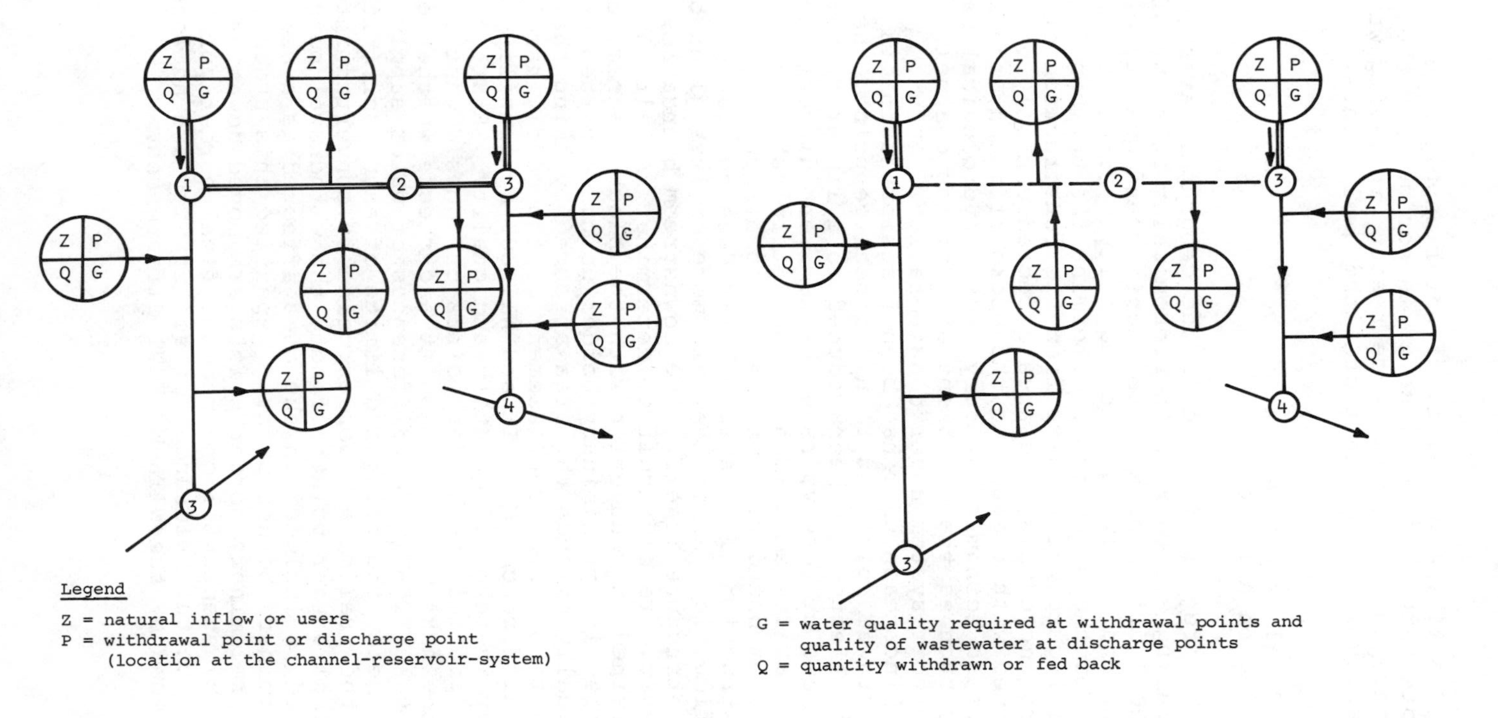


*Figure 8.4. Example for a channel-reservoir-system.*

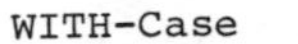

Quantity and quality of water supply is unimportant at discharge points since it is not used at these places, while, for the same reason, receiving water capacity is unimportant at withdrawal points.

## C. Final Determination of the Region Affected

The region affected is the geographic area embracing all users with economically relevant WITH-WITHOUT differences as to quantity and quality of water and receiving water capacity. Since the objective of the study is to determine the net benefit accruing to water users from the construction of weirs and canals, the area affected can only be defined with respect to the users affected; therefore it is an economic category, not a hydrological category such as the catchment area of the channel-reservoir-system under consideration.

Since each user has been transferred to his withdrawal point and to his discharge point, the area affected has been reduced to a linear system of channels and reservoirs with the users directly located at its embankment. The determination of the upstream boundaries of the region affected by the construction of a canal or weir does not cause difficulties, since they are more or less given by the structure itself. The downstream boundaries, however, are difficult to determine. As already explained in connection with the block-diagram, one starts with preliminary border points and then gradually extends the area affected by using the results of the second element of the model, namely the differences of WITH and WITHOUT availability of water in terms of quantity and quality. Starting with preliminary border points one goes downstream to the next withdrawal point and checks whether or not the WITH-WITHOUT differences of water supply fall below the tolerated limits of each user. If so, the preliminary border point is converted into a final border point; if not, the withdrawal point tested is included in the area affected as a new preliminary border point. Repeating this procedure the preliminary border points are moved downstream step by step until one reaches final border points at which the tolerances of sensitivity are no longer exceeded by the WITH-WITHOUT differences.

## D. Derivation of the Economically Relevant Supply Data Entering the WITH-WITHOUT Comparison

From the computation performed within the second element of the model one knows for both cases the quantitative supply of water and its quality for each withdrawal point and the available receiving water capacity for each discharge point. Furthermore the demand for each user is known.

Let $QA_i$ be the quantity of water available in the WITH-situation for any user i, and $QF_i$ the quantity of water required by the same user. In order to determine the economically relevant supply data for the WITH-WITHOUT comparison it is necessary to draw a distinction between two cases:

*Case 1.* The quantity available, $QA_i$, is larger than the quantity required, $QF_i$. Then the limiting factor is the quantity required so that $QF_i$ is the economically relevant supply figure ($QR_i$) entering the comparison.

*Case 2.* The quantity available, $QA_i$, is smaller than the quantity required, $QF_i$. Then the limiting factor is the quantity available so that $QA_i$ is the economically relevant supply figure ($QR_i$) entering the comparison.

The same considerations for both cases and for each of the three factors: quantity of water, quality of water and receiving water capacity leads to Table 8.1, which summarizes the derivation of the economically relevant supply data entering the comparison. The supply available in the WITHOUT-case has been marked with a dash for distinction purposes.

There is one important deviation from the procedure described above for the derivation of the economically relevant supply data. If the quantitative availability of water is smaller than the quantity required, then the quality actually available is always disregarded and the economically relevant factor is assumed to be the quality required. This procedure is the "master key" for a workable WITH-WITHOUT comparison and is admissible only because of the opportunity-cost oriented pricing concept.

As to the receiving water capacity a simplifying but realistic assumption was made concerning the supply available by supposing that at flow links or reservoir links the receiving water capacity is always sufficient to meet demand while at zero links the receiving water capacity is zero. The economically

Table 8.1

*Derivation of the Economically Relevant Supply of Water and Receiving Water Capacity for Both Cases, WITH and WITHOUT*

| | WITH-Case | | | | WITHOUT-Case | | | |
|---|---|---|---|---|---|---|---|---|
| | Supply Available | Supply Required (Demand) | Comparison: Available/ Required | Economically Relevant Supply Data | Supply Available | Supply Required (Demand) | Comparison: Available/ Required | Economically Relevant Supply Data |
| Water Quantity | | | | | | | | |
| Case I | $QA_i$ | $QF_i$ | $QA_i \geq QF_i$ | $QR_i = QF_i$ | $QA'_i$ | $QF_i$ | $QA'_i \geq QF_i$ | $QR'_i = QF_i$ |
| Case II | $QA_i$ | $QF_i$ | $QA_i < QF_i$ | $QR_i = QA_i$ | $QA'_i$ | $QF_i$ | $QA'_i < QF_i$ | $QR'_i = QA'_i$ |
| Water Quality | | | | | | | | |
| Case I | $GA_i$ | $GF_i$ | $GA_i \geq GF_i$ | $GR_i = GF_i$ | $GA'_i$ | $GF_i$ | $GA'_i \geq GF_i$ | $GR'_i = GF_i$ |
| Case II | $GA_i$ | $GF_i$ | $GA_i < GF_i$ | $GR_i = GA_i$ | $GA'_i$ | $GF_i$ | $GA'_i < GF_i$ | $GR'_i = GA'_i$ |
| Receiving Water Capacity | | | | | | | | |
| Case I | $VA_i$ | $VF_i$ | $VA_i \geq VF_i$ | $VR_i = VF_i$ | $VA'_i$ | $VF_i$ | $VA'_i \geq VF_i$ | $VR'_i = VF_i$ |
| Case II | $VA_i$ | $VF_i$ | $VA_i < VF_i$ | $VR_i = VA_i$ | $VA'_i$ | $VF_i$ | $VA'_i < VF_i$ | $VR'_i = VA'_i$ |

relevant supply data are, therefore, the receiving water capacity required at flow and reservoir sections and the receiving water capacity available at zero links, namely zero.

## E. Determination of the Net Benefits in Physical Terms

The fifth element of the simulation model is devoted to the determination of the net benefit for each user in physical terms. The net benefit in physical terms is represented by the WITH-WITHOUT differences of the economically relevant supply data in terms of quantity and quality of water and in terms of receiving water capacity. Table 8.2 explains the computation of individual costs and benefits and describes the total net benefit as a vector consisting of the three components. Table 8.3 presents an interpretation of the result of the WITH-WITHOUT comparison for any user i and explains the four basic possibilities of the outcome with x describing either quantity of water, quality of water and receiving water capacity.

*Case 1*

Supply available ($XA_i$, $XA'_i$) exceeds supply required (demand, $XF_i$) in both situations; economically relevant supply data for the comparison are the supply required in the WITH situation and the supply required in the WITHOUT situation. Since for both situations the demand was assumed to be equal, the economically relevant difference $DXR_i$ is zero, indicating that there are neither costs nor benefits accruing to user i.

*Case 2*

Supply available exceeds demand in the WITH situation but is smaller than demand in the WITHOUT situation. This means that the economically relevant difference $DXR_i$ is positive, indicating that the $i^{th}$ user benefits from the existing weir(s) or canal(s).

*Table 8.2*

*Determination of the Net-Benefit in Physical Terms Accruing to Each User Affected*

| *Net-Benefit For* | *Economically Relevant Difference of Water Supply in Terms of Quantity = Quantitative Component of the Net-Benefit* | *Economically Relevant Difference of Water Supply in Terms of Quality = Qualitative Component of the Net-Benefit* | *Economically Relevant Difference of Receiving Water Capacity = Receiving Water Capacity Component of the Net-Benefit* |
|---|---|---|---|
| 1st user | $DQR_1 = QR_1 - QR'_1$ | $DGR_1 = GR_1 - GR'_1$ | $DVR_1 = VR_1 - VR'_1$ |
| 2nd user | $DQR_2 = QR_2 - QR'_2$ | $DGR_2 = GR_2 - GR'_2$ | $DVR_2 = VR_2 - VR'_2$ |
| $i^{th}$ user | $DQR_i = QR_i - QR'_i$ | $DGR_i = GR_i - GR'_i$ | $DVR_i = VR_i - VR'_i$ |
| $n^{th}$ user | $DQR_n = QR_n - QR'_n$ | $DGR_n = GR_n - GR'_n$ | $DVR_n = VR_n - VR'_n$ |
| accumulated costs and benefits | $\sum_{i=1}^{n} DQR_i$ | $\sum_{i=1}^{n} DGR_i$ | $\sum_{i=1}^{n} DVR_i$ |

$$\overrightarrow{NNPH} = \left\{ \sum_{i=1}^{n} DQR_i,\ \sum_{i=1}^{n} DGR_i,\ \sum_{i=1}^{n} DVR_i \right\}$$

*Table 8.3*

*WITH-WITHOUT - Differences*
*With X Describing Quantity, Quality or Receiving Water Capacity*

| *Case* | *Data Entering the WITH-WITHOUT-Comparison* | *WITH-WITHOUT-Differences* | *Remarks* |
|---|---|---|---|
| I | $XA_i \geq XF_i$ and $XA'_i \geq XF_i$ | $DXR_i \equiv 0$ | neither costs nor benefits |
| II | $XA_i \geq XF_i$ and $XA'_i < XF_i$ | $DXR_i > 0$ | benefits |
| III | $XA_i < XF_i$ and $XA'_i \geq XF_i$ | $DXR_i < 0$ | costs |
| IV | $XA_i < XF_i$ and $XA'_i < XF_i$ | $DXR_i \neq 0$ | either costs or benefits |

*Case 3*

Supply available is smaller than demand in the WITH situation but exceeds demand in the WITHOUT situation. The economically relevant difference $DXR_i$ is negative, indicating that costs are imposed on user i.

*Case 4*

Supply is smaller than demand in both cases. Depending on the economically relevant difference $DXR_i$ there may be either benefits or costs for the $i^{th}$ user.

Since by definition the supply in the WITH situation exceeds (or equals) supply required in the WITH situation for both components of the net benefit, quantity of water and receiving water capacity, the effect of the weir(s) or canal(s) can either be positive or zero. The economically relevant differences for quantity and receiving water capacity fall always under the headings of cases 1 and 2 in Table 8.3, while the differences for quality can fall under the heading of all cases.

Since it was assumed that the receiving water capacity was sufficient to meet demand at flow links and at reservoir links of the channel-reservoir-system, positive differences with respect to receiving water capacity can occur at zero links, which are the canals in the WITH situation that are assumed to be nonexistent in the WITHOUT situation.

Individual costs and benefits add up to the total net benefit in physical terms accruing to water users from the construction of the weir(s) or the canal(s) under consideration.

## IV. SUMMARY

The five elements of the simulation model described above form the basic steps required to derive the net benefits in physical terms. On the basis of this model a computer program was developed that was tested for two sections of the German system of domestic waterways. The next step is to develop a general model for transforming the net benefits in physical terms into net monetary benefits. Thus far only the basic pricing concept has been developed, which is based on an opportunity cost approach.

ACKNOWLEDGMENTS

The research underlying this paper was supported by a research grant from the Ministry for Transportation of the Federal Republic of Germany.

# CHAPTER 9

# WATER RESOURCES MANAGEMENT

Y. Emsellem*

## I. INTRODUCTION

In 1964 six financial agencies were created in France, each in charge of the financial aid to the water supply and the fight against water pollution in the six basins that cover the entire country. The boards represent administration, industry, energy, agriculture, tourism and cities, and they allocate the taxes paid by consumers and polluters of water. These taxes are used for investments and partly for the common expenses of dams, treatment plants, canals and water supplies.

The main idea behind the taxes was that polluters would pay for pollution. The first taxes were allocated in 1968, the amounts being negotiated between the administration and the polluters at a market level by the council of the agencies. At the same time, the financial charges to the industry were increasing greatly.

The system of taxes in 1970 eliminated only a quarter of the pollution, and without any change in the system the situation will not improve in the future. The aim now is to reduce the gap, and then to increase the mean level of financial supply.

The main reason for the actual gap is not that the polluters are opposed to environmental improvement, for they are simultaneously polluters and citizens. As shall be seen further, it is mainly a question of a transfer problem.

---

*Pr. Y. Emsellem is the Director of the Geological Informatique Center of the Ecole Nationale Superieure des Mines de Paris, 35 rue Saint-Honore, 77 Fontainebleau, France.

If one tries to solve the problem by mathematical programming with a unique objective function, it will be very difficult to take into consideration simultaneously the technical aspects of water supply, the costs and benefits, and the influence of environmental policy on economic growth and individuals. Therefore, the problem will be posed in another way.

## II. THE FACTS AND THE IDEAS

When one manages the water supplies of a basin and is told that water is needed for irrigating an additional million acres, the question is not how to minimize the cost and optimize the benefit, but why another one million acres should be irrigated. Four politically different answers are possible:

- to produce agricultural goods for domestic consumption
- to sell these products abroad
- to create a need of labor
- to increase the standard of living of agricultural people.

It is obvious that the indirect financial participation of agricultural people in the basin will not be the same in the four examples. In the third example, the participation of the activity Agriculture in the investments is different from the participation of the group Agricultural People. Obviously, the distribution of the charges between activity and group is different from one case to another.

If the water management system for other activities and other groups is analyzed from the same viewpoint, a hierarchy in the share of expenses can be observed. It is clear that a group or an activity will not pay for and receive water if its priority is different when the indirect effects of different policies are considered. An example may be the following: For the sake of an economic policy of development and equilibrium, France and Germany increased the rate of interest by half a per cent; the weight of that general decision will not be the same for every producer, consumer or activity.

To explain further, let us choose a theoretical development policy that includes a hierarchy of the aims devoted to each group or activity. To reach the desired aims, a management scheme for water supplies is considered. Each group of activity will contribute a part to the means that are necessary to that scheme. For instance, group A will give in any currency 5, group B gives 9, group C, 2, and D, 13. Now consider the result of the planning. Suppose that five years later A receives (by the logical mechanism of economy and laws) 6, while B, C, and D receive 11, 4, and 10, respectively. It is apparent that the policy is reasonably favorable to C, which is not the biggest, while A and B are roughly on the same level of hierarchy, and D is badly considered. One can say that transfers are economically and legally organized between A, B, C and D. The first question is: Is the set of transfers and rate of growth of activities and group coherent with the policy? If not, which economical, legal and financial tools must be changed to reach the aims? If yes, what are the foreseeable difficulties? An obvious one is that D is strong and will fight against the natural evolution.

Additionally one must begin to ask, first, is the policy possible, and second, which system of transfers must be organized? Agencies, whose importance in the subject is increasing every day, are being created to assist in answering these questions. As a result many methods may be examined in order to find a solution. The method of cost-benefit analysis will be examined first.

## III. THE INADEQUACY OF COST-BENEFIT ANALYSIS

Cost-benefit analysis is highly interesting because it introduces a global objective function and then allows the examination of the problem with the aid of mathematical programming and sensitivity analysis. Yet, global cost benefit analysis is inappropriate for the following technical reasons.

1. The induced or indirect effects, which are difficult to define and model, generally are not taken into account. For instance, taxes on industry increase prices and then have an influence on the foreign trade and consequently on the level of activity.
2. To reduce the complexity of the problem, certain aspects may be neglected. For example, only a small basin or only polluter or quantity may be studied.

3. Time is a very important variable because the rate of interest gives a different importance to modern equipment, which generally uses more capital and less labor; older equipment uses less investment and more labor.
4. A bound and a variable are not taken into consideration equally in the objective function. A bound is satisfied and a production variable is not necessarily maximum; therefore, one may say that a bound is a stronger objective than the objective function itself.

These difficulties are not absolutely impossible to eliminate. Yet, from a more fundamental point of view, CBA uses one unity, the monetary value of the different elements, and only the parameters of the market may be taken into consideration. One may try to give a value to ideas, or to nonnumerical parameter such as policy, environment, life, but even when this is possible the information is not easily available.

Moreover, the desired econometric optimum is identified whatever the objective function may be and whatever the set of boundaries may represent. The method used is the optimal valorization of water. Implicitly, water is considered as the only factor of growth. This idea is true for developing arid regions, but generally not for industrialized countries because in most cases:

- for an individual user, water is often a marginal factor of production
- for a set of consumers, water may be recycled or replaced in part by other factors.

One must emphasize here the difference between scarcity and pollution. The notion of growing scarcity of water is ambiguous and dangerous because it does not take into account the reason for the situation. The correct representation of the problem is that water is a limiting factor of growth and not the only one.

In addition, the objective function is supposed to represent the preference of the collectivity. That scheme is suitable in the case of a well-identified collectivity with only one decision-maker. Even then, it is obvious that the decision-maker must know his preference perfectly. This last argument, incidentally, is against any model. Yet when the number of decision-makers increases and the partial and individual interests are divergent, the building of a unique optimum objective function is difficult and unpleasant. The observation shows many examples of unpractical optimum solutions.

It is necessary, then, to take into account a set of ideas because the water problem is part of a broader development problem. Rather than modifying the method of CBA, an appreciably different method was developed.

## IV. BASIC PRINCIPLES OF THE APPROACH

Although water is divisible through use, it is an indivisible collective item. All rational data processing of water must be inscribed in a geographical framework, allowing the interrelationships generated by this "indivisibility" to be considered. If, however, a reasonable scale is to be kept, the complete hydrological basin from the source to the sea should be accepted as the basic geographical unit because it is generally an administrative and political unit.

As a physically natural or artificially modified element, water may only be characterized by a set of nonhierarchial parameters: geographical, temporal, mechanical, and physicochemical. Whether for consumption or for use in production, water must conform to certain rules relating to the whole or part of these parameters. The distinction between quantity and quality would appear completely arbitrary. The method of data processing adopted must treat these two aspects jointly.

In temperate climate countries, water is not an element influencing the development. Determining the economic and social perspectives on the basis of an optimum valorization of water resources is possible only when water is scarce, as in the case of arid zones. In our regions, the needs that water must satisfy in the future depend on the perspectives of regional development. These data are external to all policies of water.

Because an ecological conscience has appeared in the developed countries, water may become a factor that limits development. Practically, all the technological processes controlling the degradation of the environment only change the problem from one aspect of pollution to another. In fact it is the growth itself that could be the cause. The indicators representative of this growth, which are continually used as criteria, must be complemented by ecological indicators. The objective should be not only the optimization of growth but also the conservation of natural resources.

For a theory of economic and social development of a basin there are numerous solutions of hydraulic planning. The most useful of them must be characterized by a series of indicators, including the cost, in order to clarify the choice. The optimal solution is not sought, which is equivalent to making this choice *ex ante* and which, as shown by experience, cannot please everyone completely. In contrast, the necessary means and the consequences associated with a chosen policy allow the choice between several political alternatives.

## V. THE WORK PROGRAM

Although actually all the parts are treated simultaneously, four phases may be distinguished in the work program.

### 1. *Hypothesis for the Development of the Basin*

The aim of this phase is to make the theories of regional planning more analytical. The basin is divided into zones, considered economically homogenous. For each of these zones, development projections of the principal sectors of water users are made, related among themselves by coherent rules bearing on the demography and the jobs. Then, several contrasting theories of the basin development can be obtained. Each of them will give place to a translation in terms of qualitative and quantitative water requirements, with the help of a classification of the users.

### 2. *Forecast Model*

At this stage, a principal plan is proposed for the development of the basin. This is represented by a graph in which the nodes synthesize the characteristics of a sub-basin, such as the proper resources, the requirements and the hydraulic interrelationships between sub-basins. The model works statically on horizons of discrete time *e.g.*, 1980, 1985, 2000, and for the critical periods, those with the lowest water levels. It tries to satisfy the requirements of each of these periods using the hydraulic functions. The calculations are based on a procedure of linear programming. In fact, at this level a series of local optional convexes are considered instead of "coloring" a unique convex with political options.

### 3. *Simulation Model*

This model serves to define and control the solutions of the forecast model. The basin is represented as a graph, but with many more details, and the hydraulic functions are translated constructions. The simulation works every year until the horizon 2000, allowing for the study of the construction planning. Within the same year different intervals of time can be used.

At the end of this simulation model, several planning solutions are kept for the same development theory. Each of them is characterized by several indicators other than cost.

### 4. *Finance Model*

The first part evaluates the overall cost of each of the preceding solutions. This cost may be recuperated in different ways from the users depending on political decisions. Different keys of distribution are therefore applied. The principal idea, then, is not to optimize globally, but to optimize the options of the politician. This allows the decision-makers to appreciate the monetary consequences of several possible policies. The financial problems and the resulting assignments are considered in the third phase.

The definitive results are presented in the form of files, each corresponding to a theory of economic development of a basin. Each file contains several solutions of hydraulic development, their ecological incidences, costs, and methods of possible finance and charges.

## CONCLUSION

This methodological project is actually under way in the basin of "Adour-Garonne" with an area of 115,000 $km^2$, 5,800,000 inhabitants. Eight research teams, with about 25 engineers, economists, hydrologists, geographers, sociologists, civil servants, and managers, are participating in this research which started January, 1971, and will end early 1973.

MODELS FOR ENVIRONMENTAL POLLUTION CONTROL


# CHAPTER 10

# MATHEMATICAL APPROACH TO WATER RESOURCES MANAGEMENT IN ITALY

Marcello Benedini*

## I. INTRODUCTION

The management of water resources in Italy is in many aspects completely different than in many other countries, and Italian water engineers today are called upon to deal with increasingly complicated problems. Italy has been able to cope with its own water resource problems for many years; important works have been built in the past decades by Italian water engineers. Also in the legal field and in the promulgation of appropriate regulations for water resource management, Italy has played an important role. Technological progress, on the other hand, has found Italy somewhat unprepared to face the problems arising from the modern situation, leading to a difficult situation.

The increasing demand for water for domestic, irrigation and industrial supply, strictly connected with the need for increasing the individual income, has emphasized the already existing discrepancies between the zones in which water is abundant (generally located in the center and in the north of the country) and those in which water is scarce. Even in the richest northern regions a scarcity of water is now beginning to be felt due to ever-increasing pollution. And recent disastrous uncontrollable floods caused by an improper use of the soil in the upstream regions have emphasized the need to have better control of all the existing works, and, at

*Dr. Marcello Benedini is with the Water Research Institute of the National Research Council, Via Reno 1, Rome, Italy.

the same time, for more appropriate evaluation of the natural conditions of flood generation.

The need for environmental protection has become more and more evident; aquatic and nonaquatic life must be maintained and natural amenities to cope with the increasing demand for the recreational use must be improved. With the exception of some small unused catchment areas scattered throughout the south and in the islands, the management of Italian water resources now consists mainly of settling the conflict between existing users, rather than planning and constructing new hydraulic works. New reservoirs for power production would be quite useless, not only because almost all the hydroelectric potential has been exploited but mainly because the existing reservoirs are incompatible with other uses of the water resources.

The most necessary works in the future are those for water supply and intake for domestic, agricultural and industrial uses and the design and construction of large treatment plants for the reuse of wastewaters. Italian water engineers are therefore paying much attention to all new methodologies developed and used in many other countries, especially modern computer techniques. At the same time they realize that application of these methodologies to the present Italian situation requires a radical investigation of all the basic mechanisms.

The main problem with regard to the improvement of existing situations is that of making conflicting uses compatible with one another. This often involves giving importance to aspects that are not within the field of the design and construction engineer but that require the attention of specialists of a different field. The usual design rules of hydraulics and hydrology must be expanded by the contribution of chemists, biologists and microbiologists, to study the transport of pollution, the efficiency of treatment plants and the relation of quality phenomena to the operation of the hydraulic works.

## II. Hydroelectric Power Production

The objectives pursued until a few years ago gave the most importance to the use of water for hydroelectric power. Now this use has been the first to suffer from limitations imposed by other uses.

The first series of limitations are imposed by the needs of flood control. The Ministry of Public

Works, which is now the principal authority responsible for water management, has ordered the Electricity Board (ENEL) to keep the maximum level in the reservoir below a certain predetermined level that in some seasons is considerably lower than the maximum allowable level which was established when the reservoir was built. In mathematical terms this corresponds to an "upper bound" in the following form:

$$V_j \leq VH_j$$

This constraint states that the stored volume of water at time j shall not exceed the seasonal limit $VH_j$ corresponding to the predetermined level in the same season.

Due to the convexity of the level-storage curve, the remaining volume

$$V_{max} - VH_j$$

may sometimes be very great, and the maximum power, for which the plant was originally designed, is notably reduced.

The upper bound is evaluated in a prudent manner by means of a very rough hydrological calculation. The most convenient flood storage capacity is calculated after introducing appropriate short-time hydrological input. A simulation model is available, based on the continuity equation:

$$INP_j + V_{j-1} - V_j - kP_j - X_j = 0$$

where

- $INP_j$ = hydrological input to the reservoir during period j
- $V_{j-1}$, $V_j$ = volume of the reservoir at the beginning and at the end of period j, respectively
- $P_j$ = total power produced at the plant in the period j
- $X_j$ = water quantity spilled during the period j
- k = factor converting power into water quantity.

The model can be used both with historical and synthetic hydrological input, as well as in a deterministic and probabilistic manner. The most suitable time basis for such an evaluation is an hourly one, since the effect of flood wave generation and propagation is contained in the interval of a few days.

A second order of limitations is imposed by the necessity of keeping some water in the river reaches located immediately downstream from the dam that forms the reservoir. It is very common for Italian hydroelectric plants to have the power station a considerable distance downstream from the dam, the aim being to increase the available head. Shortage of water in the river bed between the dam and the section in which water is returned from the power station is a real problem.

Withdrawal of water from a river can last for seasons, especially in the warmer months that are often characterized by low natural runoff; all the aquatic life can be seriously affected. Within this framework a certain amount of water from the reservoir ought to be released in order to restore the natural conditions.

An economic evaluation of hydroelectric production is somewhat imprecise in Italy. In fact, even if the exact market price for electricity is known and the gross benefit can be calculated, the total cost of power production is affected by many uncertainties, summarized as follows:

1. The plants were built in different periods, sometimes many years ago; therefore capital costs should take into consideration inflation and the altered value of money.
2. When electricity first became nationalized, the state board compensated the previous owners in a global manner, with no accurate distinction among the several components of the plants. The claims of private companies have very often been affected by political moves.

It should be remembered that the "Water Act" of 1933 allowed the private owners of electricity to amortize their construction capital in the "concession" time of 60 years. However some monopolistic situations, with regard to the selling of electricity within well-defined zones, have allowed some companies to amortize their own capital in a much shorter period. The calculation of the actual costs of maintenance, repair and labor is conversely easier.

Although the ecological use of water is valued by the public, the mathematical evaluation of it is difficult to show. From a purely technical view point, the effects of keeping water in the reaches of the river can be evaluated in the operation and maintenance of gates, sluice and spillways. The calculation of the corresponding benefit requires essentially intangible considerations, like the amount of money paid by people for fishing and bathing facilities or simply for enjoying the natural amenities. In the Tiber River basin a typical example is the famous Marmore Falls, which require the Electricity Board to release into the natural stream a constant value of 20 $m^3/s$ every Sunday and vacation day. Other considerations concern local agriculture, which even if it has no irrigation scheme can be affected by the existence of a certain level in the reaches, which have a regulating effect on the surrounding underground water.

When the reservoirs were built, the possibility of using them for recreational purposes was seldom considered. However, this is becoming more important now in order to meet an increasing demand for tourist facilities. The effect of recreational activity on reservoir management consists primarily of the maintenance of the water level about a pre-determined value, in order to have enough water for boating, swimming and fishing. Furthermore, the level variation should be kept to a minimum in order to permit a constant use of devices like piers, ladders and moorings.

Because of this, another reduction of the existing reservoir capacity is introduced and can be expressed as a lower bound:

$$V_j \geq VL_j$$

The term $VL_j$ varies according to the season and is expected to be higher during the summer holidays. In the reservoirs of central Italy the hydrological regime has its minimum flow during summer; therefore in order to comply with a high lower bound it might be necessary to make a considerable reduction in power production, with a notable impact on the electricity available. Fortunately such an effect can be slightly reduced by producing more electricity from the Alpine reservoirs, whose hydrological behavior is different.

## III. IRRIGATION WATER DEMANDS, WATER SUPPLY, AND POLLUTION

In the Appenine regions as well as throughout the country the demand for irrigation is particularly urgent in summer; this introduces another term, $A_j$, in the continuity equation for the reservoir:

$$INP_j + V_j - V_{j-1} - kP_j - X_j - A_j = 0$$

meaning a further reduction of water available for electricity.

Simulation models can give an evaluation of both the technical and economic effects on reservoir management by releasing water for irrigation. The location of Appenine reservoirs is preferably in narrow valleys, where the construction of high gravity and arch dams is easy and provides great storage volumes without the submersion of large areas. Conversely the areas suitable for irrigation are located far downstream, in flatter terrain, generally beside the river bed. This makes it quite inappropriate to open long channels to carry water directly to the land to be irrigated. For technical and economic reasons it is preferable to take the water for irrigation straight from the river, through small diversion works or by means of pumping stations.

The same ideas are applicable for industrial and domestic water supply, both of which are needed far from the reservoirs. All these forms of water require large amounts of water in the lower parts of the river, and the effect on the reservoir is a release of more water towards the natural downstream reaches.

Pollutants and waste discharges act in a similar way, *i.e.*, the water flow must be maintained above a certain level to provide dilution and natural pollution abatement. Simulation procedures using the Streeter-Phelps model, or those concerning the propagation of pollutants, chemicals and temperature, can help to provide some information. The degree of pollutants at a particular river cross section can also be evaluated, if the conditions existing at an upstream waste discharging section are known.

Information obtained by using the above-mentioned procedures is useful for setting appropriate lower bounds in the river reaches or some predetermined control section in the form:

$$X_j \geq XL_j$$

where $XL_j$ is a seasonal value.

At some cross sections of the rivers, especially those located immediately downstream from the waste discharging points, a great amount of water may be needed in order to dilute a large polluting load. Therefore the above $XL_j$ lower bound might reach high values, mainly during the dry season, which would mean a further demand for water from the reservoir and further reduction in power production.

The assessment of the most appropriate role for each of the above uses can be evaluated in optimization models, which have an objective function in the following form:

$$\text{maximize} \quad \sum_j \left( \sum_m c_m k_m P_{mj} + \sum_i c_i A_{cj} + \sum_r c_r F_{rj} + \sum_k c_k WIN_{kj} + \sum_n c_n R_{nj} + \sum_h c_h WS_{hj} \right)$$

in which

$P_{mj}$ = power production at the turbine m during period j

$A_{cj}$ = target value (in water units) of agricultural at district c, during period j

$F_{rj}$ = storage value for flood control at reservoir r during period j

$WIN_{kj}$ = target value (in water units) of water supply to industry k, during period j

$R_{nj}$ = recreation target value (in water units) at location n

$WS_{hj}$ = target value in water unit to urban area h.

$c_m$, $c_i$, $c_r$, $c_k$, $c_n$ and $c_h$ are "weights" for power production, irrigation, flood control, industrial water supply, recreation and drinking water, respectively, while $k_m$ denotes the factor to convert power units into water units at the plant m. The model's variables are:

$V_{rj}$ = storage of reservoir r at the end of period j

$X_{ij}$ = quantity of water flowing through a river reach i during period j

$Y_{ij}$ = quantity of water flowing through artificial penstock or canal i during period j.

$INP_{qj}$ = hydrological input to cross section (or "node") q during period j.

The evaluation of the weights (which may be considered as independent upon time) should be performed through appropriate investigation of the peculiarity of each use.

## IV. WATER QUALITY PROBLEMS

In the case of water quality problems there are again unique characteristics. Many towns in Italy have installed good treatment plants but very few of them are able to afford the operating costs, so very often the plants are not operating at all. Considering economic variables, related to the treatment costs and consequential benefits from improved water quality, a general glance at the basins reality permit the following statements:

1. The basic needs for improving water quality are required for ecological conservation.
2. Maintenance of reasonable quality standards in the natural streams contributes both to industrial and agricultural development by making possible withdrawal of water at low treatment costs.
3. Pure water in the rivers is necessary for recreational demands of the population.
4. Maintenance of high water quality in the rivers, as well as in all surface waters, is essential to prevent the pollution of underground aquifers through the natural recharge mechanisms. The development of urban water supply in Italy is based mainly on wells and springs, according to the assumption that these sources are sufficiently pure for the drinking purposes.

The substantial effect of the foregoing must be introduced in the weights of the optimization model, after detecting, by means of separate tools, the real interactions between the water quality measures and the characteristics of each of the uses. The same must be done in the constraints, after evaluating the economic characteristics of the imposed bounds.

## V. THE TIBER RIVER STUDY

In this context the work of the "Istituto di Ricerca sulle Acque" in developing methodologies appropriate for the study of integrated utilization of the resources in a large catchment area should be mentioned. With this work the Institute proposes to make known to the people in Italy who will be responsible for water management the most suitable techniques for the solution of the very complicated problems connected with the utilization of water resources. For an application of these studies, the Institute has chosen the Tiber catchment area. Reasons for this choice are many:

1. The catchment area is large and does not present excessive complexities; in fact if larger Italian catchment areas (such as the Adige or the Po basins) had been chosen, there would certainly have been greater difficulties because the elements to be examined would be more numerous. On the other hand if smaller catchment areas had been chosen, the study probably would not have been representative enough.
2. The water of the Tiber and its tributaries are utilized for numerous purposes so that the catchment area under examination makes a very appropriate example.
3. Along the banks of the Tiber and its tributaries, many industrial complexes and urban communities are developing so that it is possible to evaluate more accurately the influence of the social, economic and political elements.
4. In the catchment area there are some important sites of great interest from an historical, scenic and cultural point of view and it is very necessary to preserve these from any damage caused by an unwise use of water.

The most important aspect of this study is neither hydrologic nor hydraulic. Vast works of river hydraulics have been carried out on the Tiber; therefore, with the contribution of the available storage in the hydroelectric reservoirs, the danger of flooding seems to be averted. In fact the last damage caused by a flood was in 1937 when many of the now existing reservoirs had not yet been built.

The evaluation of the economic, political and social elements connected with water management is much more important. In the Tiber catchment area some industrial complexes are in operation (such as in the Perugia, Tivoli, Orte, and Terni areas) and considerable developments for industries with a high technological level (pharmaceutical, food, electronical) are predicted. There is also the chemical and iron and steel complex at Terni. High-level technology industrial complexes generally require small quantities of water but are very demanding with regard to quality, which can be acquired only through the use of appropriate treatment plants. However, the cost of this treatment increases with decreasing water quality. These considerations can be determining factors in the choice of location for a new industrial complex.

The hydroelectric industry is very important in the catchment area; there are numerous plants interconnected hydraulically in order to produce high quality energy. Electric energy, nevertheless, must be considered to a large extent independent of the conditions of the catchment area since the Italian electricity system is completely interconnected.

Results of a survey show very high levels of pollutants; interaction between pollution parameters and natural factors, such as climate, peak flow, and time of operation, also have been pointed out. Several mathematical models are under construction in order to represent the behavior of pollutants in the stream, according to the information given by field measurements. It is too early to discuss results, but there is a general feeling that the classic models ought to be adapted with much care to the existing situation in the basin, mainly because of the following reasons:

1. The number of dischargers is very large and the discharge points are almost continuously distributed along the river embankments. The definition of the stretches, in which the natural phenomena of dilution and degradation occur, is very difficult.
2. Statistical evaluation of the measured values should be developed in order to determine the most representative terms in time and space.
3. The interaction between discharge and intake points should be better understood.

The Institute tries to represent the Tiber system in its complexity both by means of simulation and optimization models. Simulation models are applied basically to separated subsystems, formed by the bigger reservoirs like Corbara, Salto and Turano, in order to assess some operating rules and to reconstruct the natural flow in the streams. A general simulation model for the whole system is now under construction. It will be used to test the behavior of the basin. In the meantime an optimization model is already running, with the aim of providing some appropriate working criteria for the reservoirs and for the most prominent water intakes and discharges. The following have been accomplished:

1. A very simple model was adopted to test the sensitivity of the system to the physical variables involved. In other words, in the objective function the terms concerning irrigation, flood control, industrial, recreation and urban water supply were considered as decision variables and the objective function reduced to:

   $$\text{maximize} \sum_j \sum_m c_m k_m P_{mj}$$

   subject to similarly simplified constraints. A very crude hydrological input was used as well as the assumption that the dilution phenomena in the stream could be referred to by means of appropriate minimum acceptable flow in some defined cross sections.
2. After the sensitivity is confirmed, the same tool can be used for deeper investigations, merely assessing, step by step, the most appropriate value to the involved terms both in the objective function and in the constraints, especially to the above defined weights, which shall include, at the end, the contribution of all the nonphysical impact.

The model is run using advanced techniques of linear programming and an improved software package. The way chosen by the Institute might appear somewhat cumbersome and perhaps for mere managerial purposes it should be improved and simplified, but within the scope and the mandate of the Institute it is considered the most suitable for the purpose of investigating and representing in mathematical form the complexity of water management in Italy.

# PART III

# AIR POLLUTION CONTROL

MODELS FOR ENVIRONMENTAL POLLUTION CONTROL


# CHAPTER 11

# A SURVEY OF AIR POLLUTION CONTROL MODELS

Ellison S. Burton, Edward H. Pechan, III, and William Sanjour*

## I. INTRODUCTION

The development of a methodology for analysis of air pollution abatement should be guided by several key requirements:

-- The method must have application to a wide range of urban environments with different pollution sources, fuel use and cost structures, meteorology, and different abatement objectives.
-- The method must be capable of generalization to accommodate abatement analysis of many different pollutants.
-- The method must be useful immediately with present data and become increasingly more powerful as research provides more and better data concerning air pollution.
-- The method must be useful at the federal, state, and local levels for examining the cost-effectiveness relationships implied by a given abatement policy, for judging the efficacy of competing incentives for abatement, and for indicating productive avenues of research in pollution control.

Given the large number of individual emission sources in most urban regions, the scale of any multiple-source abatement analysis is combinatorially so great that computer-assisted simulation is the only analytical method available which can meet the above requirements practically.

---

*Messrs. Burton, Pechan and Sanjour are with the Office of Planning and Evaluation, Environmental Protection Agency, Washington, D.C. 20460, U.S.A.

Before discussing the details of models now in use it will be helpful to discuss how modeling fits into an overall framework for analysis of air pollution control policy.

## II. GENERAL DECISION FRAMEWORK FOR AIR POLLUTION CONTROL

Figure 11.1 shows a macro level flow chart of a system to evaluate possible pollution control strategies within a geographic region. Such an overall model could be used either to determine appropriate standards for the region or to assist in the decision process of selecting the best strategy to meet existing standards.

The overall system works with regional and nonregional data and other inputs based on costs, benefits, possible control alternatives, etc. to produce an enumeration of all possible pollution control strategies. Each possible strategy (or only those strategies meeting standards) is characterized by costs, benefits, and nonquantitative factors. The decision process then consists of the selection of the best or most appropriate strategy.

The blocks of the model shown in Figure 11.1 are described below.

### 1. *Emission Inventory*

The emission inventory consists of data describing each pollution source in the region under study. The data must be inclusive enough to permit later phases to determine feasible control alternatives and their costs and effects on air quality in the region. Some of the data elements for these inputs would be source type (point or area), and process type (*i.e.*, SIC and process codes), geographic location, current control devices, current emissions, fuels burned, and percentage of facility currently used.

### 2. *Regional Growth Factors*

The regional growth factors are used with the emission inventory by the growth model (block 6) to update current data to the specific year under study. These data should indicate growth in emissions from present sources as well as fairly specific data on expected new sources. The accuracy of this

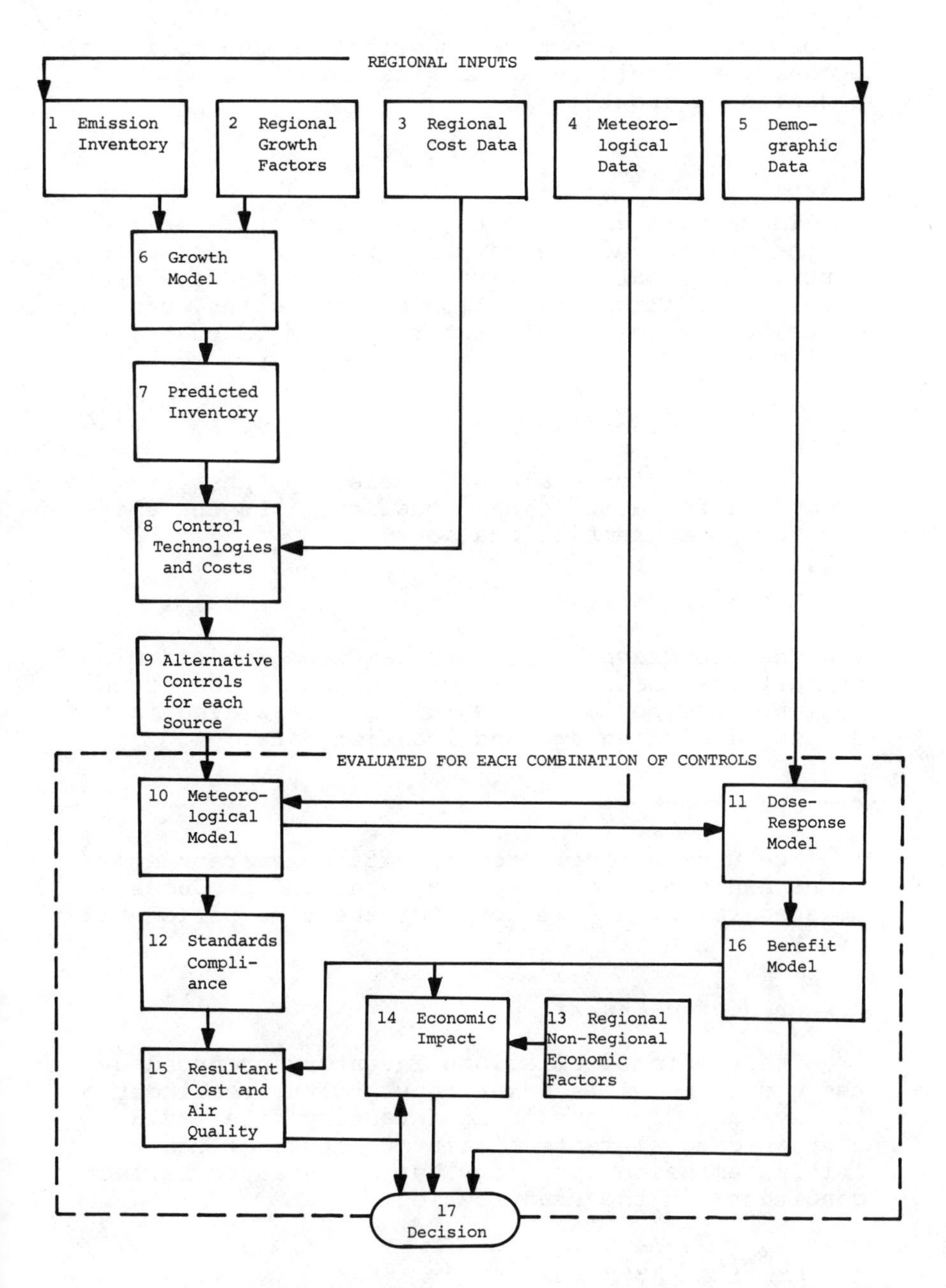


*Figure 11.1. General decision framework for air pollution control.*

information can be quite important to the decision process particularly in an area experiencing significant growth.

### 3. *Regional Cost Data*

These data are used to develop costs specific to the region under study of the various possible control alternatives. Examples of these data are labor costs for installation and operation, capital costs for devices, interest rates and tax data, and fuel costs.

### 4. *Meteorological Data*

Meteorological data are needed by the system to produce information on air quality given the emissions and their geographic locations.

### 5. *Demographic Data*

The demographic data are required to determine the effects pollution has on the population of the region. The population should be classified in some detail, including age and location data.

### 6. *Growth Model*

The growth model accepts data on current emissions and expected growth patterns and produces an updated emission inventory for the time period under study.

### 7. *Predicted Inventory*

The predicted emissions inventory produced by the growth model is a key input to the remainder of the system. The predicted inventory file would contain data elements similar to those of the initial emission inventory but adjusted to reflect conditions in the study period.

### 8. *Control Technologies and Costs*

This model uses the predicted inventory and regional cost data to produce a file containing all possible control alternatives for each pollution source. Each alternative is characterized by a specific abatement action, costs (capital, operation and maintenance, etc.), resource requirements (labor, fuels, etc.), resultant emissions, and production constraints (if any).

*9. Alternative Controls for Each Source*

This block represents the file produced by the control technologies and costs model. It is structured by source with each source having the uncontrolled state and all possible alternatives presented. The set of all possible strategies is determined by selecting one alternative for each source until all possible combinations have been selected.

*10. Meteorological Model*

For each strategy considered, the meteorological model is used to predict resultant air quality over the entire region.

*11. Dose-Response Model*

For each strategy, the dose-response model evaluates, to the extent known, the cost of the health and welfare damages caused by air pollution using the air quality developed by the meteorological model and the demographic and community data developed for the region. The outputs of this model would be the pecuniary value of lost income because of early death and illness, the medical costs of morbidity, and the value of damage done to materials, property, plants and animals, and other pollution-related damage. Since not much is known about these damage functions, this model remains more theoretical than real.

*12. Standards Compliance*

The standards compliance model determines from the resultant air quality whether or not each strategy under consideration meets the standards which have been established. If desired, strategies which fail to meet the standards can be eliminated from the remainder of the analysis. An optimization model may be a component of this model.

*13. Regional and Nonregional Economic Factors*

These data are required as an input to the economic impact model. The nonregional factors consist of data used to predict such effects as relocation of production (and jobs) to other geographic regions because of other production facilities available elsewhere. Such factors are extremely difficult to quantify. Regional factors include employment data, capital invested, etc. and are used to determine the economic impacts of each strategy considered.

*14. Economic Impact*

This model attempts to predict the economic impact of a control strategy on the region. Inputs include the regional and nonregional economic data as well as control costs. Output includes effects on regional gross product, prices, unemployment, investment, etc.

*15. Resultant Costs and Air Quality*

This model evaluates quantitatively each strategy on a cost/benefit basis using as inputs the costs of the strategy, benefits, and air quality.

*16. Benefit Model*

The benefit model evaluates the human damage reduction implied by the air quality improvements achieved and by the dose-response model. Like that model, little is quantified about benefit values.

*17. Decision*

The decision process involves the evaluation for each strategy. The inputs include costs of the strategy, air quality produced by the strategy, benefits from the strategy, and estimated economic impacts of the strategy.

## III. DISCUSSION OF MODEL TYPES

The models that have been implemented to estimate costs and air quality for regions of Air Quality

Control Region (AQCR) size can be compared in two ways. One comparison is between macro and micro level models, and the other is between analytic (or theoretical) models and heuristic models. Generally, analytic models tend to be macro models whereas most micro models resort to heuristics.

Two models illustrating the difference are Kohn's[1] model of air quality in St. Louis, and the Implementation Planning Program (IPP) model developed by TRW Systems[2] for the Environmental Protection Agency. In comparing the two models, a major emphasis will be placed on which type of model is of most help in assisting the policy decision maker.

The Kohn model is a macro level model in the sense that it measures air quality at a single point and treats sources in groups. This means that the resultant air quality computed by the model is based directly on total emissions regardless of their geographic location. An advantage of the model is that it is in linear programming format and can easily be run using existing linear programming codes.

The IPP model is a micro level heuristic model. It treats each point source in the AQCR individually, simulating the application of each feasible control alternative (of which there are 50) in turn. The effects of emissions on air quality are simulated by an atmospheric diffusion model which estimates pollutant concentrations at a number of receptor points. The original IPP did not possess an optimization capability; a heuristic integer program algorithm was later added to the system to determine near-optimal solutions. A further disadvantage of IPP was that it treated only two pollutants simultaneously.

The characteristics of the models are summarized in Table 11.1. Clearly the Kohn model offers a number of advantages. Similar approaches have been developed and documented in the literature (*e.g.*, Norsworthy and Teller[3]).

The decision maker must use some additional criteria in selecting a modeling technique. For example, if a model is used in developing standards and the standards are challenged, the agency setting the standard may be required to demonstrate convincingly its logic to nonexperts in a courtroom situation. If this occurs, the micro level model has numerous advantages.

While pollution can be viewed easily as a macro level phenomenon and strategies defined by tons of emissions reduced, an implementation of any strategy

*Table 11.1*
*Characteristics of IPP and St. Louis Models*

| *Characteristic* | *Kohn St. Louis Model* | *TRW IPP Model (with revisions)* |
|---|---|---|
| Type | analytic (LP) | heuristic (simulation) |
| Source groups | macro | micro |
| Pollutants | 5 | 2 |
| Data requirements | less | more |
| Output | general | detailed |
| Cost determination | optimal | heuristic "near optimal" |
| Computer requirements | low | high |

must involve a micro level application of each control strategy source-by-source. Thus, the aggregating of data on sources in the macro model does not really represent the physical realities of implementing controls on sources with different characteristics. For example, for some industrial processes, switching fuel types is accomplished relatively easily while for others it is close to impossible. The micro model may better reflect this difference.

In pollution control, there has been much more experience in designing and implementing models and setting standards than in actually implementing standards. Because of this there is also little experience in evaluating the accuracy of the models available. Thus, until more information is available concerning the accuracy of model predictions, care must be taken to assure that the procedures used in arriving at cost and air quality estimates are as accurate as possible. This goal can often be helped by using a more detailed level model.

## IV. DESCRIPTION OF A MICRO MODEL

One attempt that has been made to assist in the analysis of control strategies is the IPP program mentioned earlier. The stated purpose of IPP was to assist the state governments in preparing implementation plans for control of sulfur oxides and particulates

A macro flow chart of IPP is given in Figure 11.2. Each of the programs is discussed briefly below.

*1. Source Data Management Program*

This program creates, updates and lists the primary data file (defined as the Source File). The Source File will contain all sources of pollutants which may be considered by the Implementation Planning Program.

*2. Air Pollutant Concentration Program*

This program utilizes a diffusion model to transform source emission data from the source file into average, long-term, ground-level concentrations and a statistical portion to determine corresponding frequency distributions for ground-level concentrations with short-term averaging times. The output is presented on printed tables and on magnetic tape (defined as the Source Contribution File).

*3. Source Contribution File Merge Program*

This program is designed to merge the files produced by multiple (subregional) Air Pollutant Concentration Program runs into a single regional file.

*4. Control Cost Program*

The purposes of this program are to simulate the application of alternative control devices available to each point source and to determine estimates of the total annual cost and efficiency of pollutant removal for each such application. The output is presented on printed tables and on magnetic tape, defined as the Control Cost File.

*5. Control Cost File Update Program*

This program allows the user to correct or update information contained on an existing Control Cost File. In general, use of this program will reduce the number of control cost program runs required to produce a complete, error-free Control Cost File. Output is a corrected Control Cost File.

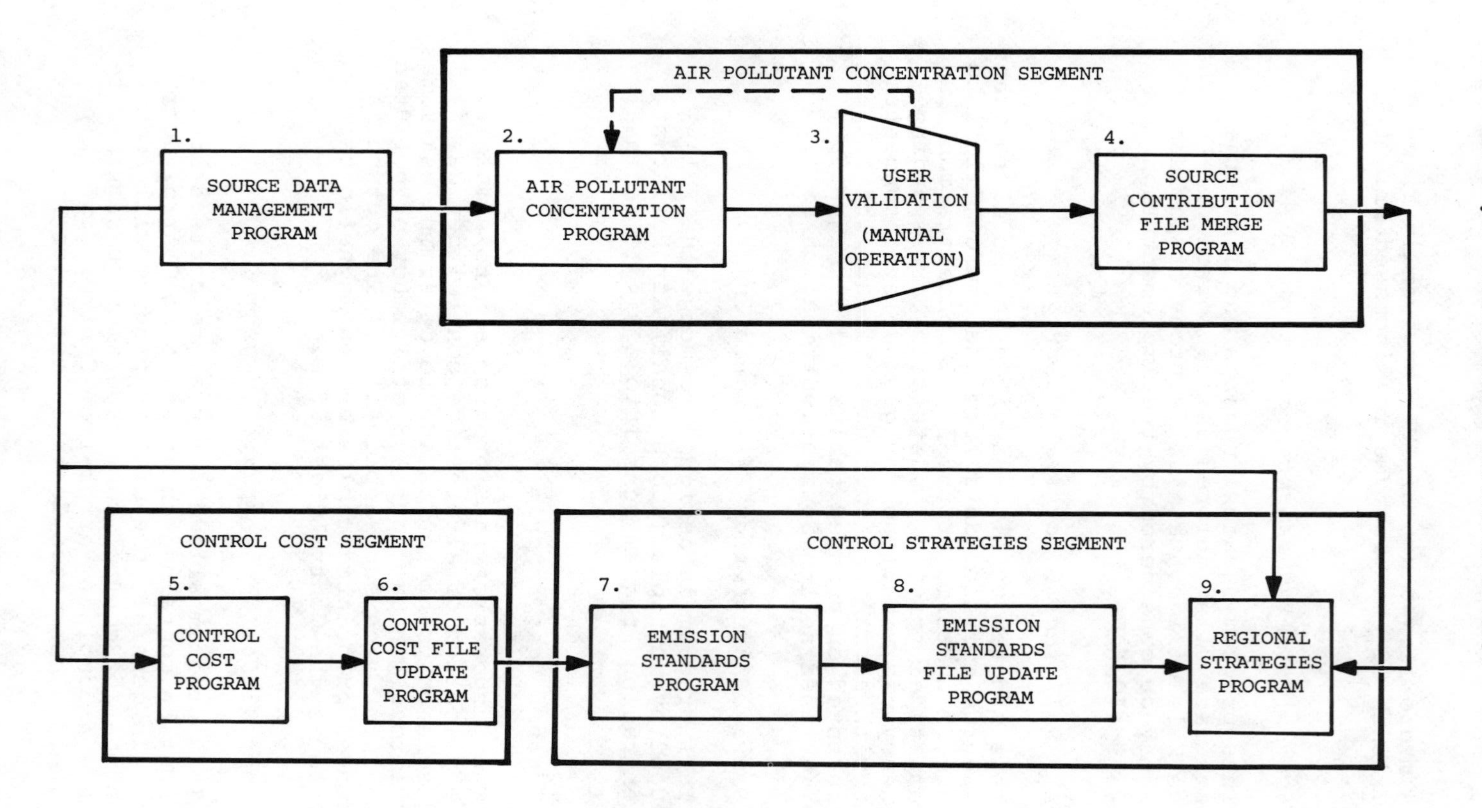


*Figure 11.2. Implementation planning program structure.*

*6. Emission Standards Program*

This program applies all candidate emission standards to the sources within each applicable political jurisdiction. The emission standard requirements are matched with the data on the Control Cost File so that the most cost-effective device for each point source, under each standard, is produced. The output is presented on printed tables and on magnetic tape, defined as the Emission Standards File.

*7. Emissions Standards File Update Program*

This program is designed to update emission standard application data contained on an existing Emission Standards File. As in Control Cost File Update Program, this program can eliminate costly reruns of the Emission Standards Program. Output is a corrected Emission Standards File.

*8. Regional Strategies Program*

This program applies specified emission control regulations and produces summary tables of the resulting emission reductions, control cost-effectiveness, and air quality values. The output from this program is the culmination of IPP. The output from a series of runs becomes input for the determination of the Control Plan for achieving ambient air quality standards.

A comparison of the IPP structure given in Figure 11.2 with the general framework of Figure 11.1 shows that the IPP system can be considered as a subset of the overall model.

Inputs required by IPP are emission inventory, regional cost, and meteorological data. The IPP Control Cost segment serves the purpose of the Control Technologies and Costs model in Figure 11.1. The Air Pollution Concentration segment deals with meteorological effects. Finally, the Control Strategies segment generates a strategy (with corresponding air quality and costs) based on user inputs to the system. The original IPP selected the strategy by applying the input criteria to each source individually. This did not permit a regional optimization capacity. Later, a heuristic optimization section was added to permit generation of a regional least cost solution given designed air quality standards at selected receptors.

The original IPP model was designed to assist the policy planner in standards evaluation. The revised model, which indicated least cost solutions and specific control alternatives selected, aided the planner in choosing near-optimal standards.

IPP did not provide for emissions growth or human effects. In addition, economic impacts were not included. However, despite these weaknesses, the model was able to show significant results.

## V. USING A MICRO LEVEL MODEL

This section describes some of the results from using the modified IPP model. The modifications made to this model permit the use of a heuristic optimization procedure described as follows.

There are N sources of emission, each source emitting k different kinds of pollution. The *n*th source has S(n) different control states (including the present state) which cost C(n,s)[n = 1, ..., N; s = 1, ..., S(n)], where C(n,s) is the cost of the *s*th state of the *n*th source. The emission from the *s*th state of the *n*th source is E(n,s,k)[n = 1, ..., N; s = 1, ..., S(n); k = 1, ..., K] for the *k*th pollutant. Furthermore, M receptor points have been specified at different locations in the area of interest. With the use of the appropriate meteorological diffusion model, we compute the matrix A(n,s,k,m)[n = 1, ..., N; s = 1, ..., S(n); k = 1, ..., K; m = 1, ..., M], which is the concentration of the *k*th pollutant at the *m*th receptor due to emissions from the *n*th source in the *s*th state.

The optimization routine is formulated mathematically as follows:

$$\text{minimize cost} \quad \sum_{n=1}^{N} \sum_{s=1}^{S(n)} C(n,s)\, X(n,s) \qquad (1)$$

subject to:

$$\sum_{n=1}^{N} \sum_{s=1}^{S(n)} E(n,s,k)\, X(n,s) \leq B(k) \quad (k=1,\ldots,N) \qquad (2)$$

$$\sum_{s=1}^{S(n)} X(n,s) = 1 \quad (n = 1,\ldots,N) \qquad (3)$$

and X(n,s) can take on values of zero or one only. This problem states that costs are to be minimized in such a way that the total emissions of each pollutant do not exceed a given amount B(k).

The set of equations (2) can be replaced by

$$\sum_{n=1}^{N} \sum_{s=1}^{S(n)} A(n,s,k,m)\ X(n,s) \leq B'(k), \quad (m=1,\ldots,M;\ k=1,\ldots,K)$$

which states that the total concentration of any pollutant at any receptor point cannot exceed a given amount B'(k). The optimization works by minimizing costs over the set of strategies while meeting air quality constraints for the two pollutants studied, sulfur dioxide and particulates.

Although the model was run for a number of AQCR's, for the sake of brevity only the results from the New York AQCR are summarized here.

The New York AQCR consists of 18 counties located in the States of New York, New Jersey, and Connecticut. The emission inventory for the region identified 678 point sources, the largest single group of which was 92 steam-electric power plants. The region was also divided into 863 area sources on a grid basis. Of these area sources, 151 were analyzed in detail manually to determine a reduction efficiency applicable to all area sources. This was done by analyzing the fuels being utilized and their alternatives. During computer runs, the air quality values reflected this reduction but no cost was applied since the cost was identical for all control strategies.

The 678 point sources averaged four feasible control alternatives, each including the base state (uncontrolled). This represents $4^{678}$ possible control strategies. In addition, air quality was modeled at 270 receptor locations.

Eleven control strategies were run for the model of which three required regional optimization. Two of the regional optimization strategies and two others requiring only individual source optimization will be discussed.

Initially, the Control Cost segment was run to determine all of the possible control alternatives for each point source in the model. (Area sources were reduced by a constant amount for air quality purposes but a corresponding cost was not computed. The area source reductions were constant over all strategies produced.) Then, a number of different control strategies were run. The most significant strategies are outlined below.

### 1. *Least Cost Subject to Primary Air Quality Standards*

In this strategy, pollutants at each receptor were constrained so as not to exceed concentrations of 80 $\mu g/m^3$ $SO_x$ and 75 $\mu g/m^3$ particulate. The heuristically obtained least-cost solution meeting these constraints was then determined.

### 2. *Least Cost Subject to Secondary Air Quality Standards*

This strategy is similar to strategy 1 but constrained both $SO_x$ and particulates to 60 $\mu g/m^3$ at each receptor. However, as indicated in the summary table, these constraints proved infeasible for the simulated New York AQCR.

### 3. *Maximum Reduction Efficiency*

This strategy maximizes the total reduction of pollutants by weight. The reduction of each alternative was determined by summing the $SO_x$ and particulate reductions. This is not an overall optimizing strategy because the optimum selection is made for each point source regardless of cost.

### 4. *Least Cost with Emission Charges*

This strategy involved the selection of the least-cost solution for each point source based on the sum of the costs of the device plus a cost for emissions produced. A uniform cost of 10¢/lb for emissions of each pollutant was used as the penalty cost.

Table 11.2 shows some of the effects of the four strategies listed above. It also presents a summary of some of the data produced by the model. Additional data not presented here included the various types of fuel and the amounts required by each strategy, the control alternatives selected for each point source by strategy, and the resultant air quality at each of 270 receptors by strategy.

The summary data leads to several interesting conclusions.

- -- A comparison of the maximum control strategy (3) with the primary standards strategy (1) shows

*Table 11.2*

*Results of Applying Strategies for the New York AQCR*

| | 1<br>*Least Cost Primary Standards* | 2<br>*Least Cost Secondary Standards* | 3<br>*Maximum Control* | 4<br>*Least Cost Emission Charges* |
|---|---|---|---|---|
| Device cost (millions $) | 54.98 | 92.16 | 406.30 | 91.29 |
| Emission penalties (millions $) | 0 | 0 | 0 | 72.21 |
| $SO_x$ reduction ($10^3$ tons) | 252.21 | 269.40 | 453.00 | 144.60 |
| Part reduction ($10^3$ tons) | 38.88 | 103.14 | 111.72 | 93.15 |
| Worst receptor $SO_x$ ($\mu g/m^3$) | 80.00 | 79.61* | 76.77 | 123.36 |
| Worst receptors particulate ($\mu g/m^3$) | 74.79 | 65.29* | 64.78 | 68.52 |

Base state emissions were 658.99 $10^3$ tons $SO_x$ and 124.75 $10^3$ tons particulate.

*Since standards could not be met, goals were set at 5% above minimum feasible for those receptors not meeting standards.

that while costs and reductions were significantly higher under maximum control resultant air quality at the worst receptor was not significantly affected.

-- The secondary standards strategy (2) was only slightly more expensive for devices than the penalties strategy (4) but had significantly better $SO_x$ air quality at the worst receptors. This indicates that the geographic location of each point source is significant with respect to the contributions to "worst case" air quality. Thus, emissions charges are not as efficient in reducing emissions as are specific alternatives selected by the optimization routine. Note also that the total cost of strategy 4 to private industry was much higher than strategy 2 because of the penalty charges.

-- A further comparison of strategies 2 and 4 indicates that the penalty cost was more effective in reducing particulate than $SO_x$ emissions.

As indicated, a variety of conclusions can be obtained from analysis of the micro models output. By analyzing the devices industry selects for a specific strategy, a series of industry-based standards can be developed. Alternatively, standards could be based on the geographic location since the model indicated that "hot spots" occurred at several receptors.

## VI. FURTHER WORK WITH A MICRO MODEL

Some of the more basic problems to be encountered with use of a micro type model are the difficulty of obtaining detailed data required, the problem of projecting growth of emissions, the difficulty of dealing with area sources at a micro level, and the difficulty arising in setting up and running a large model on a production basis.

The problem of obtaining data will probably be solved with the passage of time. Data collection efforts are underway in a number of AQCR's to define the emissions inventory. Estimating growth is difficult at a micro level. An interim solution may be to use macro level techniques until better methods are developed.

Area sources are a significant contribution to ambient air pollution but the unavailability of detailed data does not permit them to be handled as easily as point sources. Most area sources are either transportation (mobile sources) or fixed, but small, heating and incineration sources. Transportation sources will continue to be treated at an aggregate level. Heating sources can be characterized by the amounts and types of fuels burned, and control strategies would consist of fuel switching. Incineration sources are difficult to define exactly, and more work is needed in this area.

The setting up and operation of micro models is difficult and requires sophisticated personnel. Generally, manual analysis of certain parts of the data base are required and computer programs must often be modified to devise and test new strategies. For example, tests on a version of the heuristic optimization algorithm to permit constraints on fossil fuels indicated the great difficulty of modeling simultaneous constraints on fuels. Additionally, the dimensionality of the problem was further increased. These difficulties suggest the need for a large scale mixed integer LP.

While the IPP type of micro model has many weaknesses, it still provides a powerful tool to the policy planner. Work in expanding the usefulness and capability of such models will be rewarded by making strategy development and implementation more efficient and accurate.

REFERENCES

1. Kohn, R. E. "Abatement Strategy and Air Quality Standards" In *Development of Air Quality Standards*, Arthur Atkisson and Richard S. Gaines, eds. (Columbus, Ohio: Charles E. Merrill Publishing Co., 1970).
2. TRW Systems. "Air Quality Implementation Planning Program," Contract PH 22-68-60 Environmental Protection Agency (1970).
3. Norsworthy, J. R. and A. Teller. "The Evaluation of the Cost of Alternative Strategies for Air Pollution Control," APCA Paper 69-172, New York City (June 22-26, 1969).

MODELS FOR ENVIRONMENTAL POLLUTION CONTROL


CHAPTER 12

# MATHEMATICAL MODELS FOR AIR POLLUTION ABATEMENT

H. G. Fortak*

## I. INTRODUCTION

Increasing industrialization and urbanization raises the question of whether a further increase is tolerable from an environmental protection point of view. In most cases new air pollution sources are built without sufficient knowledge of possible changes in the air pollution load of a given region. Measurements cannot provide the urgently needed forecasts of anticipated changes of air pollution level. In addition, no rational planning for obtaining optimal control measures is possible. Only mathematical models for air pollution transport and diffusion together with a deep knowledge of meteorological processes involved are able to solve a wide variety of problems of this kind.

Even in view of the present difficulties in formulating and solving mathematical equations, we have achieved a state in which climatological air pollution forecast models are available and capable of solving problems of air pollution abatement and of city planning with a fairly high degree of accuracy. This has been tested for several cities since 1965, Bremen, Germany, being one of those tested. This statistical forecast of air pollution climatology is also applied to problems connected with installations of big industrial complexes and has proved to be the best method available today.

---

*Prof. H. G. Fortak is the director of the Institute for Theoretical Meteorology, Freie Universität Berlin, 1 Berlin 33, Thielallee 49, Germany.

In contrast to this, real-time forecasts of air pollution levels in regions are extremely difficult and are not feasible in the near future.

## II. THE PROBLEM FROM THE METEOROLOGICAL POINT OF VIEW

Air pollution components are produced by man at known source locations, are transported and diluted by atmospheric motion systems, and are observed at receptor locations, sometimes far away from the source areas. Air pollution abatement starts with prescribed or measured concentrations at all important receptor locations of a region, and its aim is control of the amount of emissions of all or of some of the sources. Such a control strategy is feasible only if the mechanism of atmospheric transport and turbulent diffusion is understood. Therefore, air pollution abatement is to a high degree a meteorological problem.

This is true in the case of long-range air pollution forecasting (in a climatological sense) as well as in the case of real-time short-range forecasting. In the latter case the influence of current meteorological parameters on the observed concentration field is very complex. Figure 12.1 demonstrates this complexity in a more general fashion. A dominant parameter is given by the general weather situation (*e.g.* cyclonic or anticyclonic) and therefore by the air mass and the pressure field which governs the wind field. Wind direction simply selects the receptor points with air pollution loading. Wind speed is responsible not only for the mechanism of transport but also partly for the development of atmospheric turbulence.

Turbulence depends on the properties of the air mass, mainly on the vertical temperature stratification. Here a complicated feedback mechanism takes place. The temperature lapse rate influences the kind and distribution of clouds, cloudiness interacts with the sun's radiation, and radiation finally alters the lapse rate. Cloudiness also influences photochemical reactions among various air pollution species, and, in case of precipitation, precipitation scavenging leads to another reduction of concentration

Even from a quite general meteorological point of view, then, the meteorological role in air pollution abatement is a very complex one. The problem becomes even more complicated if we consider the

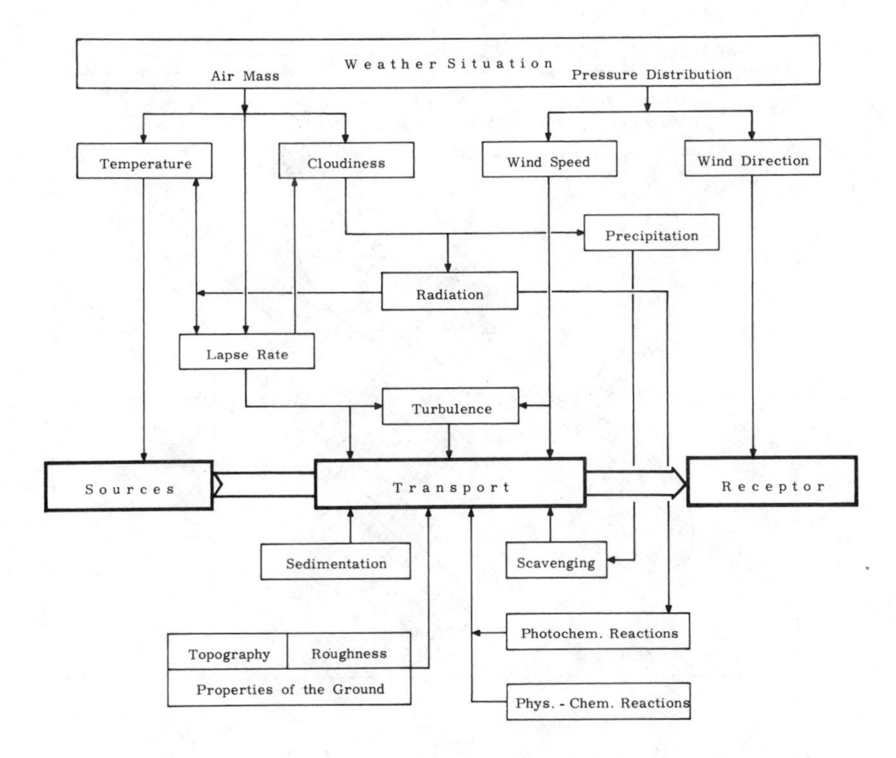


*Figure 12.1. Meteorological influences on transport and turbulent diffusion of air pollution components in the atmosphere.*

spectrum of atmospheric motions. Atmospheric processes cover a vast range in the spectrum of motions. Small scale turbulence with characteristic length scales of less than 1 cm is found at the one end and ultralong planetary waves with wave lengths of $10^4$ km are found at the other end of the spectrum (10 orders of magnitude difference). Figure 12.2 gives an overall impression of the characteristic length and time scales involved in various atmospheric processes.

The discussion of Figure 12.1 dealt with the so-called synoptic scale, *i.e.*, scales greater than $10^3$ km and characteristic time scales greater than one day. In most cases of abatement modeling, smaller scales, called subsynoptic scales, are most important. Especially under anticyclonic weather conditions a large number of subsynoptic motion systems develop. All kinds of thermal convection

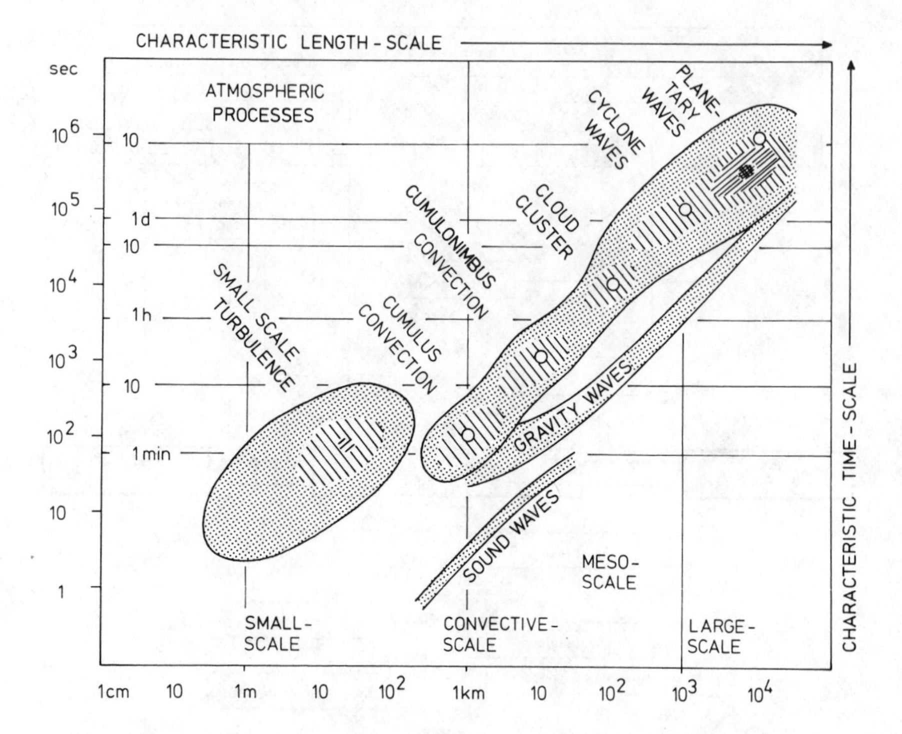


*Figure 12.2. Characteristic time- and length-scales for typical atmospheric motion systems.*

and local circulations have a strong influence on transport and dilution of air pollution.

Objective forecasting of subsynoptic atmospheric processes is not feasible in the near future. The same is true for synoptic-scale parameters, which are responsible for the development of atmospheric turbulence. Figure 12.3 shows the limitations of deterministic forecasting of meteorological fields. Barotropic and baroclinic models for large-scale atmospheric motions are useful for predictions of pressure, temperature, and wind fields up to three days ahead. The spatial resolution by a mesh of 400 km x 400 km is quite poor. Cyclones, anti-cyclones and planetary waves are close to the limits for which deterministic forecasts are possible. So far only a few models for deterministic forecasting of special subsynoptic motion systems are available. Their range of deterministic forecasting diminishes with decreasing scales. Only

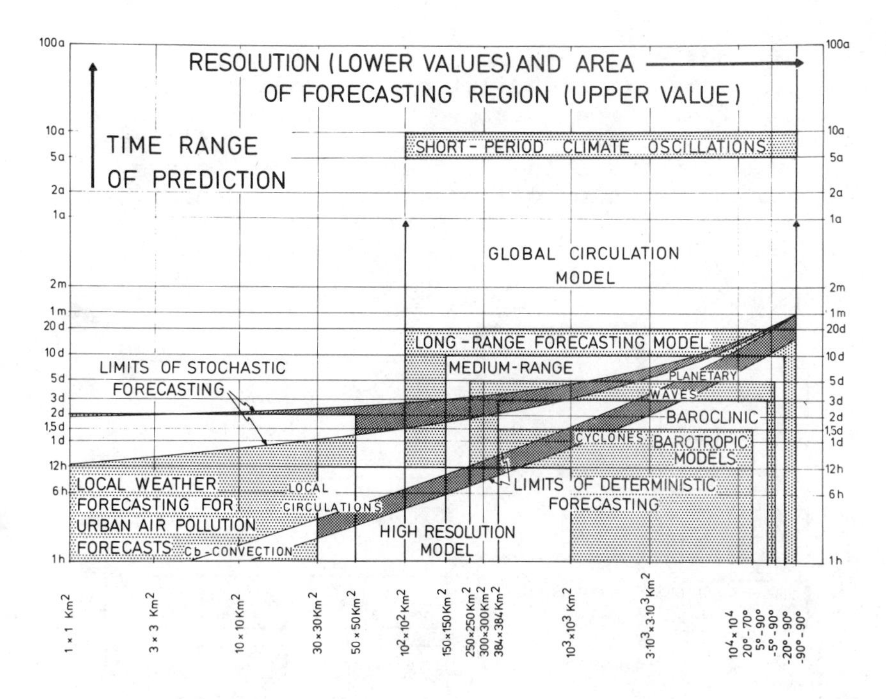


*Figure 12.3. Range of prediction of atmospheric processes, limits of deterministic and stochastic forecasting, and spatial resolution in the various models.*

newly developed models for dynamic stochastic forecasting will be able in principle to predict even the smaller scales for periods of one day or more ahead. Therefore, there is some hope of predicting the development of meteorological parameters relevant to transport and dispersion of pollutants on a real time basis in the future. Whether these stochastic forecasts are sufficient for purposes mentioned earlier is not known.

## III. GOALS OF AIR POLLUTION MODELING

Air pollution abatement strategies require predictions of concentrations in a given region. These concentrations are caused primarily by emission sources and secondarily by mechanisms of transport and dispersion in the atmosphere. Therefore, a source-orientated point of view is the natural one in this respect.

The goals of mathematical modeling of air pollution transport and dilution are demonstrated in Figure 12.4. In general, a mathematical multiple-source air pollution model (the system) will react on an input consisting of the source emissions, the regional meteorological parameters and others, and will produce an output in form of concentration fields. The output can be a real-time prediction

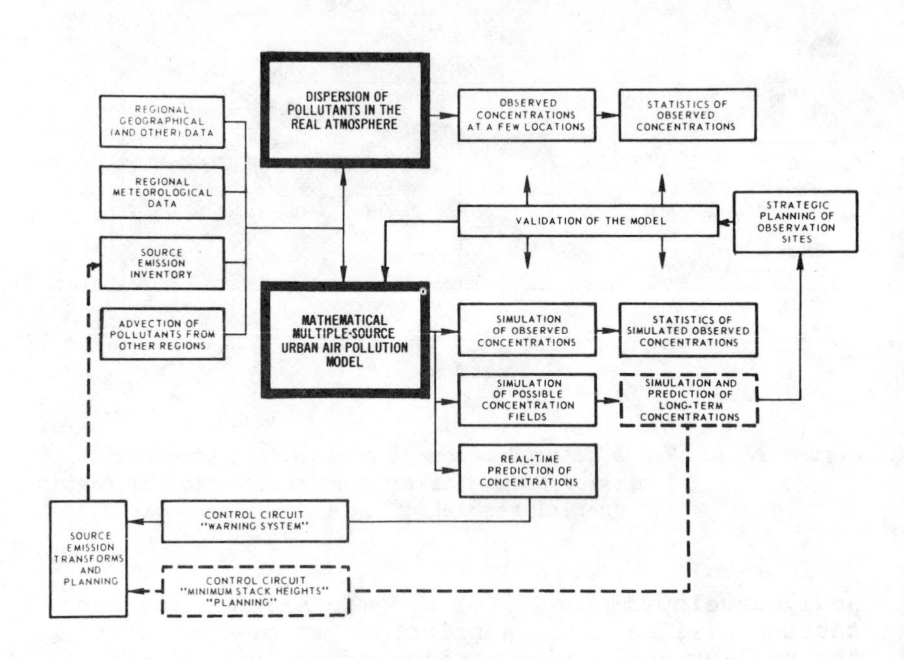


*Figure 12.4. Major elements of the urban air pollution problem connected with mathematical modeling.*

of concentrations on the basis of a real-time prediction of meteorological parameters, or it can be a prediction of all possible concentration fields for all possible combinations of meteorological parameters observed over a sufficiently long period of time. If in a real-time prediction the calculated concentrations exceed given limits, a feedback circuit can be used so that systematic variations of the source emission inventory are performed as often as necessary to obtain predicted concentrations which are tolerable. On the other hand, if all possible fields of concentration are predicted, long-term concentrations can then be predicted in a statistical sense. Such a prediction is most important for the solution of environmental planning problems. If new groups of emission sources are planned in a region, it is possible to simulate the effect of the new emissions on the already present situation and a similar feedback game as described previously will lead to optimal solutions of planning problems. Simulations of statistical properties of concentration fields are most useful for strategic planning of observation sites that are needed for model verification and, if necessary, model calibration.

It will be shown later that solutions of the problem of locating observation sites are already feasible.

## IV. MATHEMATICAL MODEL EQUATIONS IN THE SPECTRAL DOMAIN

The most frequently used model for estimating transport and spread of air pollution components is based on the simple diffusion equation

$$\frac{\partial C}{\partial t} = -\mathbf{v}\cdot\nabla C + K\nabla^2 C + q(C;\ \mathbf{r},t) \tag{1}$$

In this equation C means concentration, **v** wind vector, K turbulent diffusivity, and q the sum of all sources of the special pollutant.

In view of earlier statements, it is worthwhile to go back to the roots of this equation. The basis of Equation 1 is given by the continuity equations for air and pollutants, respectively:

$$\frac{\partial \rho}{\partial t} + \nabla\cdot\rho\mathbf{v} = 0 \tag{2}$$

$$\frac{\partial \rho^s}{\partial t} + \nabla\cdot\rho^s\mathbf{v}^s = Q^s(C^1,\ C^2,\ \ldots,\ C^s,\ \ldots;\ \mathbf{r},t); \quad s = 1,\ 2,\ \ldots,\ S \tag{3}$$

where $\rho$ means density of air, $\rho^s$ density of pollutant (substance s), $\mathbf{v}$, $\mathbf{v}^s$ velocity vectors of air and substance motion, and $Q^s$ source of substance s. Introducing the concentration $C^s = \rho^s/\rho$ into Equation 3 gives

$$\frac{\partial \rho C^s}{\partial t} + \nabla\cdot\rho C^s\mathbf{v}^s = Q^s$$

or identically

$$\frac{\partial \rho c^s}{\partial t} + \nabla \cdot \rho c^s \mathbf{v} + \nabla \cdot \rho c^s (\mathbf{v}^s - \mathbf{v}) = Q^s \qquad (4)$$

It is taken into account that the main transport is given by the air motion and that the relative motion given by the difference vector $\mathbf{v}^s - \mathbf{v}$ is generally small.

Equation 4 is valid for the micro-scale of atmospheric motions and should be averaged with respect to the scale involved. Symbolically a normal average is defined by $\overline{\alpha} = \alpha - \alpha'$, with $\overline{\alpha'} = 0$. In addition, a weighted average, weighted with the density $\rho$ of air, is quite useful: $\hat{\alpha} = \alpha - \alpha''$ with $\overline{\rho\alpha''} = o$. Averaging Equations 1 and 4 leads to

$$\frac{\partial \overline{\rho}}{\partial t} + \nabla \cdot \overline{\rho}\, \hat{\mathbf{v}} = o \qquad (5)$$

$$\frac{\partial \overline{\rho}\hat{c}^s}{\partial t} + \nabla \cdot \overline{\rho c}^s \hat{\mathbf{v}} = - \nabla \cdot \overline{\rho}\hat{c}^s (\hat{\mathbf{v}}^s - \hat{\mathbf{v}}) + \nabla \cdot (-\overline{\rho c^{s''} \mathbf{v}''}) + \nabla \cdot [-\overline{\rho c^{s''} (\mathbf{v}^{s''} - \mathbf{v}'')}] + \overline{Q}^s \qquad (6)$$

Upon introduction of Equation 5 into Equation 6 the basic equation for transport and turbulent diffusion of a substance s is obtained:

$$\frac{\partial \hat{c}^s}{\partial t} = - \hat{\mathbf{v}} \cdot \nabla \hat{c}^s + \frac{1}{\overline{\rho}} \left[ -\nabla \cdot \overline{\rho}\hat{c}^s (\hat{\mathbf{v}}^s - \hat{\mathbf{v}}) + \nabla \cdot (-\overline{\rho c^{s''} \mathbf{v}''}) + \nabla \cdot [-\overline{\rho c^{s''} (\mathbf{v}^{s''} - \mathbf{v}'')}] + \overline{Q}^s \right] \qquad (7)$$

By applying Green's theorem it is easily seen that at a boundary with outer normal vector $\mathbf{n}$ the boundary condition

$$\mathbf{n} \cdot \left[ -\overline{\rho c}^s \hat{\mathbf{v}} - \overline{\rho}\hat{c}^s (\hat{\mathbf{v}}^s - \hat{\mathbf{v}}) + (-\overline{\rho c^{s''} \mathbf{v}''}) + [-\overline{\rho c^{s''} (\mathbf{v}^{s''} - \mathbf{v}'')}] \right]$$

should be prescribed. Therefore the following must be known: transport $\hat{C}^s \hat{\mathbf{v}}^s$ from the surrounding into the region, transport of mean relative motion, turbulent transport connected with turbulence of the air, and turbulent relative transport.

In Equation 7 local time rate of change of concentration is given by advection of pollutant, $-\hat{\mathbf{v}}\cdot\nabla\hat{C}^s$, and by terms which can be interpreted as sources and sinks.

The sources $Q^s$ deserve some explanation. If there are a number of point sources fixed in space we can write

$$Q^s = \sum_m q_m(t)\delta(\mathbf{r}-\mathbf{r}_m) \tag{8}$$

where $q_m(t)$ is time dependent source strength and $\mathbf{r}_m$ location of the point source. In case of plume rise $\mathbf{r}_m$ may depend on time if a properly defined formula for plume rise is adopted. In case of a number of mobile sources we may put

$$Q^s = \sum_n p_n(t)\delta[\mathbf{r}-\mathbf{r}_n(t)] \tag{9}$$

where $\mathbf{r}_n(t)$ defines the trajectory of the mobile source. A spatial source, including an area source, is defined by a continuous function

$$Q^s = Q^s(\mathbf{r},t) \tag{10}$$

If there are chemical reactions,

$$Q^s = \rho R^s\left[\ldots\prod_{\lambda=1}^{S}\left(\frac{M}{M_\lambda}C^\lambda\right)^{\nu_{i\lambda}}\ldots\right] \tag{11}$$

where $R^s$ is the chemical production rate of species s depending on products of concentrations of substances involved in the various reactions.

Averaging Equations 8, 9 and 10 is simple. For the time being, however, it is not possible to treat the reaction rate in cases of turbulent flow.

In Equation 7 the turbulent fluxes must be parameterized. The classical way to do this is a gradient assumption

$$-\overline{\rho C^{s\prime\prime}\mathbf{v}^{\prime\prime}} = \bar{\rho}\,\mathbf{K}\cdot\nabla\hat{C}^s \tag{12}$$

which defines (an unknown) tensor of turbulent diffusivity. Something similar can be adopted for turbulent relative motions

$$-\overline{\rho C^{s\prime\prime}(\mathbf{v}^{s\prime\prime}-\mathbf{v}^{\prime\prime})} = \bar{\rho}\,\mathbf{K}_r\cdot\nabla\hat{C}^s$$

With regard to the uncertainness of $\mathbf{K}$, it is impractical to introduce a second highly unknown variable. Therefore $\mathbf{K}_r$ will be absorbed into $\mathbf{K}$. We now obtain from Equation 7 (omitting bars and dashes)

$$\frac{\partial c^s}{\partial t} = -\mathbf{v}\cdot\nabla C^s + \frac{1}{\rho}[-\nabla\cdot\rho C^s(\mathbf{v}^s-\mathbf{v}) + \nabla\cdot\rho\mathbf{K}\cdot\nabla C^s + Q^s] \tag{13}$$

an equation similar to Equation 1 if $\rho$ = const, $\mathbf{K} = K\mathbf{I}$ = const ($\mathbf{I}$: unit tensor), $Q^s/\rho = q^s$, and $\mathbf{v}^s = \mathbf{v}$. It is seen that a great number of assumptions and approximations are used to arrive at the conventional Equation 1 for transport and turbulent diffusion.

Diffusion of falling particles is influenced by the difference $\mathbf{v}^s-\mathbf{v} = (v_z^s - v_z)\mathbf{k}$ ($\mathbf{k}$ vertical unit vector) and in this case of sedimentation $\mathbf{K}$ definitely is different from the corresponding value for diffusion of gases.

In many cases chemical reactions act only like a sink for the substance s. In this case $R^s = -\lambda C^s$ and Equation 13 can be written as ($\rho$ = const)

$$\frac{\partial C}{\partial t} = -\mathbf{v}\cdot\nabla C - \nabla\cdot C(\mathbf{v}^s-\mathbf{v}) + \nabla\cdot\mathbf{K}\cdot\nabla C + q(\mathbf{r},t) - \lambda C \tag{14}$$

omitting index s except in the relative transport term.

Splitting all vectors into horizontal and vertical components according to

$$\mathbf{v} = \mathbf{v}_h + \mathbf{k}v_z; \quad \nabla = \nabla_h + \mathbf{k}\frac{\partial}{\partial z}; \quad \mathbf{K} \approx \mathbf{K}_h + K_{zz}\mathbf{kk}$$

one arrives at

$$\frac{\partial C}{\partial t} = -\mathbf{v}_h\cdot\nabla_h C - v_z\frac{\partial C}{\partial z} - \nabla_h\cdot C(\mathbf{v}_h^s-\mathbf{v}_h) - \frac{\partial}{\partial z}C(v_z^s-v_z) + \nabla_h\cdot\mathbf{K}_h\cdot\nabla_h C + \frac{\partial}{\partial z}K_{zz}\frac{\partial C}{\partial z} + q - \lambda C \tag{15}$$

In all practical cases of modeling air pollution transport and dilution involving various scales from a single plume up to intercontinental transports, a simple scale analysis is quite useful. Introducing for this purpose

$$t = Tt^*; \quad [x,y] = L[x^*,y^*]; \quad z = Hz^*$$

$$\mathbf{v}_h = U\mathbf{v}_h^*;\ \mathbf{v}_h^s = U^s\mathbf{v}_h^{s*};\ v_z = Wv_z^*;\ v_z^s = W\,v_z^{s*}$$

$$\mathbf{K}_h = K_h\mathbf{K}_h^*;\ K_{zz} = KK_{zz}^*;\ C = XC^*;\ q = Qq^*;\ \lambda = \Lambda\lambda^*$$

the dimensionless factors of the terms on the right hand side of Equation 15 are

$$\frac{U}{L/T},\ \frac{W}{H/T},\ \frac{U^s - U}{L/T};\ \frac{W^s - W}{H/T};\ \frac{K_h}{L^2/T},\ \frac{K}{H^2/T};\ \frac{QT}{X};\ \Lambda T$$

In many cases a convective time scale $T = L/U$ can be used, in which case local changes with time have the same order of magnitude as corresponding advections.

A typical example is an urban area with $L = 10$ km $= 10^4$ m, $H = 1$ km $= 10^3$ m and $U = 3$ m/sec. Here $T = 1$ h and we therefore obtain for the factors

$$1;\ \frac{10}{3}W;\ \frac{U^s-U}{U};\ \frac{10}{3}(W^s-W);\ \frac{K_h}{3\cdot 10^4};\ \frac{K}{3\cdot 10^2};\ \frac{Q\cdot 1h}{X};\ \Lambda\cdot 1h$$

Vertical motion becomes important if $W \geqq 3$ cm/sec, horizontal diffusion if $K_h \geqq 3\cdot 10^3$ m$^2$/sec $= 3\cdot 10^7$ cm$^2$/sec, vertical diffusion if $K \geqq 30$ m$^2$/sec $= 3\cdot 10^5$ cm$^2$/sec, the sources if their characteristic source strength per hour equals at least one-tenth of the characteristic concentration, and absorption becomes important if $\Lambda \geqq 0.1$ h$^{-1}$.

Many solutions of Equation 14 in simplified form have been published. It seems, however, that with regard to scales from urban scales up to higher ones a spectral approach will be most convenient for many purposes.

Assuming that there exists a three-dimensional system of ortho-normal functions that fulfills all boundary conditions, any variable can be expanded into a series like

$$C = \sum_{m=1}^{\infty} C_m(t)F_m(\mathbf{r}) \tag{16}$$

where $F_k^*F_m$ integrated over the entire domain gives $\overline{F_k^*F_m} = \delta_{km}$. It is important to note that the scale of spatial resolution can be selected by truncation of Equation 16. Introducing Equation 16 into Equation 14 and putting $\mathbf{v}^s = \mathbf{v}$ (gases), one obtains by omitting the sign for sums

$$F_m \dot{C}_m(t) = -[\mathbf{v}_m(t) \cdot F_m \nabla F_n - \mathbf{K}_m(t) \cdot\cdot \nabla(F_m \nabla F_n) + \lambda F_n] C_n(t) +$$

$$+ F_m q_m(t)$$

and after multiplication with $F_k^*$ and integration

$$\dot{C}_k(t) = -[\mathbf{v}_m(t) \cdot \overline{F_k F_m \nabla F_n} - \mathbf{K}_m(t) \cdot\cdot \overline{F_k \nabla(F_m \nabla F_n)} + \lambda \delta_{kn}] C_n(t) + q_k(t) \tag{17}$$

or

$$\dot{C}_k(t) = - M_{kn}(t) C_n(t) + q_k(t) \tag{18}$$

The matrix $M_{kn}(t)$ contains all information about the meteorological input.

Assuming the initial concentration $C(\mathbf{r},o) = C°$, *i.e.*, $C_k°$, to be known, the solution of Equation 18 describes a trajectory in phase space because the state of the system can be represented by a vector $\mathbf{C} \equiv [C_k]$ in phase space for all times. Then Equation 18 in the form

$$\dot{\mathbf{C}}(t) = - \mathbf{M}(t) \cdot \mathbf{C}(t) + \mathbf{q}(t) \tag{19}$$

is the differential equation of a trajectory in phase space.

This form of the original model equation allows a number of interesting investigations. For example, the sensitivity of the solution with respect to incomplete knowledge of meteorological input data can be studied by assuming $\mathbf{M}/\mathbf{M}+\varepsilon$ and comparing trajectories after certain times of integration. On the other hand the effect of control strategies can be evaluated by assuming $\mathbf{q}/\mathbf{q}+\delta\mathbf{q}$ and varying $\delta\mathbf{q}$ with regard to certain constraints.

A certain class of situations is connected with steady state meteorological conditions. In these cases $\mathbf{M} \neq M(t)$ and Laplace's transformation applied on Equation 19 gives

$$p\tilde{\mathbf{C}}(p) - \mathbf{C}° = - \mathbf{M} \cdot \tilde{\mathbf{C}}(p) + \tilde{\mathbf{q}}(p)$$

or

$$\tilde{\mathbf{C}}(p) = [p\mathbf{I} + \mathbf{M}]^{-1} \cdot [\mathbf{C}^{\circ} + \tilde{\mathbf{q}}(p)] \tag{20}$$

Inversion of this solution is possible and leads to a convolution integral.

The simplest type of models is the steady state model described by

$$\mathbf{C} = \mathbf{M}^{-1} \cdot \mathbf{q} \tag{21}$$

This type has been used for modeling air pollution climatologies of urban areas in the past. For all possible weather situations the corresponding vectors $\mathbf{C}_j = \mathbf{M}_j^{-1} \cdot \mathbf{q}_j$; therefore, the corresponding concentrations $C_j = \sum_m C_{j,m} F_m(\mathbf{r})$ have been calculated.

From these values $C_j$ frequency distributions can be obtained at all important locations within the urban area. An example[1] is given in Figure 12.5, in which calculated and observed frequency distributions are

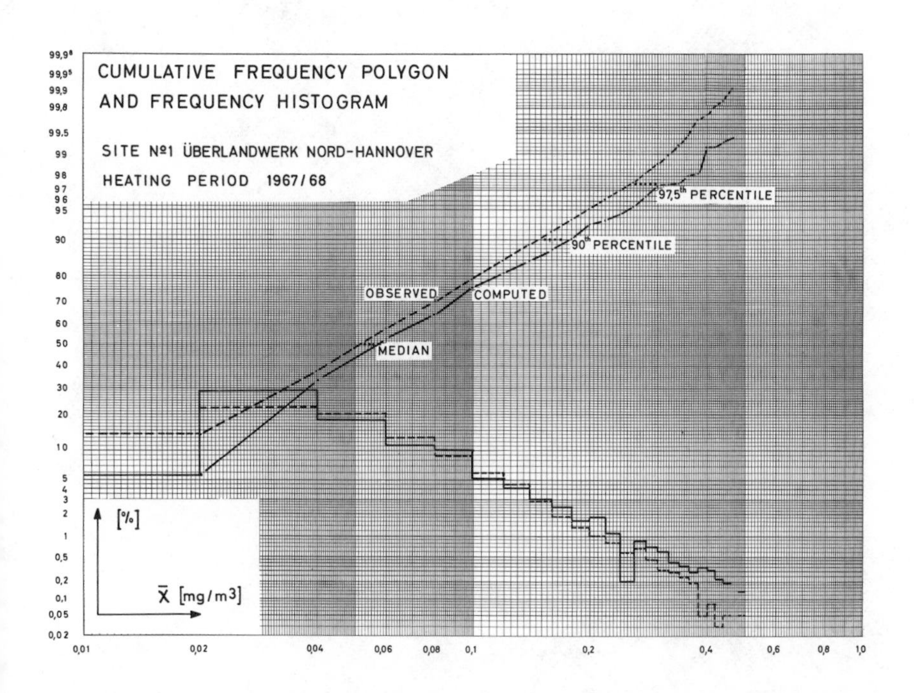


*Figure 12.5. Comparison between observed and computed frequency distributions of ground-level concentrations in downtown Bremen.*

compared with each other for one of four stations in the city of Bremen, Germany. Meanwhile other cities have been modeled in the same way (Stockholm, Düsseldorf) with equal success.

## REFERENCE

1. Fortak, H. G. "Numerical Simulation of the Temporal and Spatial Distributions of Urban Air Pollution Concentration," in *Proceedings of Symposium on Multiple-Source Urban Diffusion Models* (Washington, D.C.: U.S.E.P.A., APCO Publ. AP-86, 1969).

MODELS FOR ENVIRONMENTAL POLLUTION CONTROL


# CHAPTER 13

# MATHEMATICAL MODELS FOR AIR POLLUTION CONTROL: DETERMINATION OF OPTIMUM ABATEMENT POLICIES

S.-Å. Gustafson and K. O. Kortanek

## I. INTRODUCTION

This paper presents a mathematical model that can be used to study control strategies for air pollution abatement. The goal is to reach compliance with air quality standards for the ground level while minimizing the control costs which thereby occur. The case of a multiple-source urban air control area is treated and long-term time-averages are discussed, but the main ideas can also be applied to pollution episodes in cities or to large-scale models for vast areas.

Section II discusses various measures of air quality and gives examples of standards adopted in the U.S. Some statistical results are mentioned.

In Section III the concept of source inventory is introduced and examples of various classes of polluters encountered in an urban area are given.

In Section IV some methods of relating the sources to the concentration of pollutants in the ambient air at receptor points on the ground level are presented. There are two options: (a) statistical models correlating the observed concentration to known variables and (b) physical models based on the law of the conservation of matter. The drawback of the first approach is that the model obtained is valid for a particular city at the time of modeling

*Dr. S.-Å. Gustafson is with the Department of Numerical Analysis, Royal Institute of Technology, S-10044 Stockholm 70, Sweden. Professor K. O. Kortanek is with the School of Urban and Public Affairs, Carnegie-Mellon University, Pittsburgh, Pennsylvania, 15213, U.S.A.

and may not be used confidently for prediction of air quality in future stages of the development of the city.

In Section V the selection of control variables is discussed. It is revealed that some sources cannot be controlled independently and others cannot be regulated at all. Cost-functions associated with the control variables are also considered.

Section VI is devoted to a description of the computational solution of our optimization problem. This can be achieved by combining, in an appropriate manner, well-known numerical procedures, such as linear programming and solution of nonlinear systems of equations.

Section VII treats the question of error analysis. Applying a philosophy explained by Wilkinson,[1] one can directly relate uncertainties in input data to errors introduced during the computations (round-offs, truncation and discretization errors).

Section VIII contains a worked example. In spite of being very simple it has all the properties typical of the class of problems treated in this paper.

Section IX contains the conclusions.

## II. AIR QUALITY STANDARDS

This paper studies the appearance in the air of certain undesirable components which are termed pollutants. The composition of the air may be studied by taking samples and determining their contents. The result of this analysis is interpreted as the average over the sampling time of the composition of the air at the sampling point. If many samples are taken at the same point the results will vary with time in an irregular fashion.

Larsen[2] reports the results of an analysis of measurements taken in several American cities (Chicago, Cincinnati, Los Angeles, Philadelphia, San Francisco and Washington). The concentration of the following classes of pollutants was measured: carbon monoxide, hydrocarbons, nitric oxide, nitrogen dioxide, nitrogen oxides ($NO + NO_2$), oxidant, sulfur dioxide. Then it was found that:

1. average measured concentrations are approximately log-normally distributed (for all studied averaging times, all cities and all pollutants)
2. median concentration (50 percentile in the population of samples with a certain averaging time) is

proportional to a fixed power of the averaging time

3. the maximum concentration is proportional to a fixed power of this averaging time if the averaging time is less than a month.

Using these statistical laws it is possible to estimate how often the concentrations will surpass given levels. Thus one should be able to assess the risk that the concentration temporarily rises to a dangerously high level, which of course may happen even if the average value is satisfactory.

Generally several pollutants are present in the air at the same time. One can try to measure their effect by means of a combined pollution index.[3] Then the statistical distribution of the values of this index can be expressed by means of the corresponding information about each pollutant.

Most of the air quality standards adopted prescribe a maximum permitted value for the long-time mean concentration to protect against damage by attrition and a provision that the average concentration over certain shorter periods must not surpass given levels with more than a given frequency.

As an example of such rules the air quality standards for sulfur oxides set by Environmental Protection Agency on April 30, 1971, are summarized here.

Primary standards to protect public health:

| | |
|---|---|
| $80 \mu g/m^3$ = 0.03 ppm | annual arithmetic mean |
| $365 \mu g/m^3$ = 0.14 ppm | maximum 24-hour concentration not to be exceeded more than once a year. |

Secondary standards protect against effects on soil, water, vegetation, materials, animals, weather, visibility and personal comfort and well-being:

| | |
|---|---|
| $60 \mu g/m^3$ = 0.02 ppm | annual arithmetic mean |
| $260 \mu g/m^3$ = 0.1 ppm | maximum 24-hour concentration not to be exceeded more than once a year |
| $1300 \mu g/m^3$ = 0.5 ppm | maximum 3-hour concentration not to be exceeded more than once a year. |

Standards of the same structure apply to several other pollutants. Further, these standards must be met throughout the region, a circumstance which causes a major difficulty in the implementation. Even if compliance can be established at many sample points, it is impossible to tell a priori anything about the concentration *between* the sampling points.[4] To do this there must be a means of estimating the concentrations at all points of the air pollution control area. This is difficult since the concentrations (also the mean-values) vary irregularly from point to point. Our goal will be to split the concentration into two parts, one which is "regular" and can be estimated systematically at all points and one which is irregular and of negligible magnitude. In our pursuit of this we shall need information of the polluters affecting the air quality in the region as well as data describing its climate.

## III. SOURCE INVENTORY

In many cities a large portion of the total pollution originates from a limited number of plants emitting effluents from high stacks. Thus in the case of the Chicago metropolitan area about 65% of the sulfur oxides released into the air of the control area originate from six power plants.[5] These plants are treated as *point-sources* of negligible dimensions in space. Their contribution to the concentration at a receptor point depends on the emission rate and the height and position of the stack. The combined effect of several point-sources is found by the principle of super-position. The effect of a busy highway on the concentration of pollutants at a receptor point depends on the emission per unit length of the highway. Its combined influence is determined by integrating the contributions from each infinitesimal part that is treated as a point source. Thus the highway is called a *line-source* since we take into account its length but neglect its breadth.

A large part of the observed pollutants may emanate from a large number of scattered small sources, *e.g.*, residential heating. This is an example of an *area source*. It is described by splitting up the area in a grid and determining the emission from each mesh, which is approximated by a point-source. If the emission per area unit is a continuous function, the contribution from the area source is given by a Riemann integral. For a discussion of source inventories see Fortak,[6] Roberts *et al.*,[5] and Hanna.[7]

It should be noted here that sometimes polluters outside the control region must be taken into account as well.

## IV. CONSTRUCTION OF TRANSFER FUNCTIONS

The emission rates must be related to the observed concentrations. To do this several simplifying assumptions, which will be used throughout the paper, are introduced here.

1. A pollutant, *e.g.*, sulfur dioxide, that is chemically stable and released from a stationary source is considered.
2. Steady-state condition prevails.
3. Longer time-periods are regarded as a succession of smaller time-steps in each of which steady-state condition prevails.

Then by diffusion modeling the contribution from a point-source may be determined using the Sutton equations for unstable or neutral atmosphere while generalized Gauss formulas are used under stable conditions.[8] Fortak[6] gives more elaborate formulas, taking into account limited mixing heights--under certain rather common weather conditions the air carrying the pollutants can only rise to a specific height above the ground. The use of the formulas mentioned here requires the knowledge of parameters like wind speed and direction, stability class, effective stack-height and position of stack.[9] Consider now a point-source with an emission rate q which is a continuous function of time t. [q(t) may depend on season, weekday or time of the day, as in the case of heating.] The contribution from the point-source to the concentration of the pollutant is of the form

$$\kappa(t,x_1,x_2) = q(t)w(t,x_1,x_2)$$

where $x_1$, $x_2$ are the coordinates of the receptor point and w may be of Sutton or generalized Gauss type with the meteorological parameters expressed as functions of time.

The time-average $\overline{\kappa}$ over the period [O,T] is then given by

$$T\cdot\overline{\kappa}(x_1,x_2) = \int_0^T \kappa(t,x_1,x_2)dt = \int_0^T q(t)w(t,x_1,x_2)dt \qquad (1)$$

Weather changes irregularly with time but climate is assumed to be the same. Hence we can consider the change of the meteorological parameters with time as a realization of a stable stochastic process. Therefore if the frequencies of various values of the meteorological parameters are known for all t in the interval [0,T], we can rewrite the integrals of Equation 1 as multiple Stieltjes integrals. These can always be approximated using numerical quadrature rules with positive weights. This may prove to be an economical alternative to the computational procedures suggested by Fortak.[6] We note that in this manner the average contribution $\bar{\kappa}$ is approximated by a positive combination of a finite number of contributions occurring under definite and known conditions:

$$\bar{\kappa}(x_1,x_2) \approx \sum_{k=1}^{n} \lambda_k \; \kappa(t_k,x_1,x_2) \qquad (2)$$

$\lambda_k$ and $t_k$ are independent of $x_1,x_2$. Contributions from line- and area sources can be determined using the principle of superposition and integration. However, Hanna[7] proposes a much simpler way of handling area sources, stating that contributions from such sources are largely inversely proportional to average wind speed and directly proportional to area emission rate in a neighborhood of the receptor point.

## V. SELECTION OF CONTROL VARIABLES AND COSTFUNCTION

If the pollution is too intense, the emission rates must be reduced. It is possible to distinguish between emergency actions, like shutting down factories and closing highways, that can be taken to remedy an acute situation and long-term measures, such as installing purifying equipment in factories and prescribing the kind of fuel permitted for factories, heating and cars.

Obviously, not all sources in the source inventory can be controlled independently. Residential heating is thus considered an area source. Another example is two plants which for legal reasons must be treated identically. Let their contributions to the annual average concentration before reduction be given by the functions $\bar{\kappa}_1$ and $\bar{\kappa}_2$ and hence their combined contribution is

$$\bar{\kappa}_1 + \bar{\kappa}_2$$

After reducing the emission rates with the constant factor E, the combined contribution is

$$(1-E)u, \quad u = \bar{\kappa}_1 + \bar{\kappa}_2$$

E will be called a *control variable* and u its associated *transfer function*. After introducing all control variables, the annual average concentration κ can be written thus before reduction:

$$\kappa_0 = \Delta\kappa + u_0 + \sum_{r=1}^{n} u_r$$

after reduction:

$$\kappa_E = \Delta\kappa + u_0 + \sum_{r=1}^{n} (1-E_r)u_r \qquad (3)$$

where:

$E_r$ = control variable, real number

$u_r$ = its associated transfer function

$u_0$ = contribution from noncontrollable sources

$\Delta\kappa$ = contribution from sources not in the source inventory

$\Delta\kappa$ is thus an irregular or not explained term, $u_0$ an explained but not controlled entity. Now let $\vartheta$ be the given standard. Then we have the condition

$$\kappa_E(x) \leq \vartheta(x), \quad x \in S \qquad (4)$$

where S is the air control area and the vector x refers to the coordinates of a receptor point. The condition in Equation 4 is replaced with the simpler

$$u_0(x) + \sum_{r=1}^{n} (1-E_r)u_r(x) \leq u_{n+1}(x), \; x \in S \qquad (5)$$

where $u_{n+1}$ is chosen so that Equation 5 implies Equation 4. Hence the condition in Equation 5 is more stringent than that in Equation 4 to account for our insufficient knowledge of $\Delta\kappa$.

There are generally several different choices of $E = (E_1, E_2, \ldots, E_n)$ which meet Equation 5; hence different reduction policies can be used to meet the standard. With E we associate a cost-function G which introduces an ordering in the set of feasible reduction policies.

One must always have $0 \leq E_r \leq 1$ but there might be upper bounds $e_r < 1$ on the possible reductions. Therefore the task is:

*Program I*

Compute $\min_{E} G(E)$

subject to $\sum_{r=1}^{n} E_r u_r(x) \geq \sum_{r=0}^{n} u_r(x) - u_{n+1}(x), \; x \in S$ (6)

and $0 \leq E_r \leq e_r, \; r = 1,2,\ldots,n$

Often the cost-function is difficult to determine in practice. Then an arbitrary choice is made:

$$G(E) = \sum_{r=1}^{n} E_r^2$$

or

$$G(E) = \max_{r} E_r$$

The latter choice is equivalent to minimizing the largest reduction necessary. It seems also conceivable that in some instances G can be of the form

$$G(E) = \sum_{r=1}^{n} g_r(E_r)$$

where $g_r$, $r = 1,2, \ldots, n$ are convex one-dimensional functions. This choice indicates that the control costs increase progressively with the reduction rate.

## VI. COMPUTATIONAL DETERMINATION OF AN OPTIMAL SOLUTION

If G is convex and continuously differentiable and the $u_r$ have continuous partial derivatives of the first order then Program I is an instance of Program D in Gustafson-Kortanek.[10] As shown in Gustafson-Kortanek,[4] the $u_r$ have the regularity properties indicated above when the transfer functions are derived from commonly used diffusion models. The distinctive property of Program I is that Equation 6 defines an infinite set of linear inequalities entailing that conventional convex programming techniques cannot be used directly. A possibility is then to represent S with a finite subset T of receptor points and in Equation 6 replace $x \varepsilon S$ with $x \varepsilon T$. This task shall be called *program I-T*. It can be handled with well-known methods. An example of this is given by Kohn.[11] We point out the following result:

*Theorem 1*

There is a subset $T_0$ of S containing at most n points so that the set of optimal solutions of Program I and Program I-$T_0$ are identical. The proof is given in Gustafson-Kortanek.[12]

The points of $T_0$ may be determined as the solution of a nonlinear system with a finite number of variables and equations. An alternative is the following general algorithm of Remez type.

1. Select a subset $T_\ell$ of S.
2. Solve program I-$T_\ell$. Let $E^\ell$ be an optimal solution.
3. If condition (6) is met, then $T_\ell$ is the subset $T_0$ of Theorem 1 and we have solved program I. Otherwise proceed to 4.
4. Select a finite set $\Delta T_\ell$ of S so that there is at least one vector x* in $\Delta T_\ell$ satisfying

$$\sum_{r=1}^{n} E_r^\ell u_r(x^*) < \sum_{r=0}^{n} u_r(x^*) - u_{n+1}(x^*)$$

5. Put $T_{\ell+1} = T_\ell \cup T_\ell$ and perform step 2 with $T_{\ell+1}$ replacing $T_\ell$.

This algorithm produces a sequence $\{E^\ell\}$ whose accumulation points are optimal solutions of program I. It should be observed that the description is very general and encompasses the case when $T_\ell$ is a sequence of fine and finer grids. However, if the transition from $T_\ell$ to $T_{\ell+1}$ should count as a step in the algorithm, then $T_\ell$ must be a subset of $T_{\ell+1}$ and there must be an x* in $T_{\ell+1}$ such that Equation 6 is violated in x* for the accepted optimal solution of program I-$T_\ell$.

An application of the algorithm is the following. Assume that $u_r$ has local maxima in the finite set $M_r$, r = 1,2, ..., n and that $u_j(x)/u_r(x)$ $j \neq r$, $j \neq 0$ is negligible on $M_r$. Then take $T_\ell = M_1 \cup M_2 \cup \ldots \cup M_n$ and start the algorithm. If the maxima of $u_r$

are well separated from those of $u_j$, $j \neq r$, $j \neq 0$ then $T_\ell$ can be expected to be a good approximately of $T_0$.

## VII. SENSITIVITY ANALYSIS AND INFLUENCE OF ERRORS

In practical applications all input data will be slightly in error. The following should be noted in particular:

1. the simplifying assumptions in the beginning of section 4
2. the source inventory will not be complete and the known sources will not be described exactly
3. the climate is only known approximately; if Gauss or Sutton equations are used the constants in them are generally rough estimates
4. the transfer functions are calculated by means of approximate quadrature
5. the cost functions are generally only estimations.

The errors originating from the above sources should be compared with those arising during the computations. Let E* be a computed solution obtained by means of the algorithm in section 6 and E an optimal solution of program I. Define the following terms:

*nonoptimality* $n(E^*)$

$$n(E^*) = G(E) - G(E^*)$$

*discrepancy* $\delta(E^*,x)$,

$$\delta(E^*,x) = \min\left(0, \sum_{r=1}^{n} E^*_r u_r(x) - \sum_{r=0}^{n} u_r(x) + u_{n+1}(x)\right)$$

$\delta(E^*,x)$ is always nonpositive. Since E* is an optimal solution of program I-T for some subset T of S $G(E^*) \leq G(E)$, *i.e.*, $n(E^*)$ is nonnegative.

It is possible to find a relation between $\delta(E^*,x)$ and $n(E^*)$. Namely put

$$v = \sum_{r=1}^{n} u_r$$

and

$$c = \inf \delta(E^*,x)/v(x), \quad x \in S$$

Next define the vector E' by

$$E'_r = E^*_r - c$$

Then

$$\sum_{r+1}^{n} E'_r u_r(x) \geq \sum_{r=1}^{n} E^*_r u_r(x) - \delta(E^*,x)$$

and E' meets (6). Therefore $G(E') \geq G(E)$ giving

$$n(E^*) \leq G(E') - G(E^*)$$

and this bound is easy to compute. It is assumed that G(E') is defined ($E'_r$ could be outside the internal $[0,e_r]$). Note that E* is a feasible solution of program I if $u_{n+1}(x)$ is replaced with $u_{n+1}(x) + \delta(E^*,x)$. Hence one can compare directly $\delta(E^*,x)$ with $\Delta\kappa(x)$ in Equation 3.

Using the formulas in Gustafson-Kortanek,[4] one can relate $\delta(E^*,x)$ to the coarseness of the grid.

The set $T_0$ of theorem 1 is such that compliance in these points implies compliance in all of S. Hence in practical applications one should only need to monitor the pollution in this set.

There is a set T* that plays the same role with respect to program I-T of this section as $T_0$ with respect to program I. Generally T* is different from $T_0$ and small perturbations in input data may cause large changes in the position of T*. However, what counts is the magnitude of $\delta(E^*,x)$ and this is determined by T.

Even if the standards for the annual arithmetic mean are met, the requirements on low frequency of certain short-term very high concentrations may not be satisfied. Using the findings by Larsen,[2] one can determine if this is the case. Then the standards on the annual mean can be lowered in order to get a lower frequency of very high concentration. Another possibility is to deal with these occasional high concentrations by means of temporary emergency measures when they arise. This requires good weather forecasting. By appropriate changes of the transfer function, our model can be used for short-term studies as well, ensuring that the cost of temporary constraints is as low as possible.

Without principal changes constraints on several pollutants can be treated simultaneously.

## VIII. WORKED EXAMPLE

The following simple example was solved on computer. Consider an idealized example where the source inventory consists of three point-sources $\pi_1, \pi_2, \pi_3$ and the weather is constant. After choice of coordinate system and units the following situation prevails:

1. The wind is blowing parallel to the x-axis and in the positive direction.
2. The coordinates of $\pi_1$, $\pi_2$ and $\pi_3$ are $\pi_1$: (0,1) $\pi_2$: (0,0) $\pi_3$: (2,-1).
3. The corresponding transfer functions $u_1$, $u_2$, $u_3$ are of simplified Sutton types and given by

$$u_1(x,y) = \frac{1}{x} e^{-\frac{1}{x}[1+(y-1)^2]} \qquad x > 0$$

$$u_2(x,y) = \frac{1}{x} e^{-\frac{1}{x}[2+y^2/4]} \qquad x > 0$$

$$u_3(x,y) = \frac{1}{x-2} e^{-\frac{1}{x-2}[1+(y+1)^2]} \qquad x > 2$$

   $u_r(x,y) = 0$ in the area not covered by the definitions above (the pollution does not propagate against the wind).
4. The maximum permitted concentration is 1/2.
5. Let $E_r$ be the reduction factors. Assume the cost-function is

$$2E_1 + 4E_2 + E_3$$

6. The air quality control region is given by $|x| \leq 5$ $|y| \leq 5$.

Hence the optimization problem is:

Minimize $\quad 2E_1 + 4E_2 + E_3$

subject to $\quad (1-E_1)u_1(x,y) + (1-E_2)u_2(x,y) + (1-E_3)u_3(x,y) \leq \frac{1}{2},$

$$|x| \leq 5 \quad |y| \leq 5$$

$$0 \leq E_r \leq 1, \qquad r = 1,2,3$$

The transfer functions are unimodal. Their maximal values are:

| *function* | *maximal value* | *coordinates of maximum point* | |
|---|---|---|---|
| $u_1$ | $e^{-1} = 0.3678$ | 1 | 1 |
| $u_2$ | $\frac{1}{2}e^{-1} = 0.1839$ | 2 | 0 |
| $u_3$ | $e^{-1} = 0.3678$ | 3 | -1 |

Since the transfer functions are unimodal any function of the form

$$(1-E_1)u_1 + (1-E_2)u_2 + (1-E_3)u_3, \qquad 0 \le E_i \le 1$$

must assume its maximum value on the line connecting the two points (1,1) and (3,-1). This line is defined by the equations

$$x = 2+t \qquad y = -t, \qquad -1 \le t \le 1$$

Hence the problem takes the form:

$$\text{Minimize} \quad 2E_1 + 4E_2 + E_3$$

$$\text{subject to} \quad E_1v_1(t) + E_2v_2(t) + E_3v_3(t) \ge -\frac{1}{2} + \sum_{r=1}^{3} v_r(t),$$

$$|t| \le 1 \qquad 0 \le E_r \le 1, \qquad r = 1,2,3$$

where $v_r(t) = u_r(2+t,-t)$. Next introduce a uniform grid $T = (t_1,t_2, \ldots, t_N)$ $t_i = (i-1)\cdot h$, $h = 1/(N-1)$ and approximate the problem with the linear program:

$$\text{Minimize} \quad 2E_1 + 4E_2 + E_3$$

$$\text{subject to} \quad E_1v_1(t_i) + E_2v_2(t_i) + E_3v_3(t_i) \ge -\frac{1}{2} + \sum_{r=1}^{3} v_r(t_i)$$

$$i = 1,2, \ldots, N$$

$$0 \le E_r \le 1, \qquad r = 1,2,3.$$

This is solved for different values of the stepsize h. The following results were obtained:

| $h$ | *Optimal solution vector* | | | *Optimal value* | *Coordinate of critical point* | |
|---|---|---|---|---|---|---|
| | $E_1$ | $E_2$ | $E_3$ | | | |
| 1/10 | 0 | 00 | 0.2583 | 0.2583 | 2.9000 | -0.9000 |
| 1/160 | 0 | 0 | 0.2588 | 0.2588 | 2.8812 | -0.8812 |

Solution of the nonlinear system given in Gustafson-Kortanek[12] did not give any changes in the fourth decimal place of $E_1$, $E_2$, $E_3$. Hence compliance at the point (2.8812, -0.8812) will automatically guarantee compliance throughout the region since there is no other choice of $E_1$, $E_2$, $E_3$ giving a better value for the preference function which meets the condition at this point and violates it at others. It should be noted that the critical point *cannot* be easily found by inspection.

## IX. CONCLUSIONS

In this paper a method was described to find an air pollution control policy which guarantees that the annual arithmetic mean of the concentration of the pollutant in question is not above a given standard in any point of a control area while the control cost is minimized. As a result of the computations certain critical points can be located not numbering more than the control variables such that compliance in these points ensures compliance throughout the region.

In practice one has to compromise due to imperfections in input data which give an irregular contribution to observed concentrations. One can take this into account by entering a stricter standard into the model. It was also indicated how to relate computational errors to uncertainties in input data.

The model can easily be extended to incorporate constraints on several pollutants simultaneously or on a combined pollution index. It may also be used for short-term control after appropriate changes of transfer functions.

ACKNOWLEDGMENTS

The research reported in this paper was financially supported by NSF under grant GK-31833 and the Swedish Institute of Applied Mathematics, Stockholm. Computer programming was done by Miss Katarina Fahlander, Dept. of Optimization and Systems Theory, Royal Institute of Technology, Stockholm.

REFERENCES

1. Wilkinson, J. H. *Rounding Errors in Algebraic Processes.* (London: Her Majesty's Stationary Office, 1963).
2. Larsen, R. I. "A New Mathematical Model of Air Pollutant Concentration Averaging Time and Frequency," *J. Air. Poll. Cont. Assoc. 19:*24 (1969).
3. Babcock, L. R., Jr. "A Combined Pollution Index for Measurement of Total Air Pollution," *J. Air. Poll. Cont. Assoc. 20:*653 (1970).
4. Gustafson, S.-Å. and K. O. Kortanek. "Analytic Properties of Some Multiple Source Urban Diffusion Models," *Environment and Planning 4:*31 (1972).
5. Roberts, J. J. *et al.* "Chicago Air Pollution Systems Analysis Program. A Multiple-source Urban Atmospheric Dispersion Model," Argonne National Laboratory (1970).
6. Fortak, H. G. "Numerical Simulation of the Temporal and Spatial Distributions of Urban Air Pollution Concentration," *Proc. Symp. Multiple-Source Urban Diffusion Models,* A. C. Stern, ed. (Research Triangle Park, N.C.: U.S. Environmental Protection Agency, 1970).
7. Hanna, S. R. "A Simple Method of Calculating Dispersion from Urban Area Sources," *J. Air. Poll. Cont. Assoc. 21:* 774 (1971).
8. Islitzer, N. F. and D. H. Slade. "Diffusion and Transport Experiments," *Meteorology and Atomic Energy* (Washington, D.C.: USAEC, TID-24190, July, 1968).
9. Slade, D. H., ed. *Meteorology and Atomic Energy* (Washington, D.C.: USAEC, TID-24190, July, 1968).
10. Gustafson, S.-Å. and K. O. Kortanek. "Numerical Solution of a Class of Convex Programs," Series in Numerical Optimization and Pollution Abatement, Rep. No. 4, (Pittsburgh, Penn.: Carnegie-Mellon University, School of Urban and Public Affairs, 1972).
11. Kohn, R. E. "Linear Programming Model for Air Pollution Control: A Pilot Study of the St. Louis Airshed," *J. Air. Poll. Cont. Assoc. 20:*78 (1970).
12. Gustafson, S.-Å. and K. O. Kortanek. "Numerical Treatment of a Class of Semi-infinite Programming Problems," *Nav. Res. Log. Quart. 20,* (1973).

# PART IV

# SOLID WASTE DISPOSAL

MODELS FOR ENVIRONMENTAL POLLUTION CONTROL


# CHAPTER 14

# SOLID WASTE: MANAGEMENT AND MODELS

Robert M. Clark*

## I. INTRODUCTION

Solid waste management can be divided into two major areas: (1) collection, including storage, transfer, and transport; and (2) disposal, including any accompanying treatment. Publicly and privately about $4.5 billion per year is spent on solid waste management. Of the $1.7 billion spent publicly for municipal solid waste management, 80 percent is attributed to the collection activity.[1]

The collection operation can be subdivided into two unit operations, collection and haul. The collection operation consists of removing solid waste from the storage point at the place of generation. This operation begins when the collection vehicle leaves the garage and includes all time spent on the route whether or not it is productive. The haul operation starts when the collection vehicle departs for the disposal site from the point where the last container of solid waste is loaded and includes the time spent at the disposal site and the time after leaving the site to return to the first container on the next collection route. The haul unit operation includes, therefore, the total round trip travel time from the collection route to the disposal site.[2]

Three alternatives are normally considered for solid waste disposal: (1) direct shipment from municipalities to a sanitary landfill, (2) direct

*Robert M. Clark is an environmental engineer at the Office of Program Coordination, Environmental Protection Agency, National Environmental Research Center, Cincinnati, Ohio 45268, U.S.A.

shipment from municipalities to a transfer station where solid waste is transferred to larger vehicles and then shipped for ultimate disposal, and (3) direct shipment from municipalities to an incinerator where the solid waste is burned and the residue is shipped for ultimate disposal.

Waste collection and disposal systems are sometimes planned on the premise that they are independent operations when, in fact, the interdependencies that exist are numerous and often very significant.[3] Disposal methods can influence collection methods, and conversely, collection methods can exert strong influence on disposal practices. When these interdependencies are ignored, the cost of solid waste collection and disposal will probably be much too high.

Solid waste management planning requires an assessment of many complex interactions among transportation systems, land use patterns, urban growth and development, and public health considerations. Because of these interactions and interdependencies, much attention has been given to systems analysis and mathematical modeling techniques for sorting out various available alternatives.

Perhaps the philosophy which best characterizes the "systems" approach to solid waste management has been expressed by W. R. Lynn when he wrote:[4]

> It is obvious that satisfactory solutions to this problem (disposition of solid wastes) cannot be obtained by viewing each operation within the processes of refuse collection as an independent unrelated function. Rather, in order to achieve efficient solutions to these problems, effort must be directed toward viewing the problem in its entirety as an interconnected system of component operations and functions.
>
> Unfortunately, until comparatively recently, solutions to this class problem have been unavailable because of the difficulty in making a systematic analysis of large complex problems. However, the advent of new mathematical techniques in systems analysis and operations research coupled with the availability and facility of high speed computing devices will now permit investigation of this type.

This statement made by Lynn in 1962 is borne out by a search of the pertinent literature which reveals that a growing number of investigators have concerned themselves with the application of operations research to the study of solid waste collection systems. This work falls into three categories: broad scale attempts at outlining the nature and interactions of the entire waste management system; narrower, more specific investigations of the collection and disposal operations suited to the detailed study of specific operating systems; and, studies intended to provide an over-all economic characterization of solid waste management.

The most complete of the broad-scale studies, conducted at the University of California at Berkeley, is entitled "Comprehensive Studies of Solid Waste Management."[5] One of the study's objectives was "to explore the potential of operations research to help in the definition and solution of the solid waste disposal problem." Investigations were conducted into the development and evaluation of solutions to the total refuse disposal problem that would be applicable to a wide spectrum of alternatives and sensitive to environmental and governmental constraints as well. This involved developing conceptual models to give insight into the operation of the system and then developing mathematical models to support the conceptual models. These mathematical models included a waste generation model, a waste collection, treatment and disposal model, a regional economic model, and models for population, public health aspects, land use, and process technology.

One task in the investigation was to look at the means of determining transport of solid waste from sources through treatment and processing to disposal sinks. Anderson,[6] one of the study group, reported on the development of specific techniques for solving this problem. He was interested in modeling the optimization of flow through a given system containing existing facilities and potential new facilities, and he identified the problem as a study of a transshipment network with some peculiarities because of the nature of the processes involved. In considering a total waste management model including intermediate treatment facilities, the changes in volumes and in the state of the wastes must be considered; this is not possible using ordinary network algorithms. He suggested two solution techniques: (1) if the number of disposal sinks is small, a branch and bound model is used to generate optimal flow through the system, and (2) an out-of-kilter algorithm that includes the feature of allowing flow through a node to be proportioned in a fixed ratio. In both models, all costs are linear, and the system under investigation considers the gross shipment between points and not the routing of individual vehicles among small collection tasks. Anderson also dealt with the problem of finding flows through a known or given configuration.

To consider various combinations of new facilities each scheme must be evaluated separately. No methodology is presented for a systematic search over facility location possibilities. The assumption of linear costs omits some of the important aspects of the problem. Although nonlinear costs may be given linear approximations, this is valid only if the nonlinear costs are convex. Facilities costs are commonly concave or quasi-concave (fixed charge). Thus, linear approximation either will be unsuccessful or must be done in such a manner that one loses confidence in the results.

Other studies defining the solid waste system have been carried out by consulting organizations for governmental units but have been limited to characterization of the system. Such studies include reports by Management Technology, Inc.[7] for the State of Maryland and by Aerojet General[8] for the State of California.

The first attempt at understanding the collection process at the individual block and vehicle level was to use simulation models. Quon, Charnes, and Wersan[9] presented a model of a collection system which paid particular attention to queuing problems at the disposal point, using data from Winnetka, Illinois. Truitt, Liebman, and Kruse[10,11] constructed a simulation model using data from Baltimore, Maryland, in which the effect of a proposed transfer station could be investigated. Quon, Tanaka, and Wersan[12] continued their earlier simulation work with a model for studying changes in work rules and collection policy.

Other investigators have dealt with different aspects of the collection problem that are amenable to analytical attack. Marks and Liebman[13] have developed a number of solid waste collection models. The first model involves the location of intermediate stations where solid wastes can be transferred from collection vehicles to large trucks or trains which are more suited to long-haul transportation. The second model analyzes the best way of routing the wastes to the transfer facility, under the assumptions that all wastes are collected together and that a number of different kinds are collected separately. The third model concerns scheduling routes for individual trucks. Coyle and Martin[14] presented a heuristic method based on dynamic programming for aggregating small collection areas into work schedule assignments for crews and vehicles. Skelly[3] presented a fixed charge model for looking at the large scale problem of transportation of wastes in which the variables included the alternative location of transfer stations. He used a heuristic fixed charge algorithm developed by Walker as a solution technique.[15] Skelly's work is discussed in greater detail later in the text.

This listing of work is only a sample of the efforts that have been and are being made in an attempt to apply the "systems" approach to solid waste management. A number of specific techniques have been used to analyze various aspects of solid waste management. Locational models have been applied to the problem of selecting solid waste facilities, and both deterministic and simulation models have been applied to the collection operation.[16] Some of these techniques will be discussed now.

## II. LOCATING SOLID WASTE FACILITIES

The type of problem which one might face in locating solid waste facilities is illustrated in Figure 14.1. The problem is to select from among

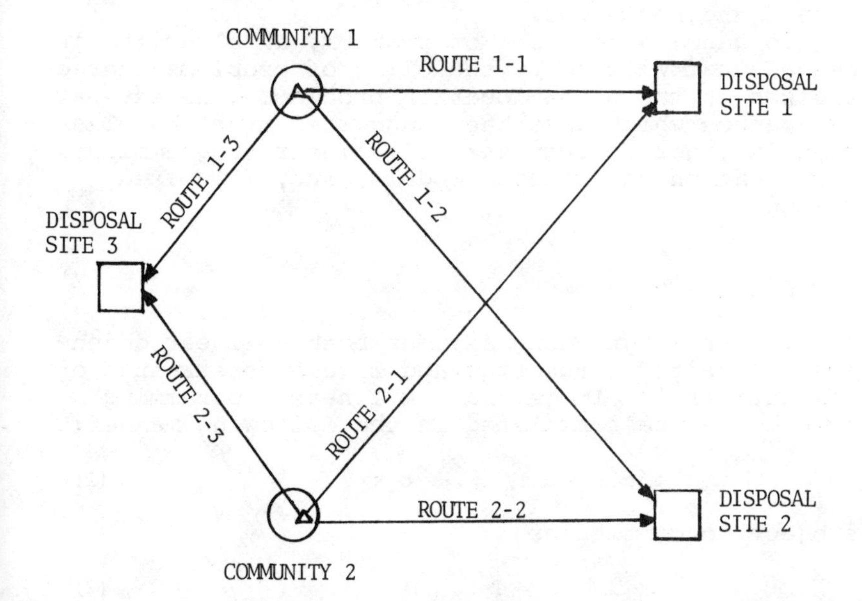


*Figure 14.1. A schematic diagram of the facility selection problem. Δ = Geographic or mass density of refuse generation area.*

three alternatives, a site to dispose of solid waste from two communities or from two generation areas in one community. To make the problem computationally feasible, the haul operation will be assumed to consist of the total round-trip time from the geographic or mass (solid waste) density center of a community to any given disposal site.[3] In effect, only one route (haul) distance from any given community or solid waste generation area to any given disposal site need be considered. Associated with each disposal site is a fixed (or acquisition) cost and an operating cost and associated with each route is a haul cost. The haul cost in this case could be based on: (1) the average round-trip time between the disposal site and the solid waste generation area; (2) the distance between generation area and site; and, (3) the average tons of solid waste shipped to disposal sites. Because each disposal site has a variable operating cost that can be lumped with the haul cost for a specific route, each disposal site plus a specific route is an alternative to be considered. There is, then, a fixed or acquisition cost and a variable operating cost for the following alternatives: Disposal Site 1 and Route 1-1, Disposal Site 1 and Route 2-1; Disposal Site 2 and Route 1-2, Disposal Site 2 and Route 2-2; and Disposal Site 3 and Route 1-3, and Disposal Site 3 and Route 2-3.

To handle this kind of problem, an algorithm is needed that will solve the class of problems characterized as the plant location problem. The several approaches which have been suggested might be classified in general terms as: (1) linear programming; (2) location equilibrium models; and, (3) fixed charge.

### 1. *Linear Programming*

Linear programming assumes both a linear objective or criteria function and linear constraints or limitations.[17] In general, a linear programming problem can be formulated in the following manner:

$$\min z = c_1 x_1 + \dots c_n x_n \tag{1}$$

subject to conditions

$$x_j \geq 0 \quad (j = 1 \dots n) \tag{2}$$

and

$$\sum_{j=1}^{n} a_{ij}x_j \leq b_i \quad (i = 1 \ldots m) \qquad (3)$$

in which $a_{ij}$, $b_i$, $c_j$ are constants. If, in fact, a linear programming formulation describes the problem adequately, then the use of this approach will provide useful answers. If, however, $c_j$ is not constant at every level of $x_j$, the linear programming approach is not valid. Unfortunately, this is the case in most realistic locational problems. Some modifications may be made to the problem by assuming specific utilization levels for the various alternatives to be considered, which fixes a value for $c_j$. After finding an optimal solution with the assumed $c_j$, a new $c_j$ can be assumed based on the previous optimal answer. By repeating this procedure until a stable solution is reached, an answer may be found but it may not be the best possible solution to the problem.

### 2. *Locational Equilibrium Models*

The locational equilibrium model consists of determining the number and location of sources or origins that will most economically supply a given set of destinations with some commodity or service.[18] A specific problem we might consider is the location of a number of landfills that are to receive solid waste from a series of waste generation sources.

The problem may be stated as follows:

Given: (1) the location of source, (2) the waste generation from each source, (3) possible landfill capacity limitations, and (4) a set of shipping costs (or a statement of the relationship between distance and cost).

Determine: (1) the number of landfills, (2) the location of each landfill, (3) the allocation of sources to landfills, and (4) amounts to be transferred or shipped.

The problem could be structured in the form of a generalized Weber problem, *i.e.*, finding a single point (origin) in two-dimensional Euclidean space, that is, the minimum transport or the cost point for any number of sources, or both. Mathematically, this problem may be formulated as:

$$\min z = \sum_{j=1}^{n} \beta_j \, [(x_{sj} - x)^2 + (y_{sj} - y)^2]^{1/2} \qquad (4)$$

in which $\beta_j$ = a weighting factor, $x_{sj}$ and $y_{sj}$ represent the coordinates of a set of n known sources, and x, y = the unknown landfill site. An iterative procedure might be used to solve this problem.

### 3. *Fixed Charge Problem*

When undertaking an activity, a fixed charge or set-up cost is commonly incurred. In such cases, the total cost of the activity is the sum of a variable cost related to the level of the activity and the set-up cost required to initiate the activity.[19]

Frequently, the variable cost will be at least approximately proportional to the level of activity j. Problems of this type take the form shown in Figure 14.2, in which $k_j$ represents the fixed cost,

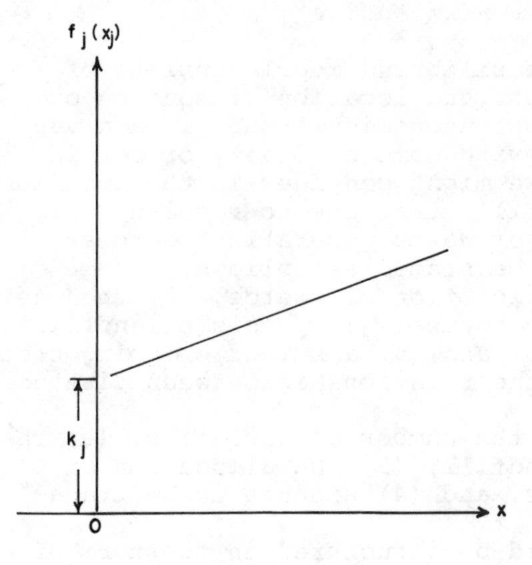


*Figure 14.2. The fixed charge problem.*

and

$$f_j\,(x_j) = k_j\delta_j + c_jx_j \qquad (5)$$

in which $c_jx_j$ = the variable cost component of the total cost $f_j(x_j)$.

If $x_j > 0$ and $\delta_j = 1$, then $f_j(x_j) > 0$, and if $x_j = 0$ and $\delta_j = 0$, then $f_j(x_j) = 0$. For example, if $c_j = 2$ and $k_j = 5$ and the function $(k_j\delta_j + c_jx_j)$ is evaluated at $x_j = 3$, the total cost is 11. If, however, $x_j = 0$, the total cost = 0.

In practice, such costs are encountered in transportation where a fixed charge is incurred regardless of the quantity shipped (provided that a positive quantity is shipped), or in the building of production facilities where a plant under construction must have a certain minimum size. The plant location problem (or rather the problem of selecting a subset of a specified set of plant locations that will minimize the cost of production and shipment of quantities to a set of destinations) is also an example of the fixed charge problem. This type of problem is closely related to the location of solid waste disposal and processing facilities.

The function $f_j(x_j)$ is concave for $x_j \geq 0$ (Figure 14.2). If the objective is to minimize a cost function that is concave, an optimal solution occurs at an extreme point of the convex set of feasible solutions. There can, however, be local optima different from the global optimum; if there is a fixed charge associated with each variable, then every extreme point of the convex set of feasible solutions yields a local optimum. This occurs because a move away from any extreme point while remaining in the convex set of feasible solutions means that more of the variables will have to be positive and more of the fixed charges will be involved. By remaining close enough to the extreme point, however, any reductions in variable costs cannot outweigh the increase in fixed costs, and thus the extreme point yields a local optimum. The existence of these local optima makes the task of solving fixed charge programming problems extremely difficult. The fixed charge problem can be formulated in the following manner:[19]

$$\min z = \sum_{j=1}^{n} f_j(x_j) \tag{6}$$

$$Ax = b \tag{7}$$

$$x \geq 0$$

in which

$$f_j(x_j) = k_j\delta_j + C_jx_j \tag{9}$$

and

$$f_j(x_j) = 0; \quad \delta_j = 0, \text{ if } x_j = 0 \tag{10}$$

$$f_j(x_j) > 0; \quad \delta_j = 1, \text{ if } x_j > 0 \tag{11}$$

Among the several available techniques that can be used to solve the fixed charge problem are Efroymson and Ray's branch and bound algorithm, Spielberg's algorithm, and algorithms developed by Gray, Kuehn and Hamburger, Kuhn and Baumol, Manne, Balinski, and Walker.

A branch-bound technique developed by Efroymson and Ray involves solving a series of linear programming problems that give progressively lower bounds (in a minimization problem) on the value of the solution to the fixed charge problem.[20] When this technique is employed, a large number of linear programs must be solved. If the computing time for each linear program is great, the method can prove prohibitively expensive. This algorithm can only handle the simple plant location problem in which the plant has no limits on its capacity.

Spielburg developed an algorithm similar to Efroymson and Ray's that can handle some side constraints but cannot handle plant location problems with capacity constraints.[21]

Gray's algorithm for the solution of the fixed charge problem incorporates a decomposition scheme.[22] Because this algorithm requires on the average more than ten minutes on a Burroughs B-5500 computer system to solve a 5 x 7 fixed charge transportation problem, this approach is expensive for larger problems.

Kuehn and Hamburger,[23] Kuhn and Baumol,[24] Manne,[25] and Balinski[26] have successfully developed heuristic or approximate methods for producing good solutions to the fixed charge transportation problem. Many of these algorithms have a high degree of computer efficiency.

Walker developed a heuristic algorithm for solving the fixed charge problem. This is an adjacent extreme point algorithm which is computationally efficient in yielding optimal solutions.[15] Walker's algorithm was tested against problems that Steinberg[27] and Gray[22] used in testing their algorithms. In most all cases, the optimal solution was obtained.

Walker reported significantly faster computer times than that of the other algorithms. This greater computational efficiency combined with its high percentage of obtaining the optimal solution makes Walker's algorithm economical and practical to use for the solution of fixed charge problems.

The main limitation of the Walker technique is it only ensures achieving a local optima, which is hopefully the global optimum or at least a good solution. From the results of extensive testing, the Walker algorithm apparently provides a good heuristic technique for solving the fixed charge problem.

## 4. *Models for Locating Solid Waste Facilities*

Several investigators have attempted to apply deterministic methods to the problem of locating solid waste disposal facilities. Wersan *et al.*[38] have used a combination locational equilibrium linear programming model to find a solution to the locational model, Schultz[29] has used a locational equilibrium model, and Baker,[30] a trial and error approach. The University of Louisville study team[31] adopted a linear programming model, and Skelly[3] used a fixed charge algorithm.

### Wersan's Algorithm

Wersan's approach to the location of solid waste disposal sites is based on minimizing travel time between solid waste generation areas and disposal sites.[28] He assumes that direct travel is not possible between arbitrary points but that travel is along streets and avenues in an L-shaped path. Travel time from $(x',y')$ to $(x,y)$ is given by $|x' - x| + |y' - y|$ where the $|\ |$ notation indicates absolute value. If the origin of the system is moved by an amount $x_o$ in the horizontal and $y_o$ in the vertical direction, the travel time becomes $|x' - x_o - (x - x_o)| + |y' - y_o - (y - y_o)|$ or $|x' - x| = |y' - y|$ because the travel time stays the same and the location of the origin is arbitrary.

The problem is to find a location for the disposal site that will minimize the total truck haul time. Finding x and y so that $\sum_{i=1}^{n} |x - x_i| + |y - y_i|$ is minimal (in which n = the number of time segments over which the truck must travel) becomes a linear programming problem. To see this, the following definitions are made:

Let

$$u_i = \begin{pmatrix} x - x_i \\ 0 \end{pmatrix} \text{ if } x - x_i \begin{pmatrix} \geq \\ < \end{pmatrix} 0 \tag{12}$$

$$- u_i = \begin{pmatrix} 0 \\ x - x_i \end{pmatrix} \text{ if } x - x_i \begin{pmatrix} \geq \\ < \end{pmatrix} 0 \tag{13}$$

A similar definition for $v_i$, $v_i'$ depending on $y - y_i$ is

$$u_i - u_i' = x - x_i \tag{14}$$

$$v_i - v_i' = y - y_i \tag{15}$$

However, if

$$x - x_i \geq 0, \; |x - x_i| = (x - x_i) = u_i \tag{16}$$

$$x - x_i < 0, \; |x - x_i| = - (x - x_i) = u_i' \tag{17}$$

and as $u_i$ and $u_i'$ are not simultaneously nonzero, it follows that $|x - x_i| = (u_i + u_i')$ and, similarly, $|y - y_i| = (v_i + v_i')$. The problem becomes:

minimize

$$z = \sum_{i=1}^{n} (u_i + u_i' + v_i + v_i') \tag{18}$$

subject to

$$x - x_i' = u_i - u_i' \tag{19}$$

$$y - y_i' = v_i - v_i' \tag{20}$$

Generally, this formulation leads to the solution $x = \text{Median}\{x_i\}$ and $y = \text{Median}\{y_i\}$.

The disadvantage of this approach is that fixed costs are not considered when various disposal alternatives are examined. For an operating model, Wersan's algorithm might prove to be very effective, but as a means of selecting disposal alternatives that require any capital investment it has many limitations.

## Schultz' Algorithm

Schultz' algorithm is based on finding a location pattern for sources that will minimize the total weighted distance from sources to sinks.[29] In this case, the sources would be solid waste generation areas and sinks would be solid waste volume reduction or disposal facilities or both. Straight line distances are measured from the center of gravity of the service area to the sink. Schultz' algorithm operates in the following manner:

1. A random pattern of initial facility location is selected; these are denoted by the coordinates $(u_r, v_r)$ in which $r = 1, 2, \ldots, s$.
2. The total area to be serviced is subdivided into compact service areas, $p_r$, each of which is associated with a facility $r = 1, 2, \ldots, s$. A solid waste generation area, i, is assigned to a service area, $p_r$, on the basis that the distance from its center of gravity to the facility, r, $d_{ir}$, is minimized over all r.
3. A new central facility location $(u'_r, v'_r)$ is located at the center of gravity of service area, $p_r$. This is determined by:

$$u'_r = \frac{\sum_{i \varepsilon p_r} w_i x_i}{\sum_{i \varepsilon p_r} w_i} \qquad (21)$$

$$v'_r = \frac{\sum_{i \varepsilon p_r} w_i y_i}{\sum_{i \varepsilon p_r} w_i} \qquad (22)$$

in which

$$d'_{ir} = [(x_i - u'_r)^2 + (y_i - v'_r)^2]^{1/2}$$

$$i = 1, 2, \ldots, n \tag{23}$$

4. If:

$$\sum_{r=1}^{s} \sum_{i \varepsilon p_r} w_i d'_{ir} \geq \sum_{r=1}^{s} \sum_{i \varepsilon p_r} w_i d_{ir} \tag{24}$$

that is, if no improvement can be made over the previous pattern, it is the optimal location pattern. Otherwise, let all $u_r = u'_r$, $v_r = v'_r$ and return to step 2.

Subareas containing households are sources of solid waste; transfer stations are considered sinks; $w_i$ = the number of households in subarea i; $x_i$, $y_i$ = coordinates of subarea i with respect to reference axes X, Y; $p_r$ = the set of subareas in the service area of facility r; $u_r$, $v_r$ = the coordinates of facility r; and $d_{ir}$ = the service distance from facility r to the central point of subarea i.

Required data input for Schultz' algorithm includes number of facilities to be located, household distributions, and number of initial patterns to be used.

The problem of finding an actual site for a transfer station near an optimally selected theoretical site would, in many cases, make the results of the algorithm impractical. Given any urban location, the downtown area would probably have to be ruled out because of zoning, capital cost, the resistance of the downtown businessmen, or traffic problems. Because the building of a transfer station near a business or housing subdivision outside the downtown area would be met with a great deal of resistance, the only available area then is an industrial district. In fact (as Schultz points out), many cities have a zoning ordinance against locating transfer stations in any area other than an industrial district. If every area other than the industrial districts is excluded, the problem can still be solved by Schultz' algorithm, but it can be solved more efficiently by other techniques described herein.

Schultz' algorithm does not consider any cost associated with the collection and disposal system. The algorithm minimizes the hauling distance between sources and transfer stations, which in many cases minimizes the haul cost. However, haul cost is more closely related to haul time than haul distance because of labor costs, which are not necessarily proportional to the distance but are directly proportional to the time. To a lesser extent, the same is true for the actual operating costs for trucks because items representing the largest percentage of total operating cost (fuel and maintenance) are most closely related to length of operations.

The different capital or operating costs associated with different facility locations cannot be considered. In many cases, operating costs for a transfer station would be similar, but differences in location costs might cause the solution to be far from the minimum. Schultz' technique could apply to incineration as well as sanitary landfill, but the solutions would probably be unrealistic as no capacity constraints can be placed on facilities.

Theoretically, landfills could be located at a selected facility site, but realistically a landfill site would very rarely be available near a theoretical site.

Schultz' algorithm would be useful only when transfer stations are being considered as an intermediate step in the disposal process. Even then, one of the more direct approaches would give more reliable results.

## Baker's Algorithm

Baker employs a trial-and-error approach to locating and assigning solid waste generation areas to disposal facilities.[30] Economies of scale are assumed for the various facilities that are considered as feasible alternatives. With this type of cost function, the unit cost varies with the level of facility utilization. Baker breaks utilization into four levels: 80-100%; 60-79%; 40-59%; and 0-39%. A unit cost is assigned for utilizing the facility at each level. Because of the assumed economies of scale, the unit cost is higher at lower utilization levels. Although Baker does not explicitly consider the fixed charge problem, he recognizes its existence by assigning a lower unit cost at high utilization levels. For example, he might assume a unit cost of $4 per ton at 0-39% utilization level and a unit cost of $2 per ton at 80-100% utilization level.

Baker's algorithm requires the following input data:

1. cost, in dollars per ton, of hauling solid waste in a collection vehicle from a source area to any disposal facility and in a transfer vehicle from a transfer station to each disposal facility
2. round-trip travel time from each source area to each disposal facility and from each transfer station to each disposal facility
3. cost of operating each facility for various ranges of facility utilization
4. population of each collection zone and
5. capacity of each disposal site in terms of population served for a specific period of time.

The Baker algorithm examines each transfer station with all possible combinations of landfills, compost plants, and incinerators. The technique assumes that incinerators and compost plants are final disposal facilities. The lowest cost per ton for each transfer station and final disposal combination is determined to permit comparison among all alternatives.

The total cost for each solid waste generation area is calculated for every possible solid waste disposal alternative on the assumption that each facility is being operated at maximum capacity (80-100%). A disposal facility representing the lowest cost for each area is then chosen.

For each disposal alternative, the total solid waste generated by the population it serves is balanced against the assumed utilization level. If the balance indicates that the total solid waste generated differs from the assumed utilization level, the utilization is changed to the actual level and the least-cost assessment begins again. When the solid waste generated is greater than the capacity of the facility, solid waste source areas are removed one at a time until generation is equal to or less than capacity. The source areas that have been deleted from the least-cost alternatives are assigned to the second-lowest cost alternatives.

Calculations are repeated until each source area employs the most economic alternative available to it. At the final solution, the assumed unit cost for the facility will match the actual utilization level and each disposal facility will be used at a level less than or equal to capacity.

Because Baker's algorithm does not explicitly consider the fixed charge problem, it has only a limited chance for reaching the true minimum cost solution. It does, however, have the basic elements of a rational approach to solve location problems for solid waste facilities.

## University of Louisville Approach

Investigators at the University of Louisville used a linear programming model to locate landfills, incinerators, and transfer stations.[31] Factors considered include the cost of landfill operation, operating costs for a transfer station, haul costs from solid waste generation sources to disposal sites, and haul costs from transfer stations or incinerators to landfill sites. Disposal is considered to be any series of events after pickup that result in the final placing of solid waste. Alternative system configurations include each source with an option of utilizing various combinations of m landfills, n transfer stations, and p incinerators, with residue from the p incinerators and the solid waste from the n transfer stations being sent to any of the m landfills for final disposal. With the use of standard linear programming procedures, the system is optimized and the minimum total cost for the entire system, based on the assumed cost for each facility, is given. The investigators, after examining the optimization results, drop any facility that is not utilized to an extent that would warrant its development. The program is then rerun until a satisfactory solution is reached. No minimum constraints are used to limit the amount of solid waste going into an incinerator or landfill.

This approach was used to verify an intuitive solution suggested by county officials for locating solid waste facilities in the Louisville area. However, the basic limitations of the linear programming approach (*i.e.*, the cost function must be assumed to be linear) make the general use of this method questionable.

## Skelly's Model

Skelly has developed a model based on Walker's solution for the fixed charge problem.[3] Because the model considers the acquisition or initial purchase cost of facilities as well as variable operating costs, flexibility in considering possible alternatives and evaluation of realistic problems are possible.

General assumptions and conditions of the model are:

1. The following parameters are constant in any time period but variable among time periods: cost per ton per hour of transporting solid waste, site operating costs, per capita refuse production for each community, population for each community, and time of travel from a community to a disposal site.
2. A landfill site is purchased in its entirety, even though all the available land at the site may not be required for disposal purposes.
3. Any site used is purchased or constructed in the first time period; *i.e.*, stage development of incinerators or transfer stations is not considered.
4. A regional authority exists with power to distribute men and machinery among the various landfill sites to ensure their efficient operation.
5. All vehicles from a solid waste generation area travel the same distance to a given disposal area.
6. Ultimate volumes or capacities of landfill sites are known.

A major shortcoming of the model is that it does not consider time variations in the cost of land or in the capital cost of facilities, the premise of this model being to design, in the present, for the future capacity of a site. Purchase of land in future time periods and stage development of incinerators and transfer stations cannot be considered with this approach.

In Skelly's model, solid waste production is considered constant during a time period. In actuality, there are seasonal variations and variations among the years of a particular period. To consider these variations, one can divide the planning period into smaller time periods. The time periods in this study are five years.

Basic data required for input into the model include: the maximum capacity of the landfill, total refuse production of each community for each time period, operating and fixed costs for each site, and haul costs for each route.

### 5. *Data Requirements*

Regardless of the method used to solve the optimal location of disposal facilities, several general types of data are required.[32] These data, in various combinations, provide input for the mathematical method to be used and include:

1. locations of landfills, incinerator, and transfer station sites
2. maximum capacities of the facilities to be considered
3. proposed haul routes from each community to each disposal site and from each incinerator and transfer station to each landfill site
4. average round-trip travel time for each route computed from the average vehicle speed from each solid waste generation source and the proposed route distances
5. cost per ton per hour for hauling solid waste in collection vehicles for current and future time periods
6. population and per capita solid waste production for each period
7. operating costs for the various disposal sites for each period and
8. purchase price of the landfill sites and capital costs for incinerators and transfer stations.

Facility location is only part of the problem of analyzing solid waste management systems. Techniques for modeling the collection operation also need to be developed. In the following discussion, some of the attempts which have been made to characterize solid waste collection will be outlined.

## III. MODELS OF SOLID WASTE COLLECTION

Both stochastic and deterministic models have been developed to describe solid waste collection policies. The use of mathematical simulation for analysis of solid waste collection systems has been reported by Quon, Tanaka, and Wersan,[12] Quon, Charnes, and Wersan,[9] and Kruse, Liebman, and Truitt.[33] A discussion of the Quon, Tanaka, and Wersan approach as a typical simulation modeling approach follows.

1. *Simulation Models*

Quon *et al.*[12] developed a simulation model to evaluate the operational characteristics of refuse collection crews employed on a constant length workday basis, usually an eight-hour workday. The solid waste collection system of Chicago, Illinois, provided the physical model on which the constant length workday model is based. Invariant parameters of the system simulated are: (1) a constant length of workday and (2) crew size consisting of three loaders and a driver. The frequency of service and the days of the week on which each collection unit is to be served may be specified. A collection unit is defined as a grouping of contiguous sources of refuse that are to be serviced. For each collection unit, the input parameters are: (1) average refuse production, in pounds per service per day, (2) standard deviation of the refuse production, (3) number of services, (4) days of the week that the collection unit is to be serviced, (5) internal distance in miles, (7) distance of the collection unit to the garage in miles, (8) truck capacity in tons, (9) number of collection units assigned per truck, and (10) constants utilized in the computation of pickup time. Frequency of service and number of working days per week are fixed by item 4.

The quantity of refuse produced per day is treated as a stochastic variable. Average pickup time per trip or truckload was taken to be a linear function of the number of services and the internal mileage traversed. Figure 14.3 shows the basic logic used for a given truck load. In Figure 14.3 CLT1, CLT2, and CLT3 are constants which can be set to control various phases of the collection operation. HIRI indicates the maximum capacity of the collection vehicle.

Most attempts at modeling solid waste collection have utilized simulation as a basic tool. There have been, however, a few investigators who used deterministic models to describe the collection operation. Two studies will be discussed: one using fixed charge programming to locate collection facilities and the other using linear programming to analyze proper fleet selection for solid waste management.

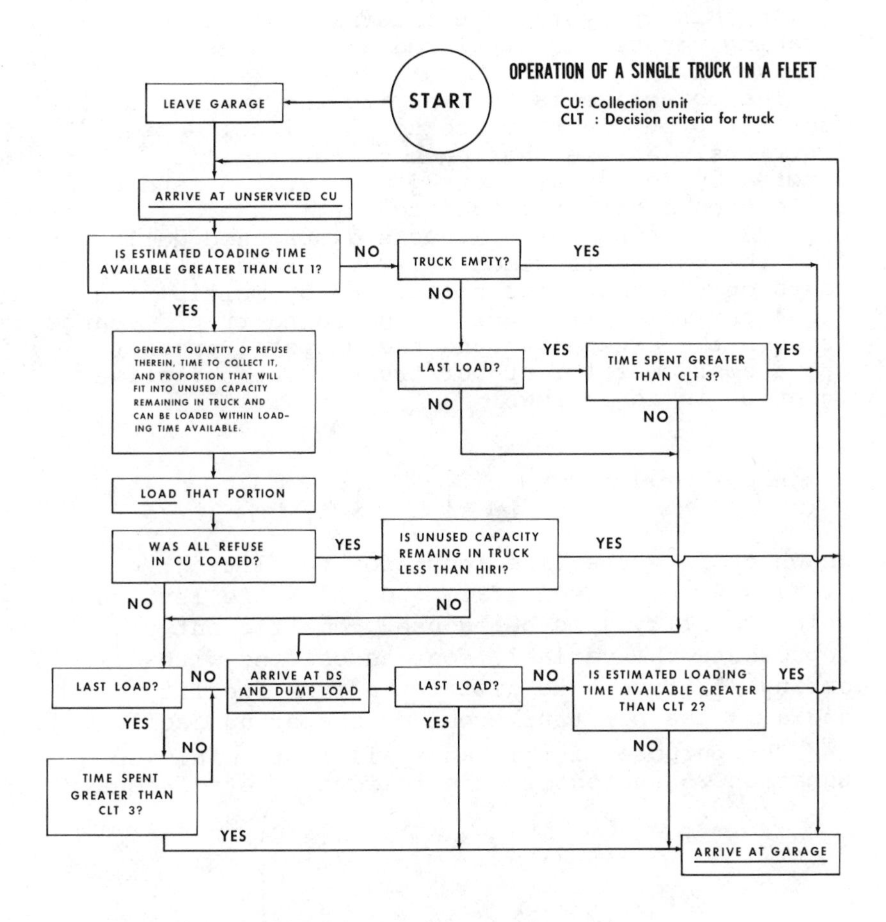


*Figure 14.3. Operation of a single truck in a fleet.*

## 2. *Deterministic Models*

### Decentralized Garaging

In this analysis, the alternatives of central garaging versus decentralized garaging were considered.[34]

The objective is to minimize the average daily cost for providing service to the 27 collection districts. Assume that each garage facility is denoted by letter j, where j = 1, ..., 4, and each collection district is denoted by i, where i = 1, ..., 27. The number of trucks dispatched daily from the garage is given by $x_{ij}$. Transportation costs on the route are calculated by multiplying $1.04 per mile per truck by the round-trip distance between the generation and the disposal areas to get a cost coefficient per truck. The objective function is as follows:

$$\text{minimize total cost} = \sum_{j=1}^{4} (f_j \delta_j + a_j z_j) + \sum_i \sum_j c_{ij} x_{ij} \qquad (25)$$

in which $f_j$ = the fixed cost for facility j, $\delta_j = 0$ if $z_j = 0$ and $\delta_j = 1$ if $z_j > 0$, $z_j$ = the level at which facility j is being used, coefficient $c_{ij}$ represents the variable cost associated with allocating $x_{ij}$ trucks to district i from facility j, and $a_j$ = the per truck cost of operating facility j.

The purpose of this analysis is to minimize Equation 25 subject to the following set of constraints

$$\left.\begin{array}{l} a_{1,1}x_{1,1} + \ldots + a_{1,4}x_{1,4} \geq T_1 \\ a_{2,1}x_{2,1} + \ldots + a_{2,4}x_{2,4} \geq T_2 \\ \ldots\ldots\ldots\ldots\ldots \\ a_{27,1}x_{27,1} + \ldots + a_{27,4}x_{27,4} \geq T_{27} \end{array}\right\} \qquad (26)$$

that represent the requirement that sufficient trucks be dispatched from the various facilities to pick up the refuse generated daily. The collection coefficients $a_{ij}$ are the number of loads which can be collected in each district and transported to a given facility, and $T_1, \ldots, T_{27}$ are the average number of truck loads generated daily in each district, as shown in Equation 26. The constraint set:

$$\left.\begin{aligned} &z_1 + z_2 + z_3 + z_4 = 100 \\ &z_1 \leq 100 \\ &z_2 \leq 50 \\ &z_3 \leq 50 \\ &z_4 \leq 50 \end{aligned}\right\} \quad (27)$$

represents the total fleet size and the capacity limitations on each facility.

To solve this problem, the Walker algorithm was used,[15] resulting in the collection districts being assigned to two garage facilities and the old garage facility abandoned (Figure 14.4). With this solution, the total cost is reduced nearly 19%: $816.34 per day as opposed to the present cost of approximately $1,012.

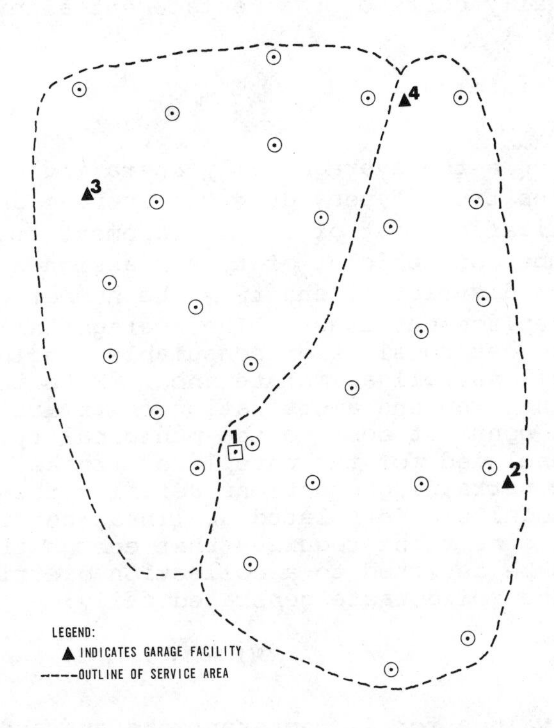


*Figure 14.4. Selection of decentralized garages.*

### 3. *Fleet Selection*

The problem to be considered is that of minimizing the average daily cost associated with the 16 cubic-yard, 20 cubic-yard, and 25 cubic-yard replacement vehicles, and the average daily operating cost of the remaining fleet.[35] A mathematical model describing the fleet selection problem considered in this study is developed. The model is structured as a linear program and is used to select the best fleet configuration from among the available alternatives and to determine the assignment of each truck to a collection district. The model is developed in general terms, and trucks of different capacity are referred to as different types: trucks of type 1 are those packer vehicles of a given capacity that currently make up the municipal fleet (16 yards). A certain percentage of these vehicles will be eligible for replacement.

The objective or criterion function to be minimized is concerned with the average daily operating cost of the existing fleet as well as the average daily costs of the replacement alternatives:

$$\text{minimize} \quad \sum_k d_k t_k + \sum_i \sum_k c_k x_{ik} \tag{28}$$

in which $c_k$ = the average daily operating cost of the various truck types, $d_k$ = the average daily crew and amortization cost of the replacement trucks, $x_{ik}$ = number of vehicles of type k assigned to collection district i, and $t_k$ = the number of each type of replacement truck. The average daily operating cost consists of consumables including gas and oil as well as maintenance. Note that because the crew and amortization costs are considered a constant cost to the municipality, they are not included for the unreplaced trucks.

The constraining equations defining this mathematical model are formulated as linear constraints. The first constraint requires that enough trucks and crews be assigned to a collection district to pick up the solid waste generated daily:

$$\sum_k a_{ik} x_{ik} \geq T_i \quad (i = 1, \ldots, I) \tag{29}$$

in which values for $T_i$ represent the average daily number of residences that must be served in each collection district i, $a_{ik}$ = the collection coefficient which represents the average number of residences that can be serviced per day in each collection district i by truck type k, and I represents the total number of districts to be served.

The second constraint is:

$$\sum_i x_{i1} = t_1 + w_1 \quad (30)$$

In Equation 30, $x_{i1}$ represents the number of 16-yard trucks assigned to district i, $w_1$ = the number of 16-yard trucks that will not be replaced, and $t_1$ = the number of new 16-yard trucks that will be added to the fleet.

The third constraint is:

$$\sum_i x_{ik} = t_1 \quad (k = 2, 3) \quad (31)$$

in which $\sum_i x_{ik}$ represents all of the 20-yard or 25-yard replacement vehicles, or both, assigned to the collection districts and $t_k$ = the number of replacement vehicles of types 2 and 3 that will be purchased.

The fourth constraint is:

$$w_1 = c_1 \quad (32)$$

giving the number of trucks in the existing fleet that will not be replaced. Note that the value of $c_1$ is given by the solid waste manager and is most often based on experience and available funds.

An example problem was used to illustrate the model's usefor selecting among the alternatives of 16-, 20-, or 25-yard packer trucks. The model suggested purchasing the largest vehicles available. Results of the model are verified based on actual municipal experience.

All of these attempts to analyze the collection systems depend on the ability of the users of the model to input correct data. Quon, Martens, and Tanaka have conducted a study to obtain adequate and reliable data on the collection operation.[36]

*4. Data Requirements*

This study[36] was conducted using City of Chicago data to examine the relationship between the efficiency of the collection and haul operations and (1) the physical and population characteristics of the area served, (2) the type and condition of the equipment used, and (3) the several modes of operation.

Data on refuse collection operation from the Department of Streets and Sanitation files are available in two forms: the original daily work sheets which each driver must fill out and a summary of time spent in each operation by each truck and crew in each of the 50 wards. To obtain the data needed for this study, it was necessary to use the original daily work sheets from which the following information was taken: the truck number, the number of loaders, the loading time for each load, the travel time and the delay time at the disposal site, the time each truck was out of service due to some malfunction, the weight of each truck load, and the numbers of all the blocks collected on each load. Information on the number of living units by block numbers was also available from Chicago, and this information, in conjunction with the block numbers, yielded the number of living units collected on each load.

The general objectives of this study were to identify certain parameters of a refuse collection system (age of equipment, number of loaders on a truck, rank of load, and population density), which may affect the work efficiency (as measured by dollars per ton, dollars per living unit per week, and loading speed).

Generally, age of truck and rank of load (first, second or third load) strongly influence work efficiency; the number of loaders on the truck and the population density have little effect. The lack of effect of population density on efficiency is contrary to the experience of other investigators and may be due to the relatively uniform collection conditions prevalent in Chicago.

The results can be separated into several major groups, each consisting of a set of parameters which exert an effect on the cost or efficiency.

## Cost and Work Efficiency

The average driver cost per load for the three wards studied was about $16.87 per load. There are only small variations from the average.

The average number of living units serviced per load versus the average number of loader hours per load is independent of population density, age of truck, or rank of load.

## Rank of Load and Age of Truck

The highest average loading speed was 1,200 pounds per man-hour and decreased with an increase in the age of truck or rank of load, or both. Both decreases were in the neighborhood of 100 pounds per man-hour.

Average cost per ton increased with the age of truck and with the rank of the load. The increase in cost due to age of truck varied between $1.43 and $3.56 per ton while the increase due to rank of load ranged between $5.59 and $8.32 per ton.

The average cost per living unit per week increased from $0.364 for the first load to $0.475 per living unit per week for the second load of the day. As truck age increased, cost ranged between $0.440 to $0.475 per living unit per week for the second load; and cost for the first load remained approximately constant at $0.364.

Collection cost expressed in terms of dollars per living unit per week is preferable to cost expressed in terms of dollars per ton. The former parameter involves a ratio of time or cost to the number of customers served while the latter involves a ratio of time to quantity of refuse collected. This quantity is affected by weather and other factors and frequently distorts comparisons.

## Labor Costs

Labor costs were found to vary significantly with rank of load: first load costs were $30.25 per load, and second load costs were $44.19 per load. This was attributed to preparation time costs of the two or three laborers while the truck was on route to and from the disposal site.

## IV. ECONOMIC CHARACTERIZATION

Another approach to analyzing the over-all solid waste management system is to examine its over-all economic characteristics. An example of this type of approach is an analysis of the capital labor ratios and the effects on productivity of capital intensification in solid waste management systems. A study of this nature was conducted on a midwestern metropolitan area.[37] In this analysis, the effect of purchasing more equipment and thereby increasing labor productivity was examined.

To measure the collection crew productivity in a given year, the total combustible solid waste collected was divided by the number of man-years expended in the collection effort (Figure 14.5). Productivity rose from 590 tons per man-year in 1960 to 675 tons per man-year by 1968. In 1969, because backyard collection was stopped and crew sizes were reduced correspondingly, productivity was approximately 800 tons per man-year.

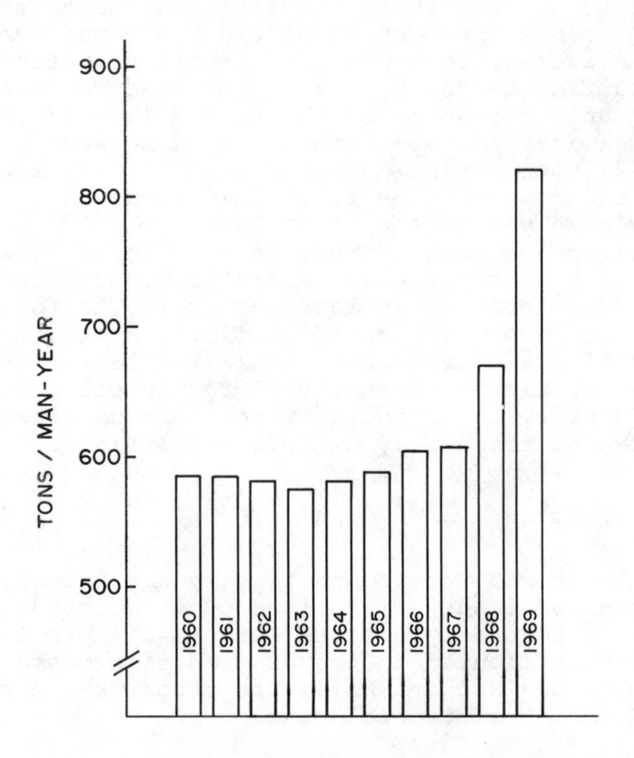


*Figure 14.5. Productivity for collection crews.*

Another factor that exerted a major impact on increasing productivity was the continuing purchase of larger vehicles. In 1965, approximately 8% of the city's fleet was made up of 16-cubic-yard packer vehicles and the remaining 92% was smaller than this size. By 1969, nearly 44% of the city's fleet was made up of 16-cubic-yard packer vehicles. Replacing small vehicles with larger vehicles reduced the number of truck loads of solid waste collected. In 1966, when 16-cubic-yard packer vehicles made up 20% of the total fleet, 65,304 truck loads of waste were collected. By 1969, when 16-cubic-yard packer vehicles made up 44% of the total fleet, 60,960 truck loads of waste were collected. In a four-year period, even with increased waste production, over 4,000 fewer truck loads were collected because of the truck replacement schedule. According to the city's annual report, "This decrease is the result of replacement of 13- and 15-cubic-yard packers with new 16-cubic-yard packers." An obvious corollary to this statement is that the increase in the number of 16-cubic-yard packers has correspondingly increased the productivity of the collection crews.

Another way to examine productivity and its effect on solid waste collection costs is to compare the change in productivity with changes in labor, equipment, and total solid waste management costs. All of these costs were calculated as dollars per man-year of effort. The per cent change from 1960 in labor cost and productivity values was calculated (Figure 14.6). For the years between 1960 and 1967, the slope of the productivity curve was less than the slope of the labor cost curve. Correspondingly, direct labor cost per ton rose during this same time span. After 1967, the slope of the productivity curve was greater than the slope of the labor curve and the direct labor cost per ton decreased. As equipment expenditures increased, productivity tended to increase (Figure 14.7). For total solid waste management costs, as for labor costs, the cost per ton tends to decrease when the slope of the change in productivity curve exceeds the slope of the change in cost curve (Figure 14.8). This analysis indicates that when the change in productivity exceeds the change in cost, the cost per ton for solid waste collected will decrease.

An examination of the ratio of direct labor cost to direct equipment cost reveals that for the period of analysis capital was substituted for labor and that when the ratio of labor to capital is decreased, productivity is increased (Figure 14.9).

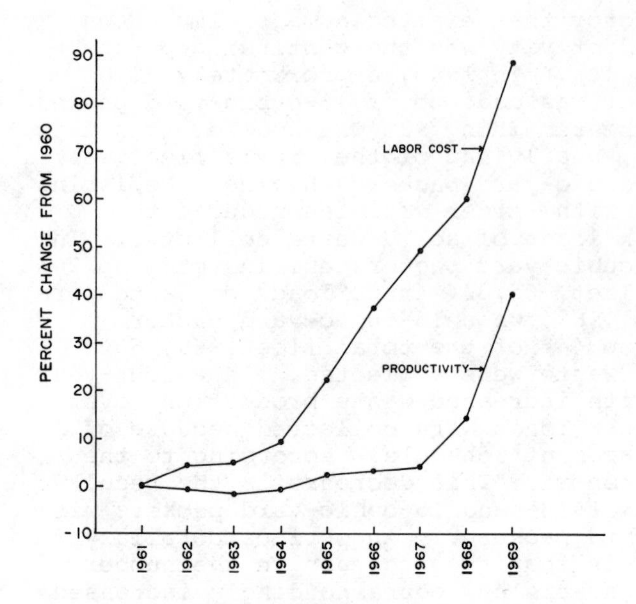


*Figure 14.6.* ***Productivity and labor*** *cost.*

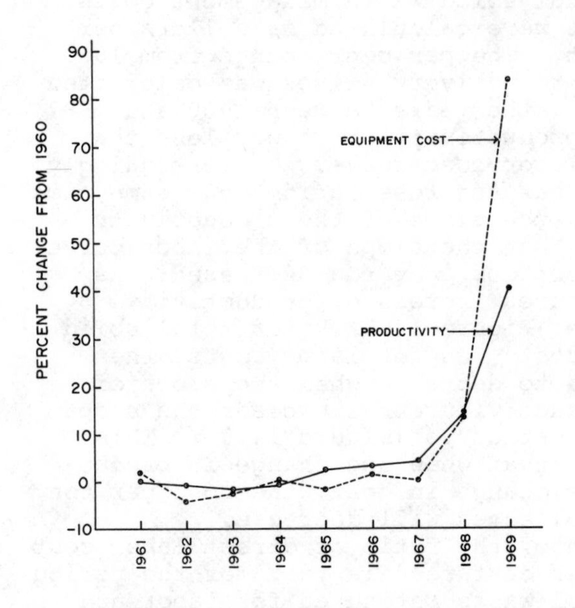


*Figure 14.7. Productivity and equipment cost.*

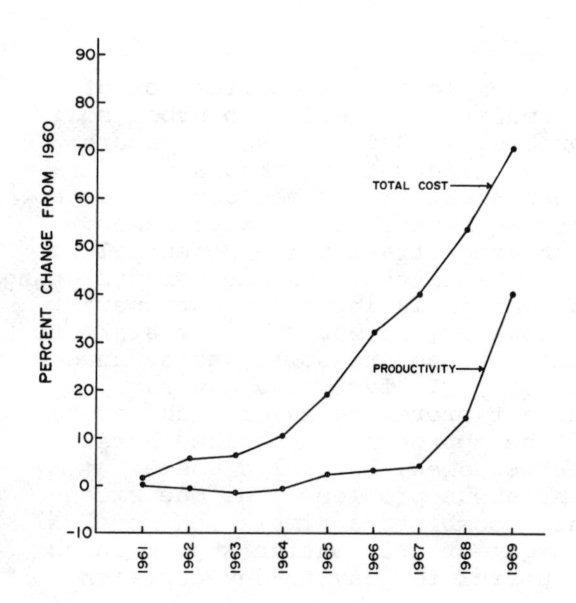


*Figure 14.8. Productivity and total cost.*

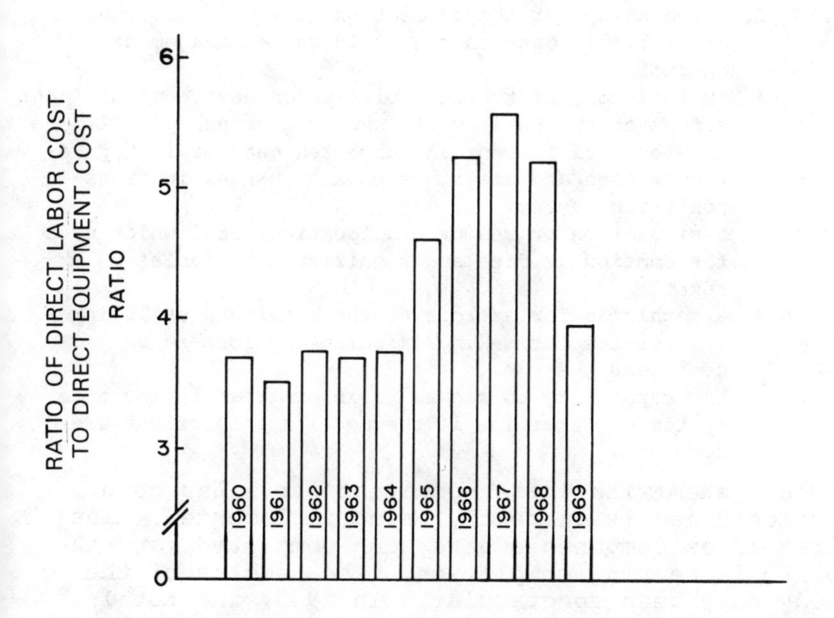


*Figure 14.9. Ratio of direct labor cost to direct equipment cost.*

## V. THE MODELING APPROACH

Many studies have explored the application of deterministic and simulation modeling to urban solid waste management problems. Several recent papers have applied these techniques to routing and facility location using either assumed data or data collected on a one-time basis. These approaches have been useful for demonstrating the potential of systems or operations research technique for assisting the solid waste manager in making important operational decisions. However, it would be a mistake to say that the systems approach to solid waste management has been a success. In fact, one can say in general that it has not proven to be of much use to managers. Most of the approaches described here have been "one-time" studies, but solid waste management is a continuous problem. The one exception is a study which combined a simulation model with an on-line management information system in an attempt to make it useful for day-to-day decision making.[38]

The results of this effort conducted in Cleveland, Ohio, indicate that five basic components are required to make a systems approach useful:

1. a mechanism for the collection of continuous, uniform, and reliable data on the solid waste management operation
2. an inventory of the variables which have a significant effect on the solid waste system; for example, a knowledge of present and expected changes in population trends and distribution and changes in transportation systems
3. a simulation or resource allocation model which uses the continuous data and significant variables as input
4. a mechanism for exercising the model and utilizing its results for making immediate or long-range decisions
5. the capability to reexamine information in the data system to determine if the model's predictions are accurate.

A system with these component parts has been developed and is currently being implemented. The first three components have been completed, and the fourth is nearing completion. The results of the study have been spectacular. In 1970, the total complement of employees of Cleveland Waste Collection and Disposal Division reached a maximum of 1,825.

Using data obtained in the study, an initial reduction was made to approximately 1,200 laborers actually collecting waste on the route. A change in collection operations was accomplished in which service was changed from backyard to curb side. Later, the total number of routes was reduced from 224 to 138. The total personnel complement now numbers approximately 600 employees, and the annual budget has declined from $14.3 million to approximately $8.0 million in 1972. It is expected that as a result of this program, Cleveland can expect its budget based on 1972 dollars to stabilize at approximately $7.2 million. Figures 14.10 and 14.11 show actual and projected changes in numbers of personnel, vehicles and routes, and in population, generation (tons), and budget for the city.

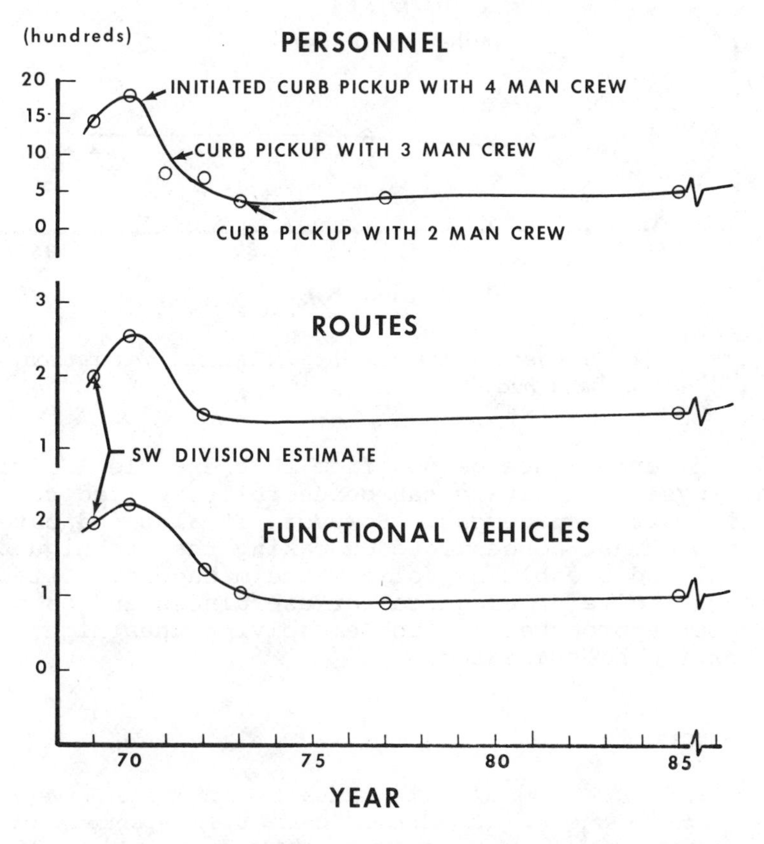


*Figure 14.10. Projected changes in personnel, vehicles, and routes.*

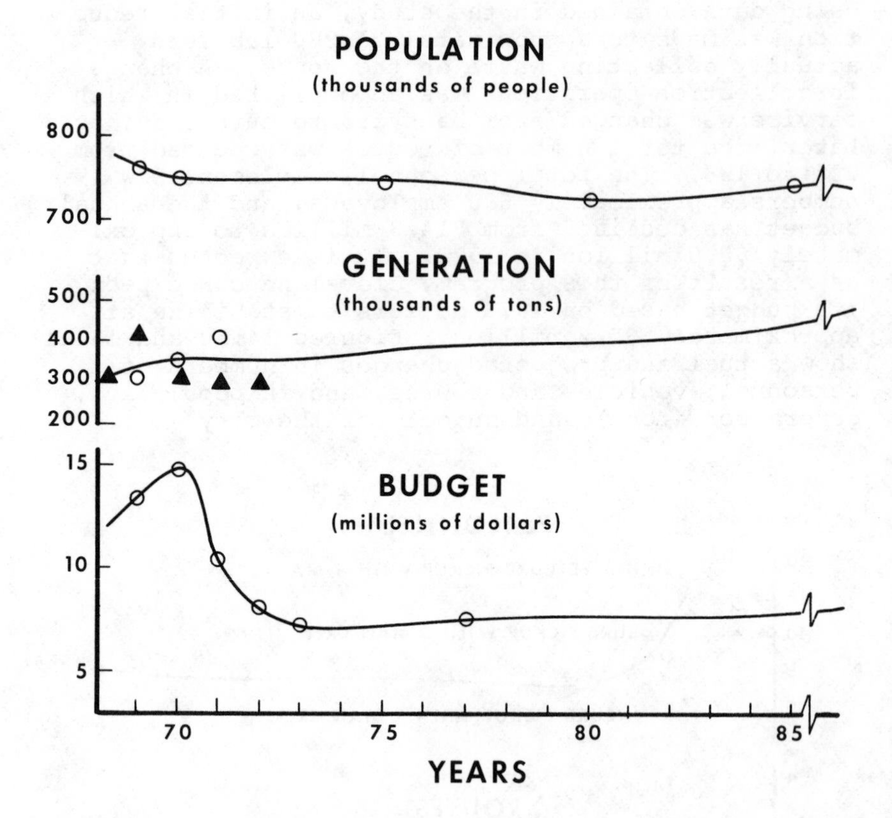


*Figure 14.11. Projected changes in population, generation, and budget.*

The experience gained from this example indicates that systems analysis can be usefully applied to solid waste management. However, it also indicates that one must concentrate on making the techniques useful and useable by solid waste managers. Often, one must give up elegance for usefulness and choose simpler approaches to problem solving than might otherwise be suggested.

REFERENCES

1. Black, R. J., *et al*. *The National Solid Waste Survey; An Interim Report*, (Cincinnati, Ohio: U.S. Department of Health, Education and Welfare, 1969).

2. Sanitary Engineering Research Project, University of California, "An Analysis of Refuse Collection and Sanitary Landfill Disposal," *Technical Bulletin No. 8, Series 37* (Berkeley, California, 1952).
3. Skelly, M. J., "Planning for Regional Refuse Disposal Systems," Ph.D. thesis, Cornell University, Ithaca, New York (1968).
4. Lynn, W. R. *Solid Wastes Research Needs,* APWA Research Foundation, Project No. 113, Appendix D, Chicago, Illinois (May, 1962).
5. Golueke, C. G. and P. H. McGauhey. "Comprehensive Studies of Solid Wastes Management, First Annual Report," Sanitary Engineering Research Laboratory, University of California, Berkeley, SERL Report No. 67-7 (May, 1967).
6. Anderson, L. E. "A Mathematical Model for the Optimization of a Wastes Management System," Sanitary Engineering Research Laboratory, University of California, Berkeley, SERL Report 68-1 (February, 1968).
7. Management Technology, Inc. "The Applicability of a Systems Approach to Solid Waste Management Problems," report presented to the Maryland State Department of Health, Division of Solid Wastes (July, 1966).
8. Aerojet General Corporation, Azuza, California. "California Waste Management Study: Report to the State of California, Department of Public Health." (1965).
9. Quon, J. E., A. Charnes, and S. J. Wersan. "Simulation and Analysis of a Refuse Collection System," *Journal of the Sanitary Engineering Division, ASCE, 91 (SA5)*:575 (1965).
10. Truitt, M. M., J. C. Liebman, and C. W. Kruse. "An Investigation of Solid Waste Policies," Department of Environmental Engineering Science, the Johns Hopkins University, Baltimore, Maryland (August, 1968).
11. Truitt, M. M., J. C. Liebman, and C. W. Kruse. "Simulation Model of Urban Refuse Collection," *Journal of the Sanitary Engineering Division, ASCE, 95 (SA2)*:289 (1969).
12. Quon, J. E., M. Tanaka, and S. J. Wersan. "Simulation Model of Refuse Collection Policies," *Journal of the Sanitary Engineering Division, ASCE, 95 (SA3)*:575 (1969).
13. Marks, David H. and J. C. Liebman. "Mathematical Analysis of Solid Waste Collection," Public Health Service Publication No. 2104 (1970).
14. Coyle, R. G. and M. J. C. Martin. "Case Study: The Cost Minimization of Refuse Collection Operation," ORSA Meeting (May, 1968).
15. Walker, W. "Adjacent Extreme Point Algorithms for the Fixed Charge Problem," *Technical Report No. 40* (Ithaca, New York: Cornell University, January 1968).

16. Marks, David H., Charles S. ReVelle, and J. C. Liebman. "Mathematical Models of Location: A Review," *Journal of the Urban Planning and Development Division, ASCE, 95 (UP1)*:81 (1970).
17. Saaty, T. L. *Mathematical Methods of Operations Research* (New York: McGraw-Hill, 1959), p. 167.
18. Cooper, L. "Heuristic Models for Location Allocation Problems," *SIAM Review 6(1)*:37 (1964).
19. Hadley, G. *Non-Linear and Dynamic Programming* (Reading, Mass.: Addison-Wesley Publishing Co., Inc., 1964). p. 136.
20. Efroymson, M. A. and T. L. Ray. "A Branch-Bound Algorithm for Plant Location," *Operations Research 14(3)*: 361 (1966).
21. Spielburg, K. *An Algorithm for the Simple Plant Location Problem with Some Side Conditions* (New York: IBM Scientific Center, May, 1967).
22. Gray, P. "Mixed Integer Programming Algorithm for Site Selection and Other Fixed Charge Problems Having Capacity Constraints," *Technical Report No. 101*, Department of Operations Research, Stanford University (November, 1957).
23. Kuehn, A. A. and M. J. Hamburger. "A Heuristic Program for Locating Warehouses," *Management Sciences 9(4)*:643 (1963).
24. Kuhn, H. and W. Baumol. "An Approximate Algorithm for the Fixed Charge Transportation Problem," *Naval Research Logistics Quarterly 9(1)*:1 (1962).
25. Manne, A. S. "Plant Location Under Economies of Scale--Decentralization and Computation," *Management Science 11(2)*:213 (1964).
26. Balinski, M. L. "Fixed Cost Transportation Problems," *Naval Research Logistics Quarterly 8(1)*:41 (1961).
27. Steinberg, D. I. "The Fixed Charge Problem," Ph.D. thesis Washington University, St. Louis, Missouri (1965).
28. Wersan, S. J., J. E. Quon, and A. Charnes. "Location of Disposal Sites," *Mathematical Simulation of Refuse Collection and Disposal Practices*, First Progress Report, Sanitary Engineering and Environmental Sciences, Northwestern University, Evanston, Illinois (May, 1963).
29. Schultz, G. P. "Managerial Decision Making in Local Government: Facility Planning for Solid Waste Collection," Ph.D. thesis, Cornell University, Ithaca, New York (1967).
30. Baker, J. S. *A Cooperative Municipal Refuse Disposal Program, Prince George's County Maryland*. Municipal Technical Advisory Service, College of Business and Public Administration, University of Maryland, College Park, Maryland (1963).
31. University of Louisville. *Solid Waste Study and Planning Grant--Jefferson County*, Louisville, Kentucky (1968).
32. Helms, Billy P. and Robert M. Clark. "Locational Models for Solid Waste Management," *Journal of the Urban Planning and Development Division, ASCE, 97 (UP1)*:1 (1971).

33. Kruse, C. W., J. C. Liebman, and M. M. Truitt. "Optimal Refuse Collection Policies," Proceedings Engineering Foundation Research Conference on Solid Waste Research and Development, University School, Milwaukee, Wisconsin (July 24-28, 1967).
34. Clark, Robert M. and Billy P. Helms. "Decentralized Solid Waste Collection Facilities," *Journal of the Sanitary Engineering Division, ASCE, 96 (SA5)*:1035 (1970).
35. Clark, Robert M. and Billy P. Helms. "Fleet Selection for Solid Waste Collection Systems," *Journal of the Sanitary Engineering Division, ASCE, 97 (SA7)*:71 (1972).
36. Quon, J. E., R. R. Martens, and M. Tanaka. "Efficiency of Refuse Collection Crews," *Journal of the Sanitary Engineering Division, ASCE, 94 (SA2)*:437 (1970).
37. Clark, Robert M. *Urban Solid Waste Management; Economic Case Study,* Environmental Protection Series EPA-R2-72-012, National Environmental Research Center, Office of Research and Monitoring, U.S. Environmental Protection Agency, Cincinnati, Ohio 45268.
38. Clark, Robert M. and James I. Gillean. *Systems Simulation and Its Application of Solid Waste Planning; A Case Study,* Joint National Meeting, AIIE-TIMS-ORSA, November 8-10, 1972.

MODELS FOR ENVIRONMENTAL POLLUTION CONTROL


# CHAPTER 15

# COMPUTER-BASED DESIGN OF REFUSE COLLECTION SYSTEMS

Robert G. Coyle*

## I. INTRODUCTION

This paper describes two approaches to the problem of providing assistance to managers faced with the task of planning refuse collection systems for modern urban communities. The work described has taken place at intervals over a period of eight years, and has been applied in practice in two English cities. It is a measure of the complexity of the planning problems involved that, even after eight years of work, there is no clearly-defined simple formula that can be applied in all circumstances. It will, however, be argued in this paper that the methods that have evolved are capable of being of real practical help to managers in dealing with actual planning situations.

The first sections of the paper are devoted to a description of the problem of designing a refuse collection system. Subsequent sections discuss the research work that has been done, concentrating mainly on managerial use of the computer programs rather than on a detailed description of the programming techniques. This is followed by a short examination of the relationships between the planning of collection and disposal systems. Finally, the last sections of the paper investigate the problem of implementation and use of computer-based planning methods in an operational situation.

---

*Robert G. Coyle is the Director of the System Dynamics Research Group at the University of Bradford, Bradford Yorkshire, BD7 1DP, United Kingdom.

## II. THE NATURE OF THE PROBLEM

It seems to be good practice for the management scientist to examine carefully the objectives of the various participants in the system before he puts pen to paper or switches on his computer. This should enable the analyst to assess the nature of the problem and to determine the criteria which his solution must satisfy. This approach may be summed up in the phrase "the optimal policy is the one that gets implemented."

In the case of refuse collection systems, management's objectives may be summarized as

1. to achieve some standard of service
2. to keep the labor force happy
3. to try to ensure that this year's costs are not too much higher than last year's.

The exceptions to this usually result from the Cleansing Department having been badly neglected or starved of funds, marked by a decline in the equipment and morale.

These objectives, however, must be achieved in an environment which is becoming more complicated because of an acute shortage of good quality labor, rising costs of labor and vehicles, improvements in collection technology and required standards of service, and a continuing process of urban construction and renewal, making plans obsolete very rapidly.

Therefore, there is a need for tools capable of planning for the short and medium-term horizons of up to three to five years ahead.

## III. REFUSE-COLLECTION SYSTEMS

In this section some of the issues and choices that arise in planning a refuse-collection system from the point of view of management will be discussed. Some conclusions about the way in which a management scientist should approach the task of providing management with assistance in planning are drawn.

First of all, it is important to recognize that there is no single refuse-collection process and, in practice, one has to allow for the simultaneous operation of up to three collection systems in the same city. These systems are domestic, semibulk, and bulk.

*Domestic Collection*

Domestic collection always involves a number of teams, each of which is comprised of a collection vehicle, its driver, and a variable number of other men. The other men, referred to as "pedestrian operators," move slowly down a street, collecting the refuse containers and placing them or their contents into the collection vehicle. The driver is nearly always required by law not to leave his vehicle, and his function is to move it in such a way that it is always in a convenient place for the pedestrian operators.

There are many variations on this simple theme. For example, some areas utilize returnable refuse containers, while others use disposable ones. The location of the container creates variations as well. It may be at the roadside or it may be near the residence, necessitating collection by pedestrian operators.

It is evident that the number of pedestrian operators will have an important effect on collection cost in a given district and that this cost will also depend on the housing composition of that district. It has, for example, been shown by Coyle[1] that the cost per house with a team of n men will be

$$C(n) = C_v\frac{t}{n} + C_v\frac{d}{r} + C_m t + C_m\frac{d}{r} n$$

where

$C(n)$ = cost per house with team of n men

$C_v$ = cost of vehicle plus driver

$C_m$ = cost of a pedestrian operator

$r$ = parameter relating to walking speed of a pedestrian operator

$t$, $d$ = parameters descriptive of the effect of housing composition in an area.

It is evident that the "optimal" team size will be given by

$$n_o = \sqrt{\frac{C_v t r}{C_m d}}$$

and

$$C(n_o' + 1) > C(n_o')$$

if

$$\frac{n_o'}{n_o' + 1} + \frac{C_m d}{C_v t r} \cdot n_o' > 1$$

where $n_o'$ is the integer part of $n_o$.

*Semibulk Collection (Paladin)*

In many cases, for schools, offices, shops, and flats, the amount of refuse exceeds the capacity of ordinary dustbins, paper sacks, etc. by such a margin as to make it worthwhile to use special containers. These usually have a capacity of about 1 cubic meter, uncompressed, and are called Paladin containers in England. Therefore, the name Paladin is used as a generic term for this type of container. They are, of course, far too heavy for man-handling, so a specially-equipped vehicle is used. This means that the crew size is not usually variable and normally comprises a driver and an assistant. There may well be more than one Paladin at a particular location; however, as the capacity of the vehicle is usually at least 30 Paladins, the problem is to design tours for the Paladin collection vehicles which meet all the various constraints on vehicle capacity, required collection frequency, etc.

Generally, one of the problems for management is that the number of Paladins tends to increase fairly rapidly and the system must be kept up to date. There are related difficulties, such as ensuring that a team does not get lost (accidentally or otherwise) and ensuring that there are unambiguous statements as to who is responsible for collecting which Paladin.

*Bulk Collection*

For locations where the output of refuse is very high and access is possible (*e.g.*, a hotel or a building site), one may use true bulk collection devices. These are large containers which are carried or towed away by a vehicle used for no other purpose. The container must be replaced immediately by an empty one, which is usually delivered to the location before the full one is removed.

Naturally, there is no question of man-handling, for these containers cannot be moved as can Paladin containers, which can be pushed on castors or moved on a trolley if the need arises and the floor is smooth. Also the technology of moving the container is such that the collection vehicle has to make a separate trip to each location, even if a container is practically empty.

Even in a large city the number of bulk containers is usually limited if only because of the difficulty of ensuring physical access. Bulk collection, then, is usually a very small aspect of the Cleansing Department's activities and will not be specifically dealt with in this paper.

## IV. MANAGEMENT PROBLEMS OF PLANNING

In order to be of any practical use to cleansing managers a planning mechanism of any kind, computer-based or not, must satisfy certain requirements. The task that the manager faces is that of dividing his city into areas, called rounds, which constitute the tasks to be assigned to the collection teams.

There are a great many constraints on the system and they must all be specifically provided for. They include

1. manpower availability
2. number of vehicles available
3. workload of individual teams must not exceed a prescribed maximum
4. workload must not vary too much from one team to another
5. the area assigned to a team must be a "reasonable" shape
6. traffic regulations (*e.g.*, one-way streets) must not be violated
7. certain areas may have to be serviced early in the morning
8. all premises in the city must be allocated to a round or tour at the proper service frequency
9. vehicle capacity must not be exceeded
10. there may be a limit on the number of loads which a team may do in a day, as well as the constraint on work duration.

Apart from these constraints, there are other factors to be borne in mind. The manager must choose which disposal site each team is to use, the task assigned to each team must be physically possible in the time available, vehicle capacity and traveling times must be allowed for, labor regulations and consideration of equity between teams may well impose constraints, and, of course, costs must be kept within reasonable bounds.

This problem would be difficult enough in a static environment but unfortunately the city undergoes continuous physical change, developments occur in vehicle technology, and relative costs of vehicles and pedestrian operators alter. Management, therefore, must face the problem of planning vehicle replacements, labor recruitment, and round redesign in such a way as to aim towards some constantly-changing ideal future.

It is necessary to acknowledge that, although most Western cities have refuse-collection services that are efficient in the sense that the refuse does get collected, hardly any have anything approaching a planning system capable of dealing with the kind of problem we have just outlined. In nearly all cases changes are made ad hoc, the principal, short-term aims being to keep the citizens and the labor force happy and to see that next year's costs are not too much higher than last. This is brought about because of the way in which local government works and partly because of the complexity of the problem--it defies analysis by pencil-and-paper calculation. The management scientist trying to deal with these problems has to develop a system which will deal with the planning problems in a reasonable manner.

The first essential for reasonableness must be that the plans produced do provide for achieving the required standard of service without violating any of the many constraints in the system. In addition the planning tool must also

1. allow for the use of local knowledge by managers, supervisors, and the work force
2. draw up tasks for the teams in a reasonable shape in order to reduce the chance of confusion or a team getting lost
3. be used and interpreted easily and be capable of producing revised plans easily, quickly and cheaply
4. be capable of updating data on the town and the collection task easily
5. produce plans requiring the absolute minimum of pencil-and-paper modification in order to make them completely practicable
6. produce work specifications that are unambiguous and usable by first-line supervisors.

## V. MANAGEMENT-SCIENCE REQUIREMENTS

The management-scientist who approaches the refuse-collection planning problem armed with the tools he would use to solve a production-planning problem in a factory is quite likely to waste his own time and everybody else's. In fact, it is necessary, if one is to have any prospect of having one's work accepted, to design a planning scheme and choose techniques that will take into account a whole range of factors; a discussion of some of these follows.

1. Because of the large number of constraints on the problem, optimization in the sense of a minimum-cost solution is likely to have a fairly limited effect on system performance. In straight cash terms it will probably be far more cost-effective to introduce good work-measurement than to indulge in mathematical analysis. The management science approach should, therefore, be geared to improved planning capability rather than to cost-reduction. However, the management science treatment should, as always, save enough to justify itself and, in any case, the long-term financial benefits of improved planning should be considerable, though hard to quantify and attribute.

2. There is a paramount need to make the planning system plausible to the managers.

3. The planning system must be easy to use or managers won't bother with it. Furthermore, it must be possible and easy for the manager to try out his own ideas, bringing to bear the local knowledge he would find it difficult to embody in a formal decision rule, and to look at the results to see if they look right.

4. It should be remembered that one of the major headaches for cleansing managers is continual change of the city. Any plan, therefore, must provide for easy incorporation of information about changes into the data file; it must be even easier to produce a synthetic data file of the city as it is expected to be in five years; otherwise the planning mechanism will be useless.

5. The planning system must also be efficient in two senses. First, it must make it so easy for the manager or planner to consider several alternatives that he will be positively encouraged to use it. This means that it must represent a substantial

saving in the time and effort he would have to expend in order to do this anyway. (Cleansing managers have been planning perfectly good collection systems for years, and are quite well aware of how to do it. They, therefore, must be able to see some appreciable benefit from using a new method proposed by an outsider.) Second, the planning process must be capable of designing a collection system efficient in its own right (*i.e.*, by Cleansing Department efficiency standards), and not requiring a lot of pencil-and-paper adjustment to make it workable in practice.

It is fairly clear from the foregoing evaluation that simple management science optimization will not meet the case and that what is required essentially is a computer simulation approach, with some optimizing overtones. The application of these precepts to the design of computer-based planning systems for domestic collection and for Paladin collection will now be considered.

## VI. DOMESTIC COLLECTION

### *The Huddersfield Study*

Early studies by this author sought to deal with collection planning in a city of about 140,000 inhabitants in the north of England. The city was particularly simple in that there were only a few types of housing and there was a weekly collection of all premises. Since it was possible to consider variations in team size from one team to the next, an approach was used based on dynamic programming. More extensive accounts have been given elsewhere[1-4] so only some of the important features of that analysis are presented here.

The first feature is that of the section. This represents a midway point between planning rounds on the basis of individual buildings and the other extreme of using great unwieldy tracts of housing. A section is simply some convenient area which common sense indicates could reasonably be subdivided. As a *rough* guide a typical section would be about two or three hours work for an average team. If the average section is too small, computer time is wasted; if it is too large, unevenness may result between team work loads. The sections form the basic framework for all subsequent development in the planning system. Deciding the section boundaries for a large city can be done in a few days and once done there is no need to update the section division.

The second idea is that of contiguity which enables one to avoid the problem of specifying the geographical layout of the city. The sections are numbered and a matrix is drawn up with one row for each section. The numbers in the rows are those of the sections which can be reached from the section in whose row the numbers are written. Thus, if two sections, A and B, are adjacent on the map, but are connected by a one-way street running from A to B, B would be recorded as being contiguous with A, but not vice-versa. The importance of this contiguity matrix is that as domestic collection rounds are built up by adding one section to another, the matrix immediately indicates the number of sections which are candidates for inclusion in the round. The computer program can then select the most satisfactory candidate section from the alternatives presented to it. This is precisely analogous to a human planner and the computer can be programmed to make a very sophisticated choice.

The Huddersfield study used the computer to do the work-measurement calculations, using data derived from a statistical analysis of time-study data. This procedure was used because it enabled the work-measurement values to be updated in a few seconds of computer time, and the values could then be kept in line with changing circumstances, which is practically impossible with pencil-and-paper methods. The values have been used successfully as the basis for an incentive scheme for about five years, but consideration is being given to making the system even easier to use and foolproof in operation.

The output from the round planning program gives a breakdown of each team's task, showing the sections to be cleared during the week, expected performances, vehicle loads, etc. The output can be modified to give a wide variety of managerial information, especially costing and bonus calculations, depending on requirements.

The program also produces an alternative plan capable of clearing the town with a minimum number of men. This is for emergency situations involving high absenteeism.

### *The Westminster Study*

The Huddersfield study[2] was completed in 1967. It was realized that the methods were not entirely satisfactory for larger cities and during the past three years work has been progressing to refine the techniques, in association with a Borough Council in London. This work has led to a computer program far more sophisticated and flexible than the earlier version and believed to be widely applicable with fairly small modifications.

The problem in London was rather different from that of the earlier work. The city was much larger and, for historical reasons, was divided into four separate collection divisions called zones. The housing types ranged from what is probably the most expensive housing in the world to something not far removed from slums. The frequency of collection could vary from daily to weekly within a small group of houses, and the city center had to be clear of collection teams before the morning rush hour started. About the only simplification was that all refuse had to be disposed of at one point, but as this was on the edge of the city, journey times could be up to 80 minutes.

Within a zone the team size could not vary, for labor-relations reasons, but it did not have to be the same for all zones.

### *Data Collection*

The first step was to divide each zone into a suitable number of sections, and to record these on a punched-card contiguity matrix. This division into sections is an error-prone job, and the computer program checks the matrix for consistency and identifies any apparent errors. These can be eliminated if they are real errors, or allowed to remain if they are deliberate distortions of the matrix to represent one-way streets, uncrossable barriers such as rivers, etc.

The second part of the data is for individual streets. A punched card is prepared for each street showing its name, the section in which it lies, a collection frequency code, and the number of houses in each of a variety of types which are to be collected at that frequency. If a street contains premises that are visited on different frequencies, it is necessary to have a card for each frequency.

Preparing this data is not difficult, although it can be rather tedious (we have used students with great success). As has been argued, this data preparation is essential if one is to have any planning system at all, and once prepared, the data file is *very* easy to update.

## *Program Operation*

The program consists of 1437 lines of FORTRAN and, therefore, cannot be described in detail here. The basis of its operation, however, can be discussed briefly. Each zone is considered separately because it may be necessary to do a replan for the whole city and the user can choose the one he wishes to deal with. For each zone, the user specifies the team size he is interested in and provides some cost data; the program prepares a collection plan for each team size for each zone.

The methods of preparing a collection plan are as follows. The program first calculates, from the housing data, the amount of refuse to be removed from each section and, taking into account the team size, the time needed to do it. This information is stored in the computer together with a code, calculated by the program, reflecting the frequency with which that section must be serviced. For example, suppose a section contains some houses which have to be serviced three times a week (Monday, Wednesday and Friday) and some which have to be done twice a week (either Monday and Thursday, or Tuesday and Friday). If the first group takes $t_1$ minutes and the second $t_2$, there are two possible work load patterns for this section.

| | *Pattern 1* | *Pattern 2* |
|---|---|---|
| Monday | $t_1$ & $t_2$ | $t_1$ |
| Tuesday | | $t_2$ |
| Wednesday | $t_1$ | $t_1$ |
| Thursday | $t_2$ | $t_1$ & $t_2$ |
| Friday | $t$ | $t_1$ |

The program will have to assign the optional, twice-weekly, part to whichever days of the week are most convenient. For domestic collection there are seven possible collection frequency combinations, each with one or more patterns, such as those exemplified here. However, the program deals with them all automatically, thus simplifying the data problem for the user.

The program starts by selecting the first section for the first round. If the user wishes, he can make the computer start at some particular section, such as the one he would have used had he been doing the calculation by hand, or, failing user intervention, the program starts from section number 1. Having assigned the section to the collection team, the program removes the section from consideration to prevent it being included twice, and checks to see if the vehicle is fully loaded or if any of the days of the week has been fully occupied. Naturally, this will not be the case after one section so the program finds, from the contiguity matrix, a series of candidate sections and chooses the most promising for allocation as the next part of the round. The choice mechanism is rather subtle, involving 200 lines of FORTRAN, but it is based on the following considerations

1. ensuring that the round is reasonably "round," *i.e.*, is not a long, thin area
2. not leaving any sections cut off
3. finding the best fit for the day's work.

After each section, the program checks the vehicle load and work load. When the vehicle is full, the program provides for emptying it and starting a new load. When a day is used up, the program starts a new round. (Clearly, the work load will inevitably be uneven for the days of the week.) When the zone is completed, the program prints out the results, showing which team has to do which work. For example:

```
TASK FOR TEAM 1

    MONDAY

WORK LOAD 419.26 MINS     PERFORMANCE 97.50%

SECTIONS TO BE CLEARED    48 53 65 50 49 64 60 24 15
```

and so on for the other days and teams. It also prints out how many teams are needed to clear the zone and the annual cost.

This specification is then followed by what is perhaps the most useful visual aid, a map of the zone showing the round boundaries. The manager can examine this and if he does not like the look of the plan can direct the computer to try again from a different starting point. In any one run the user can direct the program to examine up to four starting points, and this number could easily be increased if required.

At the end of the run the program performs a feasibility analysis to evaluate the economics of different situations in each zone. A first example is a case in which the computer has been directed to examine the effects of introducing a new, larger vehicle in a small area of the city and to determine whether the team size should be three or four men. The program produced four collection plans for each team size (which showed that it would be rather hard to fit these teams sensibly into this small area) and then printed out the fact that whether three or four men were used in a team it would still be necessary to use four teams because the vehicle was so far from the disposal point that only one load per day would be possible in such a large vehicle. Annual costs would be approximately $19,200 greater for four-man teams.

This was a very simple case where nearly all the constraints were relaxed. In a more interesting case the program was run to show the effects of trying teams of 2, 4 or 6 men in each of two zones, but with not more than 45 men or 12 vehicles total. The end result is shown in the following table.

| *Policy Number* | *Team Size in Zone* | | *Total Men* | *Total Vehicles* | *Feasible* | *Too Few* | *Annual Cost £ th.* |
|---|---|---|---|---|---|---|---|
| | *One* | *Two* | | | | | |
| 1 | 2 | 2 | 30 | 15 | No | Vehs | 135 |
| 2 | 4 | 2 | 34 | 13 | No | Vehs | 133 |
| 3 | 6 | 2 | 36 | 12 | Yes | | 132 |
| 4 | 2 | 4 | 36 | 12 | Yes | | 132 |
| 5 | 4 | 4 | 40 | 10 | Yes | | 130 |
| 6 | 6 | 4 | 42 | 9 | Yes | | 129 |
| 7 | 2 | 6 | 42 | 11 | Yes | | 139 |
| 8 | 4 | 6 | 46 | 9 | No | Men | 137 |
| 9 | 6 | 6 | 48 | 8 | No | Men | 136 |

These figures make extremely interesting food for thought. The Department currently operates at the sensible policy of four-man teams in each zone (the zones are very small, requiring four four-man teams). However, the tables show the number of vehicles needed to reduce the labor force to 30 in order to cope with a sudden epidemic of influenza, without reducing the standard of service. The end column shows that, *in this case*, the cost function is rather flat.

In both of these examples, using the computer program to set up the correction pattern is rather like using a sledge-hammer to crack a nut. The zones are very small, for purposes of illustration, and the real benefits come from using the computer program to deal with zones involving several hundred sections each, rather than a few score.

## VII. PALADIN COLLECTION

We now deal, fairly briefly, with the program which plans collection tours for Paladin containers. At first sight, there are some similarities between this problem and the celebrated traveling salesman. There are, however, more differences than similarities and one is faced, as a management scientist, with what could be called the Traveling Dustman problem.

The planning task is to assemble Paladin locations into area groups which are to be serviced by a particular team. Unfortunately, the traveling time depends on the group size, but one cannot determine the group size without knowing whether the working-plus-traveling time involved will fall within permissible limits. Therefore an algorithm was developed which assembles Paladin locations into tours, using estimates of traveling time. When a vehicle-load of Paladins has been assembled, it is put into collection order using the near-optimal Nicholson algorithm[5] and the estimated traveling time is replaced by the calculated value. This algorithm seems to give effective results, but because it takes 986 lines of FORTRAN it cannot be described in full.

### *Data Input*

The data required are essentially similar to those for the domestic planning program. The city is subdivided into sections, which need not be as small as those for domestic collection, a contiguity matrix is specified in the same way, and each Paladin location is assigned its own card, specifying the section in which it lies, the address or location number, the number of Paladins and additional information enabling the work content to be calculated by the program. Since the number of Paladin locations usually increases very rapidly, updating the data file is very easy indeed.

The collection frequency for Paladins varies considerably and in the city in which this work was done a few locations had to be served on a special Saturday shift. Management did not wish this shift to be too long but, in order to make it worthwhile for the men to turn out at all, they were prepared to have a small porportion of the twice-weeklies and thrice-weeklies done on cycles involving Saturday. One of the few controls that the user can specify is the percentage of these optional Saturday collections that he is willing to have.

### *Program Operation and Output*

The program works in much the same way as that for domestic planning. It assembles sections into rounds, and then allots the Paladin locations in each section to collection tours, taking into account work duration and vehicle load.

For each team the output shows the work to be completed on each day of the week by listing the Paladin locations to be visited in optimal collection order, showing when the vehicle is full and should return to be emptied. Vehicle utilization and driver performance are also shown. As might be expected, there are large variations in work load from one day to another, but this is inevitable given the wide variation in collection frequency for the various locations.

The program prints out a map showing the areas to be served by each team and then, as a check, lists all the Paladin locations, showing the day of the week in which they are to be visited.

Finally, the program produces an analysis of the work balancing for the Paladin system.

## VIII. PRACTICAL UTILIZATION

We have dealt at some length with the problems of planning collection systems and have discussed means by which these problems may be attacked. Lack of space precludes a full treatment of the detailed marking of the computer programs, but it is hoped that the description will convey the essence of the way in which they work. We now turn to the important question of the ways in which these programs can be applied to practical problems, and any difficulties which might be attendant on their use.

It must first be made clear that the principal benefit of using computer methods is the scope which they afford for considering many alternatives very rapidly and cheaply. All the programs described here will run in a few minutes at most even for a large city. This makes it quite possible to run through a planning study in 24 hours, a job which would take many months of work by pencil-and-paper. Since it is simply not possible to devote resources to this kind of protracted exercise, the supreme value of the computer methods is that they make possible things which could not otherwise be contemplated. In this sense they add to managerial capability, rather than supplanting it. These computer methods can be employed in a number of different modes and these will be discussed briefly.

### *Updating*

As has been observed the physical composition of a city is continually changing. The programs described here include data files that are easy to update. It is a simple matter to run the program to calculate the new work load of the *existing* tasks, based on revised data on housing types, etc. This enables one to judge where round-adjustments are necessary. Because of the way the programs (for domestic collection in particular) are written, it is easy to modify them to carry out small-scale adjustments to the round boundaries when the recalculation of work load indicated this necessity.

This is, however, a subsidiary use of the programs and they are more likely to prove fully beneficial in other, longer-range activities.

### *Environmental Changes*

The need to update the collection plan to allow for the physical changes which *have* taken place in the past has already been referred to, but it is equally valuable to study expected changes before they occur. Some examples are:

1. The proposed construction of a new housing development--how should it be incorporated into the round pattern?
2. Clearance of old houses and their replacement by apartments--what are the advantages of using Paladin at the apartments and how would the Paladin *and* the domestic services be affected?
3. Proposed construction of a major road in the city which will be difficult for teams to cross except at selected points--how will this affect the services?
4. Absorption of several separate services into one by local government reorganization--here the list of questions is almost endless.
5. If trends in housing density, refuse output and refuse composition continue, what services will be needed in, say, five years?

### *Economic Changes*

Although it is by no means easy to forecast the changes in the economic environment, the prudent manager tends to insure himself by having a contingency plan for a particular change. Thus, if the cost of labor rises more rapidly than the cost of vehicles, it is fairly evident that in five years time one would want to have relatively more vehicles and fewer men. Solving the question of number would be difficult even if other things stayed equal. Unfortunately they don't and, again, one has the kind of problem where the computer methods should help.

### *Technological Changes*

During the past few years there has been a steady progression in the carrying capacity of collection vehicles. Obviously, increasing the capacity will increase its cost, but this should be more than paid for by operating savings. Unfortunately, whether or not one can achieve the potential from a new vehicle depends on team size, the area allocated to this vehicle (and therefore the effect on adjacent teams), the distance from the disposal point, and many other factors. Deciding how useful the new vehicle would be and how many would be needed is normally a very complex, laborious, and time-consuming calculation. However, it would only take four or five minutes to prepare the data required to make the program look at this (or any other) alternative, the running time on the computer would be another few minutes, and the elapsed time would be, perhaps, overnight.

## VII. CONCLUSION

Having examined some of the situations in which the computer method could be of help one must now look at the other side of the coin. Clearly there are disadvantages, three in particular. The first is that the data on the city has to be collected and stored. This should be done in any case and whether or not it is depends on management's desire to be able to plan; it has nothing to do with the computer as such. However, a good deal of thought must be given to the exact type of planning desired because this affects the fineness of detail and hence the quantity of data. The second disadvantage is that few cleansing managements have much acquaintance with computers and management scientists. This creates a problem of training and attitudes. It is important that some training in computing and related matters be given right at the start of the development work in order that the managers involved can take part in and fully understand the development process. It is also important to liaize frequently and effectively with top management, while a junior manager works closely with the management science team.

The third disadvantage is the most serious. Round design, in the sense of working out a scheme which will operate in detail in actual practice, is a very subtle process involving chances which experienced planners apparently are unable to describe. Under the circumstances it is impossible to write a computer program which will produce a perfect daily operating scheme. It is highly unlikely that two experienced cleansing superintendents would agree with each other over an operating scheme, let alone with a computer. The computer program can be written to incorporate all the objective factors and to reflect many of the subjective factors which the cleansing superintendent would consider. However, there still remains the problem of convincing them that producing an operating plan for adoption next Monday is *not* the same as medium-term planning of the six-months-to-five-year time horizon.

In conclusion it is contended that the methods described in this paper are valid and valuable for the medium-term planning function. It is hoped that during the next few years more use will be made of them and more applications will be reported.

REFERENCES

1. Coyle, R. G. "The Economic Optimization of Public Cleansing Systems," Ph.D. Thesis, University of Bradford, (1967).
2. Austin, E. "Computer Planning of Refuse Collection in Huddersfield. The Manager's View," *Public Cleansing* (1969).
3. Coyle, R. G. "Computer Study of Some Refuse Collection Problems," *Public Cleansing* (1966).
4. Coyle, R. G. and M. J. C. Martin. "Operations," *Operational Research Quarterly 20:*43 (1969).
5. Nicholson, R. A. J. "A Boundary Method for Planar Traveling Salesman Problems," *Operational Research Quarterly 19(4):*28 (1968).

MODELS FOR ENVIRONMENTAL POLLUTION CONTROL


# CHAPTER 16

# REGIONAL PLANNING MODELS FOR SOLID WASTE MANAGEMENT

Jochen Kühner and Bernard Heiler*

## I. INTRODUCTION

Solid waste processing and disposal has become a fairly significant item in the municipalities' budget due to increased costs associated with tightened landfill and air and water pollution-control standards. These expenses can usually be reduced when several municipalities join together to operate as a region. Where available economies of scale can be utilized (see Figure 16.2) and when many facility site locations can be made available for eventual use, regionalization is a logical consideration. If there is no regional scheme better than the individual schemes, then the optimal regional scheme would be to continue operating on an individual basis.

The region's size is usually determined by the level of government that has (or can obtain) ultimate jurisdiction over zoning. The level of government which has power to zone is important since this power can be invoked to deny the use of land for disposal by other governmental units. This zoning power is exercised when the municipality does not want to be a net-importer of another municipality's refuse because the land suitable for sanitary land fill is a "scarce" resource. If available land is acquired by "outsiders," there is obviously less available for the "insiders," and consequently

---

*Jochen Kühner and Bernard Heiler are environmental engineers with the Environmental Systems Program, Harvard University, Cambridge, Massachusetts 02138, U.S.A.

available land will become more expensive. In New England the towns have this zoning power. In other parts of the United States the jurisdiction can usually rest with the counties. The decision of a municipality having ultimate zoning power to enter into a regional agreement is usually based on its desire to improve its condition. The question of what actually constitutes an improvement of condition is complicated by nonmonetary factors such as land-use policies and environmental pollution. Despite the political problems of obtaining agreement to submit planning and operation responsibility to a regional authority, planning can proceed on a regional basis to determine the advantages that could be expected. In the United States funds for solid waste planning are granted under the conditions that the planning will be done on a regional basis.

*What is meant when we talk about modeling regional solid waste management (SWM)?* SWM is conventionally partitioned into the categories of *collection* and *disposal* and subdivided into the elements of waste generation, storage, pick up, hauling, transfer and/or processing (with or without sale of the residuals), transport, and ultimate disposal. Although all of the elements could be considered in a regional scheme, we will limit our approach to the free-body-cut that includes the elements hauling, transfer and/or processing, transport and ultimate disposal. This limitation is needed to keep the models general enough to be useful in practical applications.

The parameters of waste generation and pick up are exogenous. The area's waste generation is introduced as a particular amount at a point source which is usually the population center of gravity of that area. Originating from this point transport takes place, either directly or via transfer station and/or processing plant, to disposal at a sanitary landfill. If no residuals remain after a process (*e.g.*, the residuals are sold and not disposed of), the transport cost to the landfill is zero. In the U.S. the sale of residuals is not expected to contribute much monetary benefit and is neglected.

Before a literature review on the regional modeling is done, some significant factors of regional solid waste management should be considered:

- general data considerations
- quality of generated solid waste is nonhomogenous and variable
- quantity is variable

- environmental impact
- financial versus economical considerations.

Planning implies looking into the future; data indicate only what has occurred. It is on the basis of this data that future inferences are drawn. To gain insights through the use of mathematical models, thoughtful considerations must be given to the quality of the inputs.

Good waste-quantity and -quality data upon which to base a planning model are desirable. Sadly, however, there are few good data sets available because there has been little incentive to record quantities. If one does find quantities well-documented, the quality is usually not noted. Another problem is that the quantities recorded may be only those received for processing while the portion that went to a land disposal site unprocessed was not recorded. Where all records are kept faithfully they may only reflect the quantities collected, not the amount generated.

Cost and operational capability data is not always certain, but it usually is reliable when it concerns equipment and operations that are in common usage. Costs and capabilities cannot always be reliably predicted for new technologies.

Predictions of refuse quantities and composition and costs and capabilities of equipment and operations increase in uncertainty with the length of time. They are influenced very much by unpredictable changes in regulations, education of the population and advancing technology. Despite these facts, data are treated as if a stationary condition exists.

The composition and the density of solid waste is characterized by wide fluctuations. These have a strong influence on the performance of processing plants. Solid waste is assumed to be a homogeneous material despite what we know about it. This simplifies the models but it also lowers their accuracy.

In the same manner as quality, the quantity of the solid waste generation and processing exhibits strong daily, weekly and seasonal variations. One of the few investigations that studies the statistics was that done by Harrington and Partridge.[1] They indicate that a strong cost-bias could be introduced if only the average values of waste collection per time unit would be considered. Because of the various time-dependent variations, the question of the choice of time units for the SWM set-up seems also to be highly significant and sensitive.

Transporting, processing and disposing of the solid waste creates an environmental impact. To make a good decision on the regional scope we must find ways to include this impact into the investigation of the alternatives. This task seems to be very difficult since there is no professional agreement on quality indices related to SWM. This is unlike indices of air quality and water quality, where $SO_x$ and CO, and DO, BOD, and SS are commonly, if uncritically, used. There are two provisional possibilities for including this impact: one is an *indirect approach* of estimating how much people are willing to pay to avoid noise, air and water pollution and then taking this cost into consideration. The other is the *direct approach* of analyzing every single impact, such as leachate from landfills and particulates from incineration, with available measurements. To avoid the problem of comparing the nonbeneficial impacts, standards could be introduced as constraints which have to be met. The difference in the costs to meet the different levels of standards could also give the planner an indication of the otherwise immeasurable costs associated with personal assessments of environmental impact.

Very often a planning pattern for a regional approach is found from economic considerations only. Unfortunately, the financing aspects are omitted; these actually play an important role in implementing a regional decision. The difference between an *economically optimal decision* and a *financially feasible decision* is too often forgotten. The consideration of financing should not be limited to the first capital investment, but also extended to the operation of the activities. Consideration of financing should be included to add realism.

In order to avoid imposing a large special tax assessment, large projects are usually financed by bonds. The type of bond financing available presents differences in interest rates which may be applicable as a result of capital requirements. Municipalities in the U.S. finance mainly by general obligation bonds or by revenue bonds. The financing aspect in the model could be handled, for example, by working with the general obligation bond interest rate and a constraint on capital cost. This result would be compared with that obtained from revenue bond interest rate use (Note: the revenue bond interest rate is usually higher, but the money to be raised is not limited by statutory debt ceilings).

Other forms of financing, such as assessments based on expected service or effluent contribution, should be compared also and the least costly form chosen.

The discussion that follows deals with existing optimization models but not with simulation models for regional solid waste problems.

## II. LITERATURE REVIEW

The following three groups define the criteria chosen to classify the models in a systematic way for the review of the literature.

1. *solution algorithms:* linear programming models (LP) and their extensions, or integer and mixed integer programming models (MIP)
2. *time horizon considerations:* static models (only 1 point in time is considered) or dynamic models (inclusion of a time horizon to trade-off decisions in time)
3. *location and distance description:* average distance between locations independent of their geographic points or "Lp-Metric" (distance (Lp) = $[|x - x'|^p + |y - y'|^p]^{1/p}$) where the location is given by the coordinates (x,y) and the distance is described according to p: p = 1 gives the $L_1$ metric: distance ($L_1$) = $|x - x'| + |y - y'|$; and p = 2 is the Euclidean distance $[(x - x')^2 + (y - y')^2]^{1/2}$. The $L_1$-metric seems to be justified in cities laid out in rectangular grids and providing L-shaped paths.

Stochastic regional planning models are not considered. It is not intended to imply that such models do not exist, nor is it intended to imply that such models are not important. Table 16.1 is a matrix of the previously described classification, filled in with citations of some of the representative literature which will be briefly reviewed.

A more detailed review of the static-average-location-LP models plus recent advancements in the average-location MIP models will be presented. The detailed description will include a real world problem.

The static (LP) model in the form of an expanded transportation model is the most common type discussed in solid waste literature and most often used in preliminary planning practice. A typical example is the set-up in Figure 16.1 (Kühner[2]). The objective is to minimize the transportation and

*Table 16.1*

*Deterministic Models of Regional SWM*

| | | *L.P.* | *M.I.P.* |
|---|---|---|---|
| Static | average location | expanded transportation models | fixed charge models: Helms and Clark,[3] Marks,[4] Marks and Liebman[5] |
| | $L_1$-metric | Wersan *et al.*[6] | Wersan *et al.*[6] |
| | Euclidean ($L_2$-metric) | | Golueke and McGauhey[7] |
| Dynamic | average location | Harrington[8] | Skelly and Lynn[9] Kühner[2] |
| | Lp-metric | Nothing Found in Literature | |

operations costs subject to constraints like mass balance, continuity, and maximum and minimum capacity. This type of model could be called an *operational* model since the solid wastes collected from point sources are directed to the existing disposal facilities in such a manner that the costs are minimized. A lower bound is introduced on the capacity utilization of the existing plants. This provides dual variables for testing the sensitivity of the system where facilities operate in a range that in reality is characterized by increasing average costs with decreasing capacity utilization. As is seen in Figure 16.2, the total cost function is approximately exponential in form (exponent less than unity). The average cost function is relatively constant for larger usage rates, but the slope increases as the usage rate goes back to zero.

Systematic changes of the lower bound and corresponding constant average cost will give some plants dual-variable-values from which inferences can be drawn about continuation or abandonment of plants.

The University of Louisville[10] approached the problem of *developing* a regional system by minimizing the total costs for the entire system and dropping any facility not utilized to an extent that would

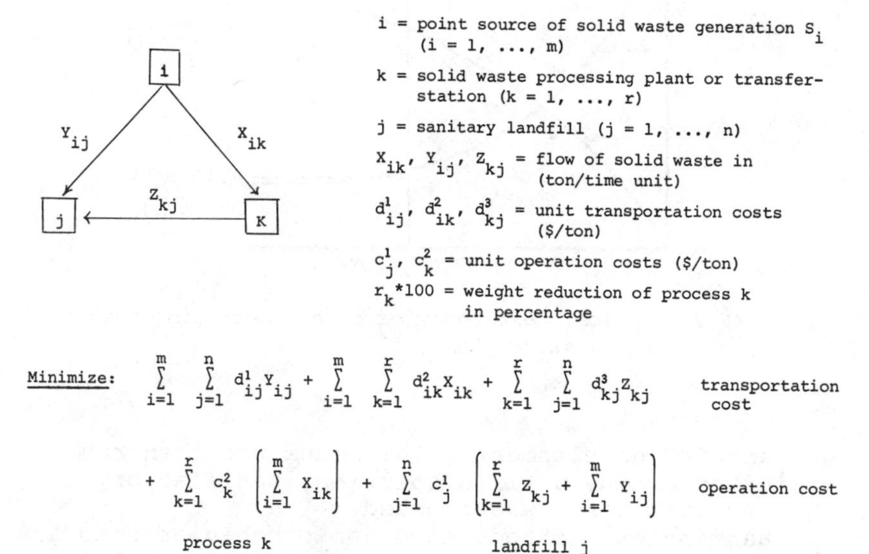


$i$ = point source of solid waste generation $S_i$ ($i = 1, \ldots, m$)

$k$ = solid waste processing plant or transfer-station ($k = 1, \ldots, r$)

$j$ = sanitary landfill ($j = 1, \ldots, n$)

$X_{ik}, Y_{ij}, Z_{kj}$ = flow of solid waste in (ton/time unit)

$d^1_{ij}, d^2_{ik}, d^3_{kj}$ = unit transportation costs (\$/ton)

$c^1_j, c^2_k$ = unit operation costs (\$/ton)

$r_k*100$ = weight reduction of process k in percentage

<u>Minimize:</u>

$$\sum_{i=1}^{m}\sum_{j=1}^{n} d^1_{ij}Y_{ij} + \sum_{i=1}^{m}\sum_{k=1}^{r} d^2_{ik}X_{ik} + \sum_{k=1}^{r}\sum_{j=1}^{n} d^3_{kj}Z_{kj} \qquad \text{transportation cost}$$

$$+ \underbrace{\sum_{k=1}^{r} c^2_k \left(\sum_{i=1}^{m} X_{ik}\right)}_{\text{process } k} + \underbrace{\sum_{j=1}^{n} c^1_j \left(\sum_{k=1}^{r} Z_{kj} + \sum_{i=1}^{m} Y_{ij}\right)}_{\text{landfill } j} \qquad \text{operation cost}$$

<u>Subject to:</u>

(1) $$\sum_{j=1}^{n} Y_{ij} + \sum_{k=1}^{r} X_{ik} = S_i$$ $i = 1, \ldots, m$
mass balance of $S_i$ in area i

(2) $$(1 - r_k) \sum_{i=1}^{m} X_{ij} - \sum_{j=1}^{n} Z_{kj} = 0$$ $k = 1, \ldots, r$
modified "Kirchhoff-Node-Law" for process k

(3) $$\sum_{i=1}^{m} X_{ik} \leq CP_k$$ $k = 1, \ldots, r$
maximum capacity of process k

(4) $$\sum_{i=1}^{m} X_{ik} \geq MCP_k$$ $k = 1, \ldots, r$
minimum load (process k)

(5) $$\sum_{i=1}^{m} Y_{ij} + \sum_{k=1}^{r} Z_{kj} \leq L_j$$ $j = 1, \ldots, n$
maximum capacity of landfill j

(6) $$X_{ik}, Y_{ij}, Z_{kj} \geq 0$$ nonnegativity of variables

*Figure 16.1. Static average-location LP-model.*[2]

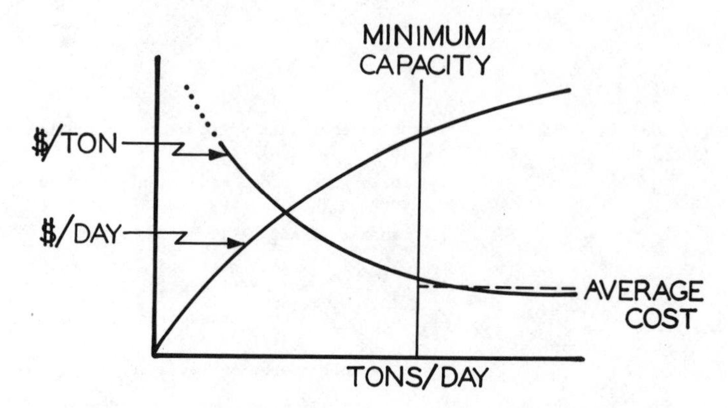


*Figure 16.2. Typical cost-behavior of the operation-cost of an existing plant.*

warrant its development. The model was then run again and again, until a solution, satisfactory to the investigators, was reached.

Harrington[8] introduced dynamic behavior into a one-source model when formulating the LP model in a way to minimize the *present value* of transport-, processing-, disposal and capacity expansion costs (of an incinerator) over a time horizon of T time units (see Figure 16.3). This model is an investment model since in every time period a decision has to be made whether the incinerator should be expanded to take account of the increased amount of the solid waste or if this amount should be distributed to available landfills.

Harrington assumes an exponential growth rate for the solid waste generation. The time is handled in a discrete way, so that for every time period mass balance, continuity, and capacity constraints must be considered. Although the incinerator cost can be thought of as a concave function, a linear function is used since the economies of scale are very small, especially when replication is used for capacity expansion (*e.g.*, a 450 tpd incinerator would more likely be 3-150 tpd furnaces than only 1-450 tpd furnace and the expansion would be of increasing to 4-150 furnaces rather than modifying 3-150 tpd furnace to become 3-200 tpd furnaces). Land policy is assumed fixed for the time horizon of T time units, although this seems to be one of the critical points in long term considerations of solid waste in addition to the question of sharply

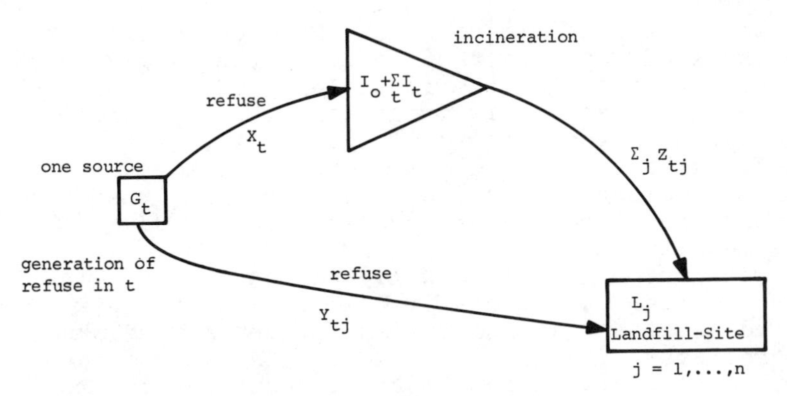


t = 1, ..., T (number of time-periods)

d = interest rate

variables have the dimension (ton/time unit)

<u>Minimize</u>: $PV = \sum_t (1 + d)^{-t} a_t I_t$ — linearized costs for incinerator increments $I_t$

cost for transport and process where the $d_t$, $d^1_{tj}$ and $d^{11}_{tj}$ are the unit costs:

$$+ \sum_t (1 + d)^{-t} d_t X_t$$

$$+ \sum_t (1 + d)^{-t} \sum_j d^1_{tj} Y_{tj}$$

$$+ \sum_t (1 + d)^{-t} \sum_j d^{11}_{tj} Z_{tj}$$

<u>Subject to</u>:

$$\sum_j Z_{jt} - \cdot 15 X_t = 0 \qquad t = 1, \ldots, T$$

(reduction of weight is assumed 85%)

$$\sum_j Y_{tj} - (P_t - X_t) = 0 \qquad t = 1, \ldots, T$$

$$X_t \leq I_o + \sum_{k=1}^{t} I_k \qquad t = 1, \ldots, T$$

$$\sum_t (Y_{tj} + Z_{tj}) \leq L_j \qquad j = 1, \ldots, n$$

All variables $\geq 0$

*Figure 16.3. Dynamic 1-source model.*[8]

advancing technologic process. The dual variables are the shadow prices that indicate the land-value of the landfills for the fixed land use policy.

Before attention is focused on the inclusion of a fixed charge into the above models, we should briefly investigate the $L_1$- and $L_2$-(Euclidean)-metric models that are used for describing the distance and location of sources, sinks, and intermediate facilities.

With the $L_1$-metric model, the basic set-up is that a number of sources (i = 1, ..., M) are given (coordinates are $a_i$ and $b_i$) and a number of sinks (coordinates $x_j$ and $y_j$)(j = 1, ..., N) are sought which will minimize the distance under the assumption that the load from each source is indivisible:

$$\text{Minimize} \quad \sum_{i=1}^{M} \sum_{j=1}^{N} \alpha_{ij}(|x_j - a_i| + |y_j - b_i|)$$

$$\text{Subject to} \quad \sum_{j=1}^{N} \alpha_{ij} = 1 \quad \forall_i,\ i = 1, \ldots, M;$$

$$\alpha_{ij} = \{0,1\}$$

If one particular pair of source and sink are chosen to act together, $\alpha_{ij} = 1$; otherwise 0. This means that we have an integer programming problem. The constraint insures that the quantity originating at a source is directed in its entirety to a specific sink.

Wersan *et al.*[6] dealt with single- and multiple-sink problems in such a way that the coordinates of the sinks became variables also. They defined absolute value variables ($u_i^+, u_i^-$) and $v_i^+, v_i^-$) as the positive and negative parts respectively of the functions $x - a_i$ and $y - b_i$ (in the one sink problem) to get rid of the difficulties arising from the absolute value of the distance in the $L_1$-formulation. To make the set up even more realistic, constraints are introduced that exclude certain areas k from being chosen as sinks. Thus a *one-sink* problem (x,y) can be set up as an *LP*-problem:

$$\text{Minimize} \quad \sum_{i=1}^{M} (u_i^+ + u_i^- + v_i^+ + v_i^-)$$

$$\text{Subject to}\quad \left.\begin{array}{l} u_i^+ - u_i^- = x - a_i \\ v_i^+ - v_i^- = y - b_i \end{array}\right\} \quad i = 1, \ldots, M$$

$$c_k x + d_k y \le e_k \qquad k = 1, 2, \ldots, K$$

$$\text{all variables} \ge 0$$

The set-up neglects operation costs; the transportation unit-costs are assumed to be constant for *all choices.*

Once one returns to a two or more sinks problem, the integer variable $\alpha_{ij}\{0,1\}$ as allocation variable of load i to sink j has to reenter the problem. Because of the combined use of the integer and the continuous variables, we have a nonlinear objective function and linear constraints:

$$\text{Minimize}\quad \sum_{i=1}^{M} \sum_{j=1}^{N} \alpha_{ij}(u_{ij}^+ + u_{ij}^- + v_{ij}^+ + v_{ij}^-)$$

$$\text{Subject to}\quad \left.\begin{array}{l} u_{ij}^+ - u_{ij}^- = x_j - a_i \\ v_{ij}^+ - v_{ij}^- = y_j - b_i \end{array}\right\} \quad \begin{array}{l} \forall_i, \forall_j \\ i = 1, \ldots, M \\ j = 1, \ldots, N \end{array}$$

$$\sum_{j=1}^{N} \alpha_{ij} = 1, \ \forall_i, \qquad i = 1, \ldots, M$$

$$\text{all variables} > 0 \qquad \alpha_{ij} = \{0,1\}$$

Wersan *et al.*[6] found a process resting on the nonlinear requirement $\alpha_{ij} = 0$ or 1 only, to solve this nonlinear problem by quasilinearization.

In summary, the criterion for locating the disposal sites j and allocating the indivisible load i to one of the j's is the minimization of the transport-distances; no transportation cost depending on the actual load and no operation costs and no capacity constraints are considered.

The Berkeley Group[7] calls the problem "Optimal Activity Locations": locating a treatment plant site or disposal site or a group of "M" sites to serve "N" sources (waste generators) in the most economical manner. The mathematical form of the problem is:

Given a set of generation points N, their locations, and their loads $w_i$ ($i = 1, \ldots, N$).
Find the disposal sites $(x_j, y_j)$ ($j = 1, \ldots, M$) by

Minimizing $$F(x,y) = \sum_{i=1}^{N} \sum_{j=1}^{M} \alpha_{ij} w_i \ell_{ij}$$

Subject to $$\ell_{ij} = [(a_i - x_j)^2 + (b_i - y_j)^2]^{k/2}$$

$$\alpha_{ij} \{0,1\}$$

with $\alpha = 1$ if generation point i is assigned to site j; otherwise 0. A general power k considered in the distance (or cost) expression could allow treatment of all possible cases where the power of cost is not unity. Two general assumptions have been made finally:

1. the unit transportation costs are independent of the amount transported
2. there is no capacity restriction on the sinks. We should note that the sinks are given through their coordinates while in the above models the coordinates were variables.

A branch-and-bound algorithm was developed to find optimal results for reasonably sized problems without the necessity for investigating every possible assignment of sources to destinations. The solution procedure can be modified to include the fact that some disposal sites are already existing and operating and that only a few additional ones may be required. The solution may also be modified to account for capacity constraints at some sites. The problem of locating intermediate sites can only be solved after the optimal location of the sinks are found. This program has been developed for use in a study of the nine San Francisco Bay area counties, a region containing 120 cities, 75 operating disposal sites, and 15 potential disposal sites. Results of the application of the model are not yet available.

To find the logical transition to the fixed-charge models at this point of the review, it is necessary to return briefly to the description of the $L_1$-metric models of Wersan *et al.*[6] Their first set up for one sink included no restriction for the location of the sink on the plane; this set-up was extended by eliminating certain areas from consideration, first for the one-sink problem and then for the multi-sink problems. By recognizing the fact that only a few parcels are available for solid waste use, the next logical step in changing the procedure would be to set up the problem in such a way that the best sites would be chosen from a set of sites eligible for landfilling or processing solid waste. Hence, the original problem is extended: some locations are eligible for landfilling or processing solid waste and some point-sources are given. The question: which sites should be picked for the activities out of the set of eligible sites and how should the loads be distributed in order to minimize cost? This final set-up was the form investigated by several persons, among whom were Marks,[4] Marks and Liebman,[5] and Helms and Clark.[11] This form has now evolved to the Fixed Charge Model by including the costs that are fixed but independent of the level of activity as well as the costs that depend on the amounts transported and processed. These fixed charges represent capital costs such as site acquisition and construction. The fixed costs are completely independent of design capacity and thus inflexible to design capacity changes.

Marks[4,5] made an important contribution in his work on fixed charge models which he set up as location allocation models to establish the location of transfer-stations between sources of solid waste and the sinks. This problem is represented by a model for determining the intermediate nodes that should exist in a transshipment network that has fixed charge functions and capacity constraints at each intermediate node. Linear cost relations with respect to the amount transported and processed are used.

Choice of locations depends on minimization of an objective function which considers all costs (including the fixed charge). These are the charges relating to transporting solid waste, constructing and operating transfer-stations, and disposal operations. This mathematical set up is shown in Figure 16.4. It should be noted that sufficient dummy facilities with $F_i = 0$ can be established to guarantee alternate paths, where transshipments are not utilized. Marks' basic assumption was that all refuse was transshipped. A branch-and-bound algorithm is used to find the optimal solution.

Objective: Minimize the total cost of facilities and transshipment

Minimize:

$$\sum_{i=1}^{m} F_i y_i + \sum_{i=1}^{m} \sum_{j=1}^{n} c^*_{ij} X^*_{ij} + \sum_{i=1}^{m} \sum_{k=1}^{p} c^{**}_{ki} X^{**}_{ki}$$

Subject to:

(1) $\sum_{i=1}^{m} X^{**}_{ki} \geq S_k$ $\quad k = 1,\ldots,p$: mass-balance at all *p sources*

(2) $\sum_{j=1}^{n} X^*_{ij} = \sum_{k=1}^{p} X^{**}_{ki}$ $\quad i = 1,\ldots,m$: mass-balance at *intermediate sites*

(3) $\sum_{k=1}^{p} X^{**}_{ki} \leq Q_i y_i$ $\quad i = 1,\ldots,m$: upper bound on the capacity of the *intermediate site*

(4) $D^u_j \geq \sum_{i=1}^{m} X^*_{ij} \geq D^l_j$ $\quad j = 1,\ldots,n$: upper and lower bounds on the *n sinks*

(5) $X^*_{ij}, X^{**}_{ki} \geq 0$ (flow of the material); $\quad y_i = \begin{cases} 1, & \text{if the } i\text{th facility is built;} \\ 0 & \text{otherwise} \end{cases}$

where:

$c^*_{ij} = c_{ij} + R_j$ = unit cost (\$/unit) associated with a transfer of solid waste from facility i to sink j

$c_{ij}$ = unit shipping cost ($i \rightarrow j$) (\$/unit)

$R_j$ = unit variable cost (\$/unit) associated with using sink j

$c^{**}_{ki} = c'_{ki} + T_k + V_i$ = unit cost (\$/unit) associated with transfer of material from source k to facility i

$c'_{ki}$ = unit shipping cost ($k \rightarrow i$)(\$/unit)

$T_k$ = unit variable cost associated with using source k (\$/unit)

$V_i$ = unit variable cost associated with using facility i (\$/unit)

$F_i$ = *fixed charge* for establishing facility i (\$)

$S_k$ = solid waste (tons/time unit) generated at point source k

m = number of proposed intermediate sites

*Figure 16.4. Marks[4] fixed-charge, transshipment, facility location model.*

Marks' models are interpreted as being static since they do not include a time horizon. Trade-offs over time are not considered.

Skelly and Lynn[9] introduced a time horizon T with discrete time units (t = 1, ..., T) into their fixed charge model for considering dynamic behavior. All parameters can be described relative to time as well. As discussed before, this is an advantage since most aspects of solid waste involve dynamic changes; solid waste generation usually increases with time, while certain mechanical operations could have decreasing costs because of improved technology.

This can result because trade-offs with reference to the combined use of various processing and disposal facilities occur over time. The total capital cost for a processing facility is set up as a fixed charge plus a cost linearly depending on the actual design capacity of the plant.

Skelly and Lynn also use the fixed charge model to incorporate the concave character of the cost functions of operations such as incineration and landfilling. The cost function can be approximated by two linear segments as shown in Figure 16.5.

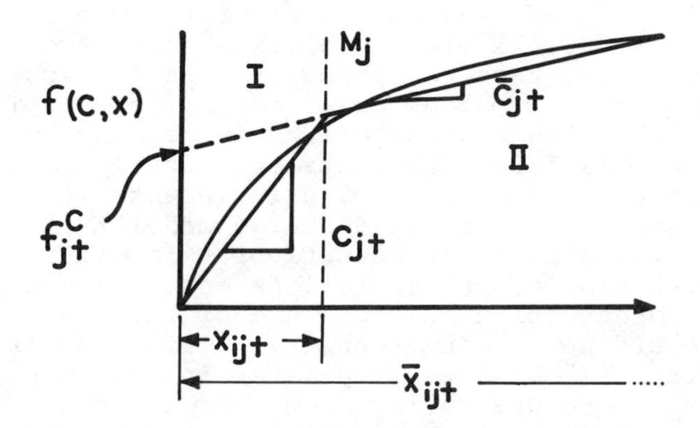


*Figure 16.5. Use of the fixed charge model for concave functions.*[9]

This technique requires that each segment be treated as a separate facility. The two segment example of Figure 16.5 would be considered as either a small site with a linear cost function ($c_{jt}$) and maximum size constraints of up to $M_j$ or as a larger site with fixed cost plus a linear cost function ($\bar{c}_{jt}$) and a minimum size constraint of $M_j$. The two sites are really only one since "one or the other, but not both" would exist. The existence of either will depend on whether the assigned load is greater than or less than $M_j$.

The objective function to be minimized combines all fixed charges and variable charges of all activities at all time periods. This set-up gives us the flexibility to include the dynamics of solid waste generation rates but it does not permit dynamic investment choices. If a facility is to be built, it would have to be built at a predetermined time, usually at the initial time period.

Skelly and Lynn used the Walker algorithm[12] to solve this fixed charge problem. Unfortunately, this algorithm is only a heuristic algorithm and therefore gives solutions that quite often may be different from the true optimal solution. Helms and Clark[3] applied the Walker algorithm to a cost minimization problem, but their solution was not optimal. Liebman[13] compared the nonoptimal solution with the optimal solution obtained by total enumeration. The Walker algorithm was found to have given a value significantly above the optimal solution.

In larger problems, total enumeration is not possible; a branch-and-bound algorithm can be used. It screens out all of the solutions that are dominated (and thus nonoptimal in the particular model set and the accompanied assumptions). The added cost of such an algorithm is usually justified since its added running costs are usually less than the expected saving upon plan implementation.

Kühner[2] set up a fixed charge model that permits dynamic investment considerations. In order to keep the model simple, the operations costs were kept linear; otherwise a multiple use of the fixed charge model would have been necessary. This model's extension will be discussed later in this paper.

The literature review shows the factors of regional solid waste management which are considered important and which have been modeled. Of note is that almost all of the models are of network type and constructed deterministically. No attention is paid to the variability of refuse or the processes. The waste flow is described in flow per time unit with the time units set at years or greater which avoids recognizing the shorter term variations.

There was no recognition for environmental impact analysis, even for the indirect approach for including the environmental impact. Most authors considered only the more readily quantifiable business-cost factors.

There are additional data uncertainties. The quality of the solid waste was considered to be both consistent and homogeneous, whereas in reality it is not of any consistent quality and is heterogeneous. Compounding the data problems is the question of what measure should be used. There is nearly universal agreement that weights or volumes or both should be used. It is common practice to assume that these figures can be nearly interchangeable at certain degrees of compaction. The flow of solid waste is modeled almost exclusively as weight-per-time-unit but never as volume-per-time-unit; why this is the case is a question that should be investigated.

Transport is dependent on compacted volume. Weight could be a factor where axle-load restrictions prevent full utilization of the equipment. New equipment on the market promises 600 $kg/M^3$ with American household refuse. Old equipment provides only 300 $kg/M^3$ of compacted density. The assumption that weight is a more consistent measure is true in a sense, but it does not take into account *moisture*, a factor that increases specific weight but usually decreases volume.

Processes other than transfer depend more on quantities of particular types of refuse than on total quantities. Incineration depends on thermodynamic properties such as moisture content, BTU content, and heat release rates. Composting depends on the types of organic materials available.

Final disposal depends on compacted volumes and ultimate volumes. The raw refuse usually has a compacted density of over 300 $kg/M^3$ but can settle to over 600 $kg/M^3$. The incinerator residue has a compacted density of over 600 $kg/M^3$ and settles less.

Average relations are calculated from time to time for the conversion factors. Actually, weighted refuse quality factors are needed. Measurable indexes that relate to the refuse composition would have to be determined. A set of regression relations for each element could be used: several factors will influence the output at different elements in different ways. However, this has the unfavorable property of changing the problem into a nonlinear set up.

Because of these difficulties refuse weight will have to continue to be the characteristic that describes flow. It is hoped that a method of using volumes and weights interchangeably will be found.

## III. STATIC LP-MODEL

The static average distance LP-model can be used for today's usual regional transportation and processing problems. It (Figure 16.1) is further generalized to include more of today's recognized Processing and Disposal alternatives, some of which are:

1. direct delivery from source i to landfill j (solve by program in Figure 16.1 with K = 0).
2. transfer-stations are put into the flow between the sources i and the landfills j (the exact set-up of Figure 16.1).
3. a processing plant, *e.g.*, an incinerator, exists. This is the same problem as (2) if the incinerator replaces the transfer-station. If the landfill site is adjacent to the processing plant, either the amount of land filled to maximum capacity of the processing plant can be limited (then the costs of two operations can be included into one cost-term) or the landfilling operation can be larger than the processing plant (then the two activities are considered as two nodes and the model cited above would be applicable).
4. The transfer-station (node between source i to landfill j) includes an on-site process. We can use the same set-up as above if the two processes have the same capacity. If we assume that the flow through the transfer-station can be higher than through the on-site process we have to introduce a new node. Thus the basic set-up is changed.
5. The Three-Haul-System consists of transport from the collection area source to a transfer (and/or processing) site, then to a processing plant (*e.g.*, incinerator or composting plant) and then to a landfill. Alternatives 4 and 5 give rise to the kind of flow diagram in a regional scheme depicted in Figure 16.6.

To model this set-up is a simple extension of the model in Figure 16.1 (see formulation in Figure 16.7).

Engineers and planners, with the aid of Operations Research people, should be able to extend the models to include all the alternatives available in regional solid waste management. These models can also be extended to include more constraints; for example, transportation limitation can be put on certain routes in order to confine the frequency of the traffic on a small road to a technically and socially accepted value. Dummy transfer-stations

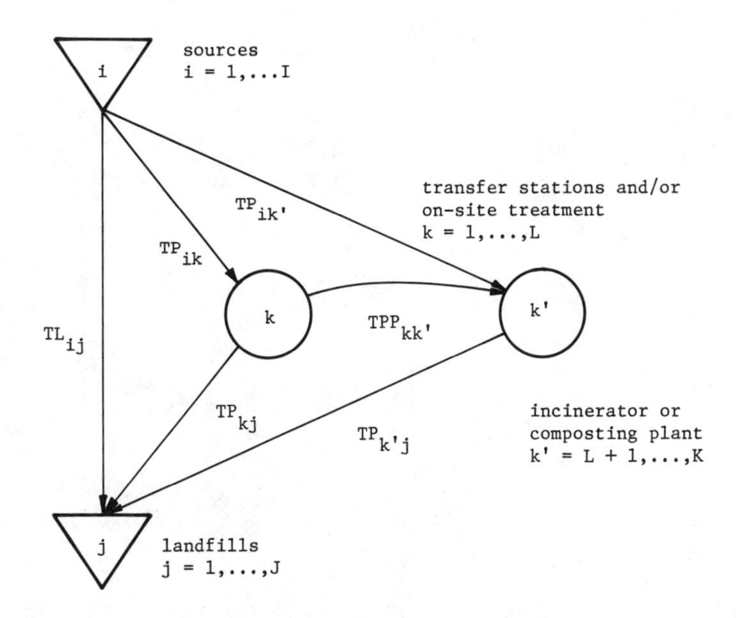


*Figure 16.6. Flow diagram of Three-Haul System.*

can be set up as a way for expressing alternative paths which may be more costly than the main path. These alternatives might be worthwhile to consider since the main path has a traffic constraint (*e.g.* pipeline capacity or socially acceptable traffic level). The static LP-model also can be extended and serve as an introduction to fixed charge problems and mixed integer programming.

A problem arises when regionalization is contemplated to take over the existing facilities of many independent operations. Some of the existing activities may need to be phased out as economically infeasible, and associated with the phasing out will be costs such as dismantling an incinerator or recultivating over a dump. Since most discontinued operations are expected to be old dumps where the costs are recultivation fees, the designation "CF" is used.

The objective function is supplemented by the term $\sum_{j=1}^{P} CF_j * y_j$, the fee for all dumps P under consideration of being closed. The $y_j$ term is an

Minimize:

$$\sum_i \sum_j TL_{ij} \; C^{T1}_{ij} + \sum_i \sum_k TP_{ik} * C^{T2}_{ik} + \sum_k^{L} \sum_{k'=L+1}^{K} TPP_{kk'} * C^{T3}_{kk'} + \sum_k \sum_j TPL_{kj} * C^{T4}_{kj} \quad \text{transport}$$

$$\sum_j \left(\sum_i TL_{ij} \; C^{OL1}_{ij} + \sum_k TPL_{kj} * C^{OL2}_{kj}\right) \quad \text{landfill operation}$$

$$\sum_i \left(\sum_k^{L} TP_{ik} * C^{OP}_k\right) + \sum_{k'=L+1}^{K} \left(\sum_i TP_{ik'} * C^{OP1}_{ik'} + \sum_{k=1}^{L} TPP_{kk'} \; C^{OP2}_{kk'}\right)$$

operation cost

Subject to:

(1) $$\sum_{k=1}^{K} TP_{ik} + \sum_j TL_{ij} = SW_i \qquad i = 1, \ldots, I$$

(2) $$\sum_i TP_{ik} (1 - r_k) = \sum_{k'=L+1}^{K} TPP_{k'k} + \sum_{j=1}^{J} TPL_{kj}$$

$k = 1, \ldots, L$

(3) $$\left(\sum_i TP_{ik'} + \sum_{k=1}^{L} TPP_{kk'}\right)(1 - r_{k'}) = \sum_{j=1}^{J} TPL_{k'j}$$

$k' = L+1, \ldots, K$

Further: Capacity constraints (similar to Figure 16.1)

Nonnegativity of all variables

*Figure 16.7. LP-formulation of the Three-Haul-System.*

integer variable: {0,1} with $y_j = 1$ if recultivation fee is applicable (facility is to be closed) and $y_j = 0$ if the fee is not applicable. Thus the LP-set-up became a linear MIP-problem. The model must have the following additional constraint:

$V_j - (1 - y_j)\, L\, MAX_j \leq 0$ to prevent inflows $V_j$ to the facility j with a maximum capacity L $MAX_j$ if it is in fact incurring a closing cost.

Further constraints could be introduced as

$\sum_{j=1}^{N} y_j \leq (N - 1),$ that at least one of the N facilities remains in use or be upgraded or

$\sum_{j \varepsilon e_i} y_j \leq 1,$ that only one facility per municipality or other subdivision will remain in operation (to prevent concentrating many facilities in only some of the towns).

Consideration of the problems concerning regional solid waste management showed that extensions to the simpler static average location model would be strongly desirable. Unfortunately many of the desired extensions cannot be solved by an LP but would also require solution by mixed integer programming.

## IV. DYNAMIC INVESTMENT MODEL

The dynamic investment model (see also Appendix, p. 356) is in contrast to the static model because it introduces time as a state variable. It can be set up to include a time horizon and thus allows possible trade-offs over time to be considered because the solid waste facility investment's timing is included in the Present Value (PV) criterion. Some ideas on how to model decisions affected by a changing waste load with respect to time will be illustrated.

The basic set-up is known also as a Mixed Integer Programming (MIP) Model. Actual computational work could be solved by the MPSX-MIP of the IBM-360 system (a linear mixed-integer programming program). The basic assumptions for the model presented are:

-- Future information on the existing and future projects can be transformed into a cash-flow of costs and revenues. Utility values associated with the projects do not change. The information transformations are deterministic.
-- Interactions between other type public works' projects and their investments are omitted.
-- The technologic future that is assessible is represented in the cost terms.
-- The interest rate used for Present Value calculations is determined by the prevailing present rate of interest for bonds of the type that would be sold. These rates are deterministic.
-- Replacements are not considered. The question of durability of process technologies is not relevant and thus neglected.
-- Land use policy is fixed: assumption that land available at t = 0 is also available at t = T.
-- There is no "warehousing" of "disposal-site supply" allowed.
-- No storage is allowed within a city.

The scope of our investment decisions can be identified and will be listed below. A number of landfills ($e_j$) and plants ($e_k$) are presently (at time = 0) in operation.

-- At what time $t_k^*$ should we invest in which process k and at what time $t_j^*$ in which landfill j?
-- If a maximum potential capacity can be designed for a location, what should the actual capacity be? Should this capacity be installed immediately or part initially ($DC_{kt}$) and part as capacity expansions ($EXP_{kt}$) at later times?
-- What benefits are foregone by the use of a location for solid waste disposal?

The last question can be answered by considering these benefits foregone as a rent $R_{kt}$. These $R_{kt}$ can be positive or they may be negative in the sense that the use of the land would be an improvement, *i.e.*, an "attractive nuisance" such as a quarry getting filled or a flat area becoming a landscaped hill. This rent $R_{kt}$ can also be used for the indirect approach of assessing environmental impact. The rent implied by $R_{kt}$ that is needed to justify an *a priori* decision not to use the site is an indirectly assessed impact cost.

The two earlier questions will be answered in the basic model description.

An investment made can be viewed as consisting of two parts, the first of which is a fixed charge independent of the chosen capacity. The second part is a capital cost which is linearly dependent on the initial design capacity $DC_{kt}$ chosen.

The design capacity, $DC_{kt}$, at a location is a decision variable limited by a maximum capacity at the location. If the design capacity is less than the maximum allowable capacity, then there is still an option for future capacity expansion. This increment of expansion of capacity at any time will be limited by the site size restriction. Expansion should occur conditional to the appropriate initial site development. Initial site development done prior to the start of the planning period (*i.e.*, existing facilities) is recognized.

The objective function is to minimize the present value (PV) of time path costs, time-operations costs and time-fixed-charge costs minus the rental income of the sites up to a solid waste management development. The source for sanitary landfill costs for each time period consists of the path amounts multiplied by the linear path costs and the present value multiplier.

Landfill costs are made up first of the amount landfilled multiplied by the landfill operation cost and the present value operator, and second, of the rental income up to the year $t_j^* - 1$ and the fixed charge at the year $t_j^*$ the operation is established.

The intermediate processing plant's costs consist of the elements similar to those of landfill costs: transport, fixed costs and operations costs, plus an additional capital investment-charge that depends on the design capacity DC. Expansion is allowed in the first time period after the first investment and at every time for plants already in operation. The path costs of the residue to the landfill are also represented.

The format of only two sets of alternatives of transport from the source [(1) to the landfill (TL) or (2) to the intermediate processing site (TP) with residue continuing to the landfill (TPL)] was chosen for simplicity. Any weight reduction due to intermediate processing is accounted for by r, and inflow-outflow balance of an incinerator node considers this factor (see constraint 10).

Time dependence is shown by the subscript t. The factor $\beta_t$ is the present value factor (1 + rate of interest)$^{-t}$. The first capacity DC is decided upon only once per location and thus is time dependent in the model formulation through $DC_{kt}$ and by its cost coefficient $c_{kt}^{cp}$. If the construction of a new processing plant at the site k is more appropriate than the expansion of the previously developed plant, the site k should be artificially divided for the model formulation in order to include more than one fixed charge for each site.

On the following pages the model is set up in such a way that it can be programmed for the mixed integer program of IBM (MPSX-MIP). Note that because of simplicity, the rental income from a site not used for solid waste activities is combined with the initial fixed investment FC, in t to the fixed charge $F_t$; thus

$$F^L_{jt} = \beta_t * FC^L_{jt} - \sum_{e=1}^{t-1} \beta_e'^{-1} * R_{je} \qquad \text{(landfill j)}$$

and

$$F^P_{kt} = \beta_t * FC^P_{kt} - \sum_{e=1}^{t-1} \beta_e'^{-1} * R_{ke} \qquad \text{(processing site k)}$$

Hence:

<u>Minimize P.V.</u>

$$\sum_{t=1}^{T} \beta_t \ (\sum_i \sum_k C^{T^1}_{ikt} \ TP_{ikt} \qquad \text{transport cost: source to process}$$

$$+ \sum_i \sum_j C^{T^2}_{ijt} \ TL_{ijt} \qquad \text{source to landfill}$$

$$+ \sum_k \sum_j C^{T^3}_{kjt} \ TPL_{kjt}) \qquad \text{process to landfill}$$

$$+ \sum_{t=1}^{T} \beta_t \sum_j C^{OL}_{jt} \ (\sum_i TL_{ijt} + \sum_k TPL_{kjt}) \qquad \textit{operation cost} \text{ (landfill)}$$

$$+ \sum_{t=1}^{T} \sum_{j=1}^{n-e_j} F^L_{jt} * y^L_{jt} \qquad \text{fixed charge (landfill)}$$

$$+ \sum_{t=1}^{T} \beta_t \Sigma_i \Sigma_k C^{OP}_{kt} \ TP_{ikt} \qquad \textit{operation cost} \text{ (plant)}$$

$$+ \sum_{t=1}^{T} \sum_{k=1}^{r-e_k} F_{kt}^{P} * y_{kt}^{P}$$ fixed charge (plant)

$$+ \sum_{t=1}^{T} \sum_{k=1}^{r-e_k} \beta_t \, (C_{kt}^{CP}) \, DC_{kt}$$ capital investment depending on the decision variable "design capacity, DC" in the investment period $t_k^*$ (plant); *note:* $DC_{kt}$ in (ton/day)

$$+ \sum_{t=1}^{T} \beta_t \sum_{k=1}^{r} C_{kt}^{exp} \, EXP_{kt}$$ linear cost for capacity expansion of k that occurs after the initial capacity is installed. *note*: $EXP_{kt}$ in (ton/day)

<u>Subject to</u>:

(1) $$\sum_k T \, P_{ikt} + \sum_j T \, L_{ijt} = S_{it} \qquad k = 1, \ldots, m; \quad t = 1, \ldots, T$$

mass-balance at the point sources

(2) $$\sum_{t=1}^{T} y_{kt}^{P} \leq 1; \quad \sum_{t=1}^{T} y_{jt}^{L} \leq 1 \qquad k = 1, \ldots, r-e_k \quad j = 1, \ldots, n-e_j$$

only 1 fixed charge has to be paid in T time-units at $t_j^*$, $t_k^*$ (only for $n-e_j$ landfills and $r-e_k$ plants, not in operation at t=1)

(3) $$DC_{kt_k^*} + \sum_{t=t_k^*+1}^{T} EXP'_{kt_k^*t} \leq MC_k * y'_{kt_k^*} \qquad k = 1,\ldots,r-e_k \quad t_k^* = 1,\ldots,T$$

the design capacities before $t_k^*$ and the expansion before $t_k^* + 1$ have to be zero; the prime denotes the time adjustment of the variables [ton/period]

(4) $$\sum_{t=1}^{T} EXP'_{kt} \leq MC_k - DC_k \qquad k = r-e_k+1, \ldots, r$$

the existing capacity and their expansions have to be less than the maximum capacity of the location

$$(5)\quad -\sum_i TP_{ikt} + \sum_{t'=1}^{t} (DC'_{kt} + EXP'_{kt}) \geq 0 \qquad t = 1, \ldots, T \quad k = 1, \ldots, r$$

the inflow into the plant has to be less than the existing capacity (for existing plants DC is an exogeneous variable; for the others a decision variable)

$$(6)\quad EXP_{kt} = \sum_{t_k^*=1}^{t-1} EXP_{kt_k^*t} \qquad k = 1, \ldots, r-e_k \quad t = 2, \ldots, T$$

Since the expansions are indexed according to the period of the site's initial development and the remaining periods up to T, only one term of the sum on the RHS will be different from zero due to the constraint 3.

$$(7)\quad \sum_{t=1}^{T} DC_{kt_k^*} \geq MINDC_k * \sum_{t=1}^{T} y_{kt_k^*} \qquad k = 1, \ldots, r-e_k$$

the first capacity has to be greater than a minimum value

$$(8)\quad \sum_{t=1}^{t_j^*} (\sum_k TPL_{kjt} + \sum_i TL_{ijt}) - MCL_j * \sum_{t=1}^{t_j^*} y_{jt} \qquad j = 1, \ldots, n-e_j \quad t_j^* = 1, \ldots, T$$

the sum of all material landfilled between $t_j^*$ and T has to be less than the maximum capacity of the landfill (estimated at t = 1)

$$(9)\quad \sum_{t=1}^{T} (\sum_k TPL_{kjt} + \sum_i TL_{ijt}) \leq MCL_j \qquad j = n-e_j+1, \ldots, n$$

for the landfills already in operation at t=1

$$(10)\quad \sum_i TP_{ikt} (1 - r_k) - \sum_j TPL_{kjt} = 0 \qquad k = 1, \ldots, r \quad t = 1, \ldots, T$$

modified Kirchhoff-Node-Law for the plants, whereby $r_k$ describes the weight-reduction

(11) all variables $\geq 0$; $y = \{0,1\}$

This model shows how inferior investment decisions in a regional solid waste management system can be screened out according to economic criteria if a mathematical optimization algorithm is used with certain simplifying assumptions. This economic approach still lacks considerations about the financing situation of the regional agency or the participating municipalities.

The usual financial alternatives for capital-investments have been mentioned; an attempt to formalize the financing in a mathematical sense is shown. These considerations will not be included in the dynamic investment model itself, since some inconsistencies have yet to be eliminated.

Bonds, the main financing means, can be issued at an initial time and the capital kept available until it is needed. Retaining the proceeds of the bond sale in reserve can be done with no real monetary risk since the unspent proceeds of the bond sale could be deposited in a bank and draw interest to pay the interest due to the bond holders. The unfortunate feature is that the amount of bonds outstanding (Bonded Indebtedness) could be high enough to affect the ability of the governmental unit to float other bond issues that would be ear-marked for other needs. We should rather assume that the solid waste management activity will have a limited allowable Bonded Indebtedness at an attractive interest rate.

Another assumption is that each capital project will be funded by issuance of the exact amounts of bonds in a series that is paid off in equal annual increments (any pay-off schedule is actually possible). One also assumes unlimited borrowing power at increased interest rates.

The basic condition just described is shown below.

$$\text{Bonds outstanding at } t = \sum_{\tau=0}^{t} I_\tau \frac{\gamma_\tau - (t-\tau)}{\gamma_\tau} \delta(\tau)[1 - \delta(\tau + \gamma_\tau)]$$

where:

$\tau$ is time that investment is made

$t$ is time of observation

$\gamma_\tau$ bond life of investment $CD_\tau$

$I_\tau$ is lump sum raised by sale of bonds at year $\tau$. Assume that the bonds are redeemed at a constant rate during life $\gamma_\tau$

$$\delta(\tau) \begin{cases} 1 \text{ if } t \geq \tau \\ 0 \text{ if } t < \tau \end{cases}$$

$$\delta(\tau+\gamma_\tau) \begin{cases} 1 \text{ if } t \geq \tau + \gamma_\tau \\ 0 \text{ if } t < \tau + \gamma_\tau \end{cases}$$

The amount of capital outstanding is represented for every time period and can affect the interest rates. This nonlinear reality complicates situations somewhat, placing dependent, nonconstant present-value-factors in the objective function.

A link to the investment model could be set up so that in each time period the capital expenditure for all costs must be less than or equal to the bond sum available for investment (we assume that operation costs and scheduled bond redemptions are covered by user charges or tax assessment):

$$\sum_{j,k} F_t y_t + \sum_k DC_{kt} C^{CP}_{kt} + \sum_k EXP_{kt} C^{exp}_{kt} \leq \text{[maximum bonded indebtedness} - \text{(capital bonds outstanding at time} = t)] \; \forall_t$$

This formulation has to be studied further to be compatible with our original investment model. It is hoped that this will provide a link between economic and financial modeling.

Lynn[14] and Clark[15] may be consulted for additional comments on the financing aspect in relation to economic aspects of public work projects.

## V. CONCLUSIONS

This paper has attempted to give an overview of the literature, an interpretation of existing models, and some insight into shortcomings in the state of regional solid waste management. The most advanced modeling level presently achieved appears to be the use of mixed integer programming in static models. Areas requiring considerable future work to truly advance the state of modeling are:

- -- description of indexes of environmental impact
- -- trade-off structure relating to interaction with air and water quality
- -- generalized inclusion of financing considerations

-- models to provide choice of various processes at each potential site
-- Investigate the merits of processing the separate components at different locations. More data about components in the refuse would have to be obtained.
-- formulation to take into consideration the stochastic nature of the inputs and the processes (use of stochastic simulation and stochastic programming ? synthetic generation of solid waste flows ? )
-- controlled economic models to give insights into waste disposal practices and requirements that could change as a result of public policy changes. These models could also help show areas where technological progress should be forced.
-- political gaming models and models which consider only solutions providing benefits to all rational parties affected (see Dorfman and Jacoby[16]).

Real progress toward achieving solid waste management success will require a cooperative spirit between many different professions. The problems are not always easily defined by any one single professional group, and the interdisciplinary approach is advocated for this interdisciplinary problem.

REFERENCES

1. Harrington, J. J. and L. J. Partridge, Jr. "Analysis of Variability in Loadings on a Municipal Refuse Incinerator," *APCA Annual Meeting*, Paper No. 72-175, Miami Beach, Florida, 1972.
2. Kühner, J. "Mathematische Modelle im Müll- und Abfallwesen" (Mathematical Modelling in Solid Waste), *Müll und Abfall* (1972).
3. Helms, B. P. and R. M. Clark. "Selecting Solid Waste Disposal Facilities," *J. San. Eng. Div., ASCE 97 (SA4):* 443 (1971).
4. Marks, D. H. "Facility Location and Routing Models in Solid Waste Collection," Ph.D. thesis, Johns Hopkins University, Baltimore, Md. (1969).
5. Marks, D. H. and J. C. Liebman. "Location Models: Solid Waste Collection Example," *J. Urban Plan. Devel. Div., ASCE 97 (UPI):*15 (1971).
6. Wersan, J., J. Quon, and A. Charnes. "Mathematical Modelling and Computer Simulation for Designing Municipal Refuse Collection and Haul Services," U.S. Environmental Protection Agency, EPA-SW-6RG-71 (1971).
7. Golueke, C. G. and P. H. McGauhey. "Comprehensive Studies of Solid Waste Management, First and Second Annual Reports,"

U.S. Department of Health, Education, and Welfare, Bureau of Solid Waste Management, UI-00547 and SW-00003 (1970).

8. Harrington, J. J. "Systems Analysis: A Key Aspect of Modern Management and Planning," *Mod. Gov. Natl. Devel.* 57 (1969).
9. Skelly, M. J. and W. R. Lynn. "Planning for Regional Refuse Disposal Systems," *Second National Symposium on Sanitary Engineering Research, Development and Design* (Ithaca, New York, 1969).
10. University of Louisville, Institute of Industrial Research, Louisville, Ky.-"Ind. Metropolitan Region Solid Waste Disposal Study," U.S. Department of Health, Education, and Welfare, Bureau of Solid Waste Management (1970).
11. Helms, B. P. and R. M. Clark. "Locational Models: Solid Waste Management," *J. Urban Plan. Devel. Div., ASCE 97 (UP1)*:1 (1971).
12. Walker, W. "Adjacent Extreme Point Algorithms for the Fixed Charge Problem," Center for Environmental Quality Management and Department of Operations Research, Cornell University, Ithaca, New York (1968).
13. Liebman, J. C. Discussion paper on B. P. Helms and R. M. Clark, "Selecting Solid Waste Disposal Facilities," *J. San. Eng. Div., ASCI 98 (SA1)* (1972).
14. Lynn, W. R. "Stage Development of Wastewater Treatment Works," *J. Water Poll. Cont. Fed. 36(6)*:722 (1964).
15. Clark, R. M. "Economics of Solid Waste Investment Decisions," *J. Urban Plan. Devel. Div., ASCE 96 (UP1)*:65 (1970).
16. Dorfman, R. and H. D. Jacoby. "A Model of Public Decisions Illustrated by a Water Pollution Policy Problem," in *Public Expenditures and Policy Analysis*, Robert H. Haveman and Julius Margolis, eds. (Chicago: Markham Publishing Co., 1970); also: Discussion Paper No. 70-1, Environmental Systems Program, Harvard University.

## APPENDIX

The dynamic investment model (p. 347) is illustrated by an example. The scenario consists of three cities ($C_1$, $C_2$, and $C_3$), three sanitary landfill sites (two potential sites: $L_1$ and $L_2$; and one operating site: $L_3$) and three processing sites ($P_1$ is a potential transfer-station, $P_2$ is a potential incinerator, and $P_3$ is an operating transfer-station). Three time periods of about four years each are considered (see values in Table 16.2). The optimal investment and operating decisions are shown in Table 16.3.(variables with zero values are not depicted). The decision sequence illustrated in Figures 16.8 16.9, and 16.10 refers to landfill site $L_3$ being limited to disposing 500,000 tons of refuse.

*Table 16.2*

*Informations on Solid Waste Generation* and Capacity Limitation*

| *City* | *Period 1* | *Period 2* | *Period 3* |
|---|---|---|---|
| $C_1$ | 585,000 | 655,000 | 730,000 |
| $C_2$ | 1,750,000 | 1,900,000 | 2,010,000 |
| $C_3$ | 1,070,000 | 1,250,000 | 1,300,000 |

| *Process* | *max. capacity [tons/day]*† | *min. capacity [tons/day]* |
|---|---|---|
| $P_1$ (transfer) | 2920 | 220 |
| $P_2$ (incinerator | 2920 | 440 |
| $P_3$ (transfer) | 876 | 438 (in operation) |

| *Landfill* | *Tons* | |
|---|---|---|
| $L_1$ | 1,000,000 | |
| $L_2$ | 5,000,000 | |
| $L_3$ | 500,000 | level ground |
| (in operation) | 1,300,000 | build a hill |

*Solid waste generation in tons/period.

†1000 workdays per four years period are assumed.

*Table 16.3*

*Results of Two Computer Runs with Different Capacities at Landfill 3*

| | $L_3 \leq 500{,}000$ | $L_3 \leq 1300{,}000$ | |
|---|---|---|---|
| *Variable* | *Amount* | *Amount* | *Unit Cost* |
| TP131 | 438,000 | 438,000 | 1.84 |
| TP132 | 438,000 | 438,000 | 1.38 |
| TP133 | 438,000 | 438,000 | 1.12 |
| TP211 | 507,000 | | 1.69 |
| TP212 | 1,900,000 | 1,753,500 | 1.25 |
| TP213 | 1,900,000 | 1,753,500 | 1.03 |
| TP321 | 438,000 | 438,000 | 5.40 |
| TP322 | 1,250,000 | 1,250,000 | 4.05 |
| TP323 | 675,117 | 351,029 | 3.03 |
| TL122 | 217,000 | 217,000 | 3.33 |
| TL123 | 292,000 | 292,000 | 2.62 |
| TL131 | 147,000 | 147,000 | 2.30 |
| TL211 | 1,000,000 | 1,000,000 | 2.50 |
| TL231 | 243,000 | 750,000 | 2.60 |
| TL232 | | 146,500 | 1.95 |
| TL233 | 110,000 | 256,500 | 1.58 |
| TL321 | 632,000 | 632,000 | 3.30 |
| TL323 | 624,882 | 948,970 | 2.00 |
| TPL321 | 438,000 | 438,000 | 2.10 |
| TPL322 | 438,000 | 438,000 | 1.57 |
| TPL323 | 438,000 | 438,000 | 1.28 |
| TPL121 | 507,000 | | 2.50 |
| TPL122 | 574,882 | 898,970 | 1.88 |
| TPL221 | 65,700 | 65,700 | 1.32 |
| TPL222 | 386,268 | 315,679 | 0.99 |
| TPL223 | 386,268 | 315,679 | 0.81 |
| TPP122 | 1,325,118 | 854,529 | 4.20 |
| TPP123 | 1,900,000 | 1,753,500 | 3.40 |
| DC11 | 507 | | 1,000 |
| DC12 | | 1,753 | 750 |
| DC13 | | | 600 |
| DC21 | 438 | 438 | 10,500 |
| DC22 | | | 7,600 |
| DC23 | | | 6,200 |
| EXP12 | 1,393 | | 700 |
| EXP13 | | | 570 |
| EXP22 | 2,137 | 1,666 | 6,000 |
| YL11 | 1 | 1 | 200,000 |
| YL21 | 1 | 1 | 445,000 |
| YP11 | 1 | | 500,000 |
| YP12 | 1 | 1 | 280,000 |
| YP21 | 1 | 1 | 2,450,000 |

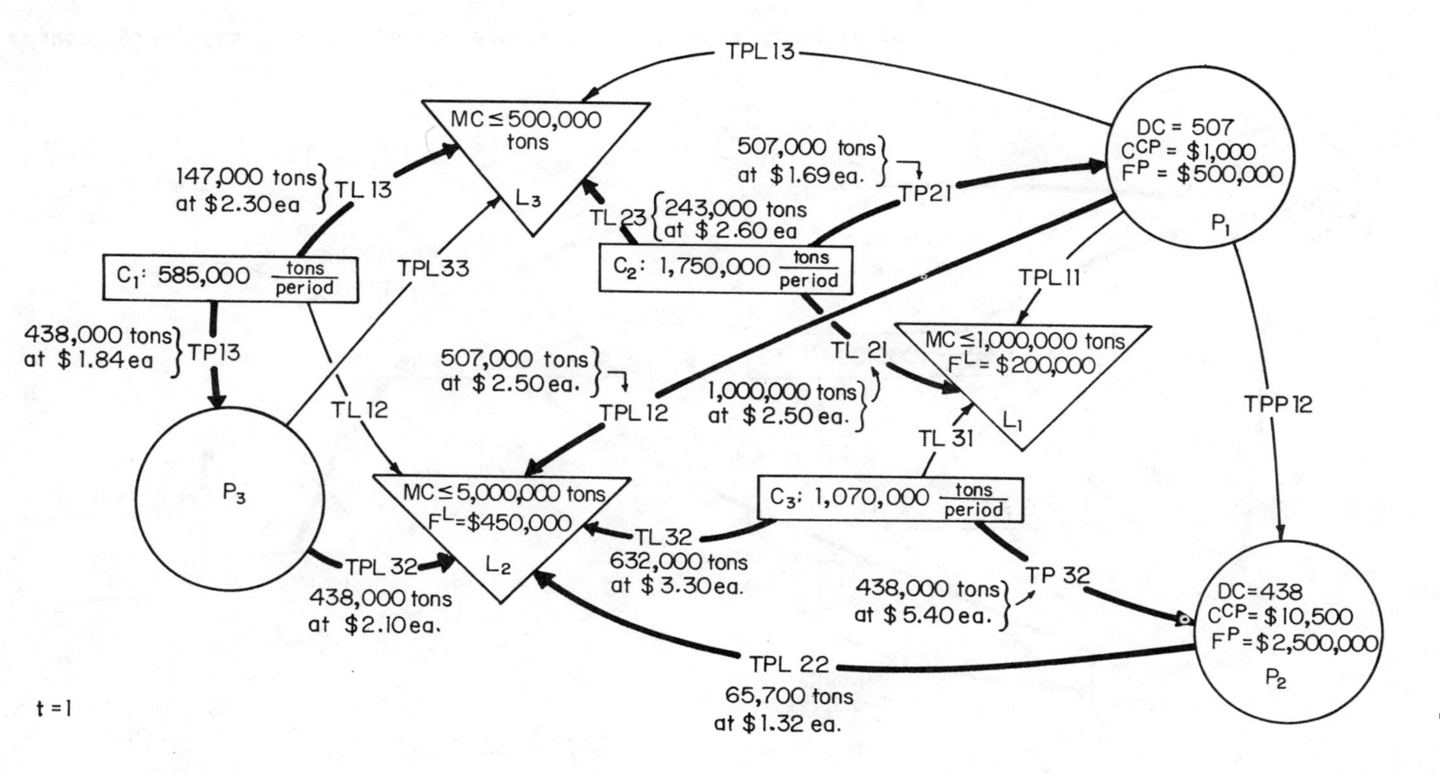


*Figure 16.8. Landfill 3* ***capacity*** *equals 500,000 tons: flows in period 1.*

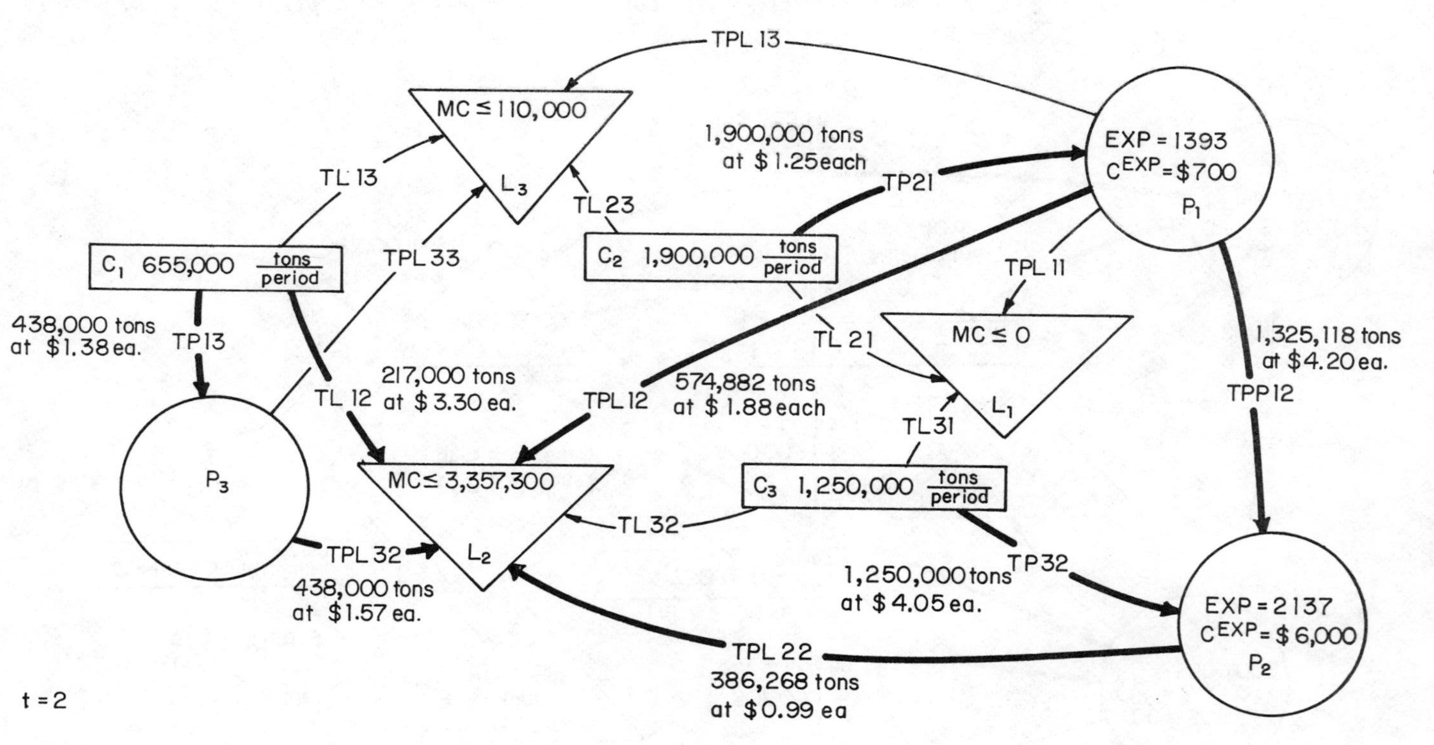


*Figure 16.9. Landfill 3 capacity equals 500,000 tons: flows in period 2.*

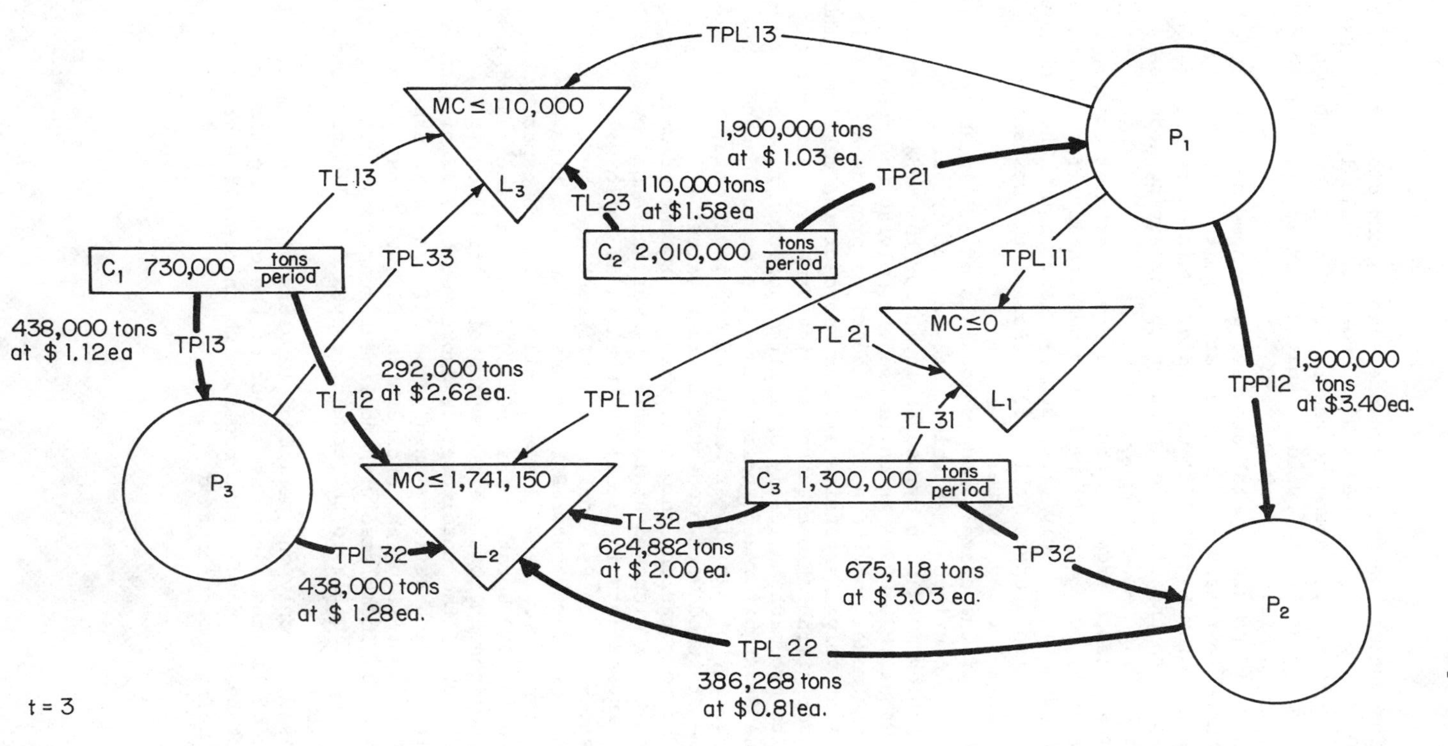


*Figure 16.10. Landfill 3 capacity equals 500,000 tons: flows in period 3.*

The optimal decision sequence when $L_3$ can dispose of 1,300,000 tons of refuse is also shown in Table 16.3 but is not illustrated. Expanding the capacity of $L_3$ by 800,000 tons is equivalent to building up the land level, such as providing a hill.

Not all possible paths are used with some being excluded by heuristic reasons. Also the path from transfer station $P_1$ to incinerator $P_2$ (TPP12t) was added. Unit costs include transport to and operating cost of the receiving node (see Table 16.3).

The fixed charges at the nodes are investment costs plus opportunity costs in all periods minus receipts from nonsolid waste management activities, such as land rent. These charges are also displayed in Table 16.3. All of the costs are adjusted to present value.

Each of the example models consists of 58 linear programming rows, 70 continuous variables, and 12 integer variables. This information must be provided as input for the MPSX-MIP-system (IBM Manual H20-0908) which is used in solving the model.

This model and the solution technique provide an orderly and relatively inexpensive aid for the decision maker who must consider timing of operations and investments. Although the results are mathematically optimal, discretion must be used in the straightforward application. The output depends on the quality of the linear model and the input data; the parameters used were projected average values.

The computations show that when the amount of waste permitted to be disposed at landfill $L_3$ was increased by 800,000 tons, investments in facilities changed as follows: transfer-station $P_1$: only one investment at time t=2, no expansions; incinerator $P_2$: expansion done at time t=2 is 470 T/day less. Landfills are still filled to capacity ($L_3$ is now 1,300,000 rather than 500,000 tons) at the end of the time horizon. Flows are appropriately varied.

The PV of the total cost is decreased by $6.2 million, a magnitude which reflects mainly the difference in expenditures between landfilling and incinerating 800,000 tons. This model may also be formulated with nonlinear functions. The algorithm used is applicable to the case of convex and slightly concave functions.

Assuming only one period of 12 years gives an answer, but it is of limited use. The effect of the changing flow rate cannot be included in the model, with an overall average flow being chosen instead.

MODELS FOR ENVIRONMENTAL POLLUTION CONTROL


# CHAPTER 17

# ROUTING PROBLEMS FOR STREET CLEANING AND SNOW REMOVAL

Thomas M. Liebling*

## I. INTRODUCTION

The purpose of this study was to aid municipal authorities in the planning of the following tasks:

- choice of suitable equipment for street cleaning
- determination of sites for depots
- planning of individual tours and routes.

Different streets had to be cleaned according to need, and therefore frequencies varied from street to street, from once weekly to four times a day. Obviously it was desirable to find reasonable cost solutions. Analogous problems were posed by snow-removal and salting of the streets during the winter season. In these cases the object was not primarily to satisfy given wishes at a minimal cost but to use all means available to clean snow or to salt as quickly as possible. In all cases, one restriction was binding: in order to formulate the problems mathematically and to organize the large amounts of required input data, the city (Zurich) was represented as a graph. On the graph each intersection corresponded to a vertex and each street segment to an arc.

For each of the over 8000 segments, the following information was collected: length, width of street and sidewalks (if any), trees, parked cars, cleaning frequency, and priority during the winter.

---

*Dr. T. M. Liebling is with the Institut für Operations Research, Eidg. Technische Hochschule Zürich, Clausiusstr. 55, 8006 Zürich, Switzerland.

Figure 17.1 shows a portion of a map of Zurich from which the topology and measurements were taken. Figure 17.2 shows the corresponding graph, and Table 17.1 shows the representation of the data in the computer. All further computations are based on this data-bank.

These computations were carried out in two phases--a rough- and a fine-planning phase. An example of a fine-planning model, which is an application of the "Chinese Postman's Problem," is shown in Section II, and some ideas about the rough-planning problems are sketched in Section III.

## II. A FINE-PLANNING MODEL

The object of this model is to find a tour with minimal length that covers a given subset of connected segments of a city's street network. This is sometimes known as the Chinese postman's problem.

### *Problem Formulation*

*Given* a strongly connected graph $G(X,U)$ (street network) whose arcs (one-way street segments) $u \in U$ have the nonnegative lengths

$$d(u_j) := d_j, \quad u_j \in U$$

Further a merely connected subgraph $G_1(X_1,U_1)$ of $G(X,U)$, that is $X_1 \subset X$; $U_1 \subset U$.

*Find* a tour on $G(X,U)$ that has the following properties

(a) it covers all of the arcs of $G_1(X_1,U_1)$
(b) the sum of the lengths of its arcs is minimal.

### *Solution*

First, a Euler-graph must be found with the following properties:

1. it contains all arcs of $G_1(X_1,U_1)$ in at least simple multiplicity
2. all its elements are taken (the arcs in arbitrary multiplicity) from the graph $G(X,U)$
3. the sum of its arc-lengths is minimal.

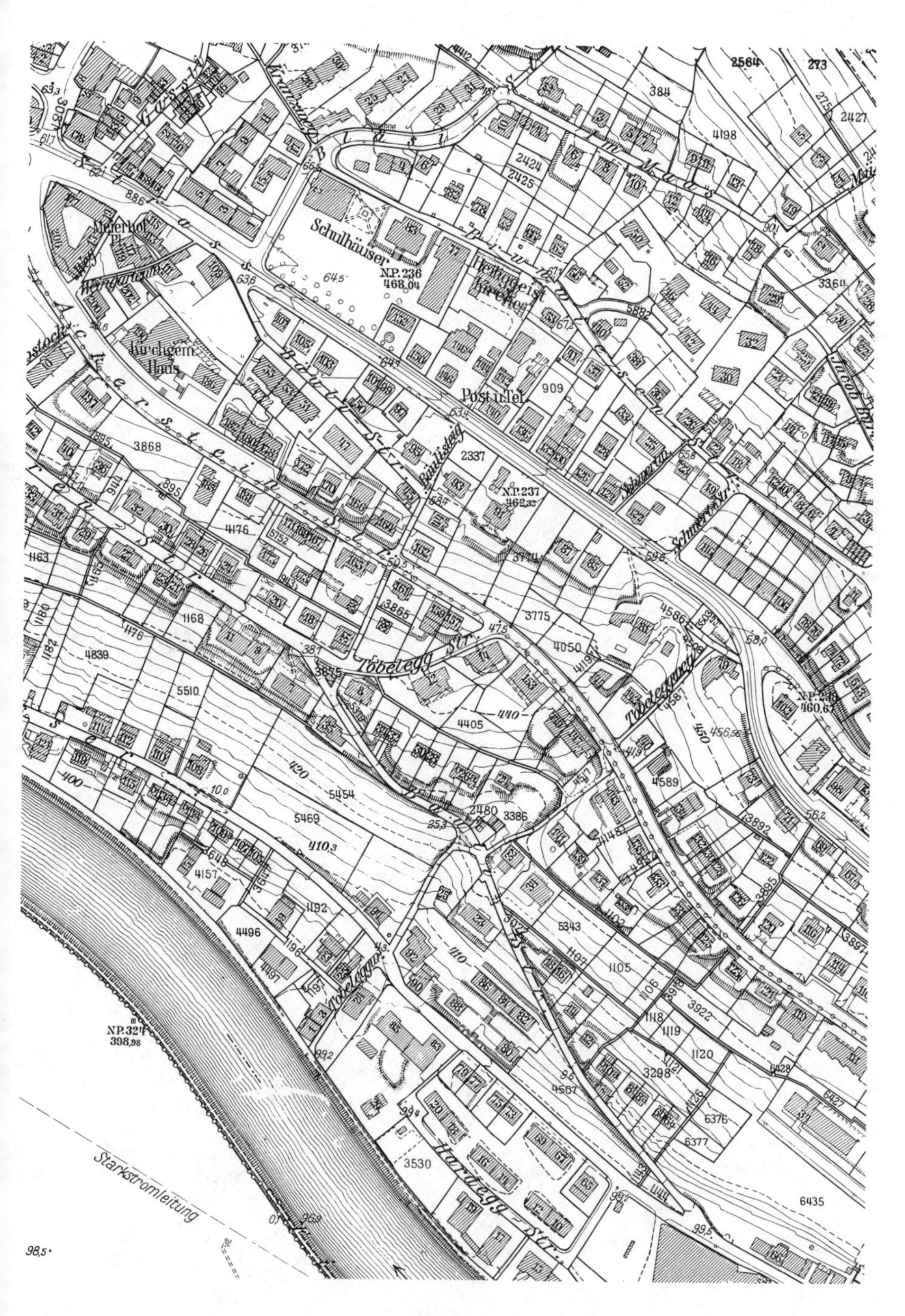


Figure 17.1. Part of a map of Zurich.

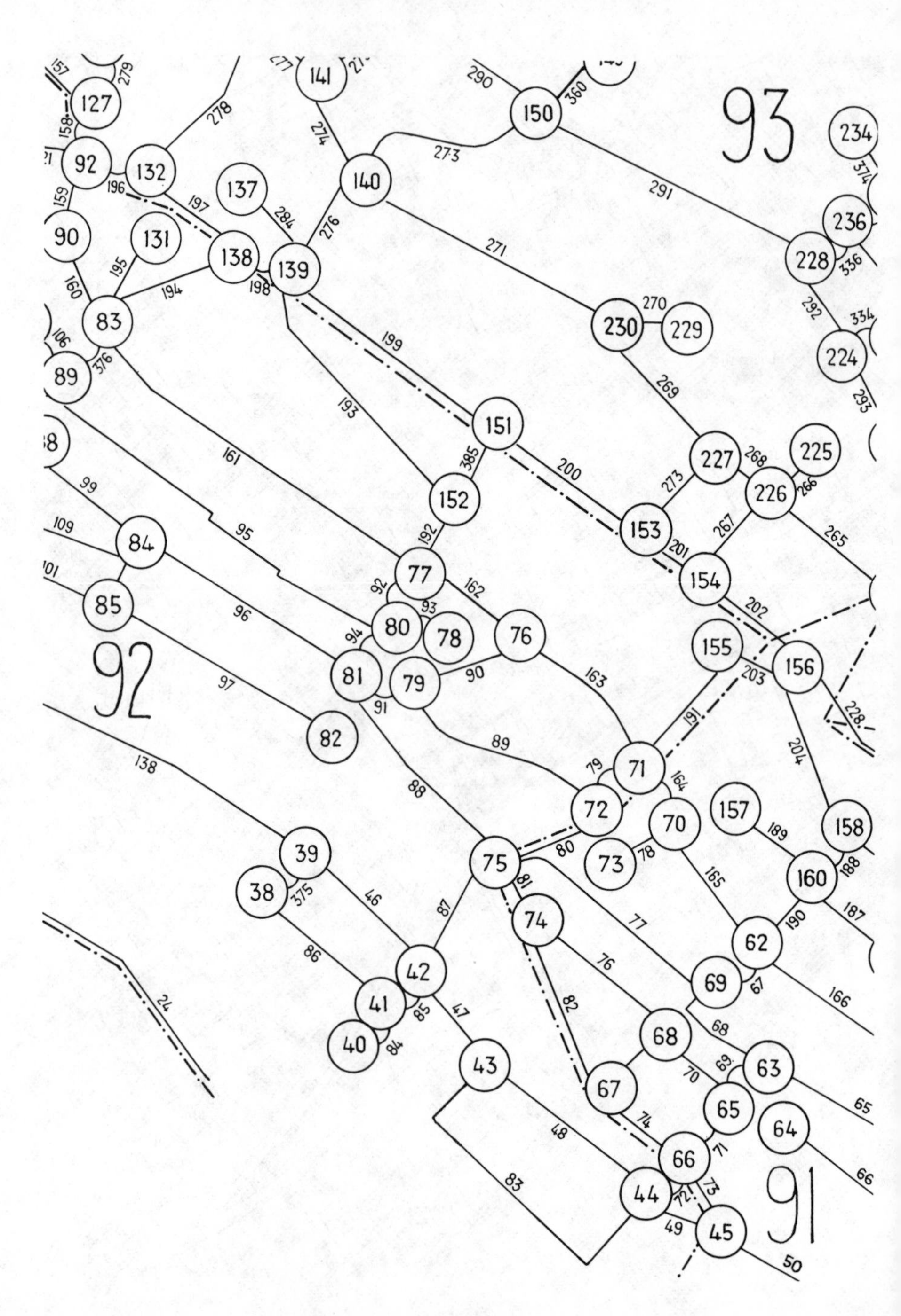


Figure 17.2. Graph corresponding to Figure 17.1.

*Table 17.1*

*Data of Some of the Streets Shown in Figures 17.1 and 17.2*

| Street No. | Begin. Point | End Point | Street Name | Code | Pavement Length | Pavement Width | Sidewalks Left Length | Sidewalks Left Width | Sidewalks Right Length | Sidewalks Right Width | Steps Trees | Cleaning Freq. |
|---|---|---|---|---|---|---|---|---|---|---|---|---|
| 18158 | 18092 | 18127 | Ackersteinstr | 25 | 40 | 9.0 | 20 | 2.5 | 20 | 2.5 | T | 2.0 |
| 18159 | 18092 | 18090 | Ackersteinstr | 25 | 50 | 11.0 | 30 | 3.0 | 40 | 2.5 | T | 2.0 |
| 18160 | 18090 | 18083 | Ackersteinstr | 25 | 60 | 12.0 | 60 | 3.0 | 60 | 2.0 | T | 2.0 |
| 18161 | 18077 | 18083 | Ackersteinstr | 25 | 260 | 8.5 | 260 | 3.0 | 170 | 1.5 | T | 2.0 |
| 18162 | 18076 | 18077 | Ackersteinstr | 25 | 70 | 6.0 | 70 | 3.0 | | | T | 2.0 |
| 18163 | 18071 | 18076 | Ackersteinstr | 25 | 110 | 6.0 | 110 | 2.0 | | | T | 2.0 |
| 17073 | 17066 | 17067 | Am Giessen | 230 | 130 | 6.0 | | | | | | 1.0 |
| 18042 | 18032 | 18034 | Am Giessen | 230 | 30 | 7.0 | 25 | 2.2 | | | | 2.0 |
| 18046 | 18039 | 18042 | Am Wasser | 270 | 100 | 6.0 | 100 | 2.5 | | | | 2.0 |
| 18047 | 18042 | 18043 | Am Wasser | 270 | 80 | 7.0 | 80 | 2.5 | | | | 2.0 |
| 18048 | 18044 | 18043 | Am Wasser | 270 | 140 | 6.5 | | | | | | 2.0 |
| 18049 | 18045 | 18044 | Am Wasser | 270 | 60 | 6.0 | 45 | 2.0 | | | | 2.0 |
| 18132 | 18036 | 18039 | Am Wasser | 270 | 90 | 10.0 | 90 | 2.5 | 90 | 2.0 | | 2.0 |
| 18135 | 18037 | 18036 | Am Wasser | 270 | 100 | 10.0 | 80 | 2.5 | 100 | 2.0 | | 2.0 |
| 18136 | 18039 | 18037 | Am Wasser | 270 | 270 | 8.0 | | | | | | 2.0 |
| 17057 | 17077 | 17098 | Am Wettingertobel | 275 | 70 | 8.0 | 70 | 2.0 | 100 | 2.0 | S | 1.0 |
| 17060 | 17098 | 17106 | Am Wettingertobel | 275 | 50 | 4.0 | | | | | | 1.0 |
| 18128 | 18102 | 18103 | Am Wettingertobel | 275 | 100 | 4.5 | | | | | | 2.0 |
| 18143 | 18104 | 18111 | Am Wettingertobel | 275 | 110 | 2.5 | | | | | | 1.0 |
| 18146 | 18103 | 18104 | Am Wettingertobel | 275 | 50 | 8.5 | 20 | 1.5 | 20 | 1.5 | | 2.0 |

The Euler-graph $G_E(X_E,U_E)$ is found as follows:

(1) *Classify* the vertices of $G_1(X_1,U_1)$ according to their degrees, *i.e.*, call a vertex $x_i$ a $q_i$-source if the number of arcs entering $x_i$ exceeds the number of arcs leaving $x_i$ by $q_i (q_i \varepsilon \{1,2,...\})$. Call a vertex $x_j$ an $s_j$-sink if the number of arcs leaving $x_j$ exceeds the number of arcs entering $x_j$ by $s_j (s_j \varepsilon \{1,2,...\})$. It is easily shown that

$$\sum_{\text{(sources)}} q_i = \sum_{\text{(sinks)}} s_j$$

(2) *Add* arcs of $G(X,U)$ to $G_1(X_1,U_1)$ so that no vertex of the new graph is either a source or a sink and the sum of the lengths of the arcs added is minimal. These conditions are both satisfied if the arcs added are those determined by the solution of a simple transportation problem, in which the sources are viewed as warehouses with $q_i$ units in stock and the sinks as customers with demand of $s_j$ units, and the transportation costs per unit $c_{ij}$ are equal to the length of the shortest path from $x_i$ to $x_j$ on the graph $G(X,U)$.

Using the arcs of $U_E$ exactly once, any closed circuit on $G_E(X_E,U_E)$ satisfies (a) and (b) so that finding any particular circuit is satisfactory. An algorithm, best explained by the following, accomplishes this.

> Taking a roll of red and of white string, walk through a maze in an arbitrary sequence, unwinding the red string, observing one-way segments and using no segment more than once, until the entry is reached and no unused segments leave from it. Next backtrack on the "red" path while unwinding the white string and rewinding the red one, until a yet unused and permitted segment is found. Deposit the white roll and go on unwinding the red one. Continue this procedure, until at one backtracking phase, no further unused segment can be found: the unrolled string then describes the desired route.

A computer code that makes use of this idea has the advantage of requiring only "local" information at each step, that is, it can be programmed with low storage and computation time.

Example 1

Consider a street network which yields the graph G(X,U) shown in Figures 17.3 and 17.4.

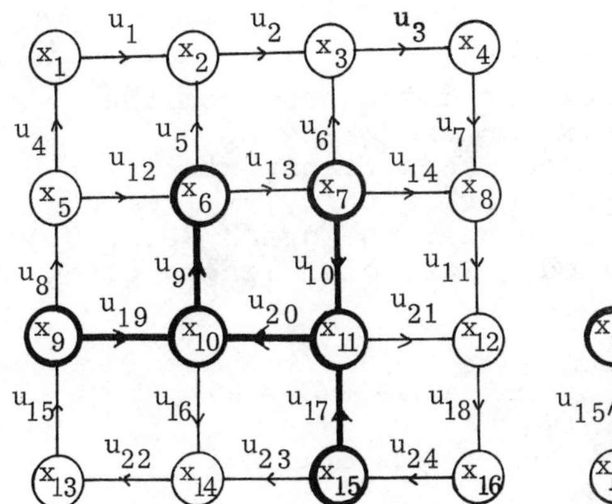


*Figure 17.3.* ***Street network graph.***

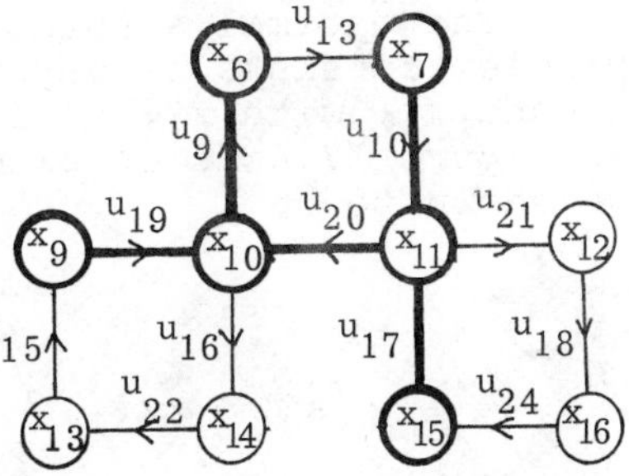

*Figure 17.4.* ***Street network graph.***

With

$$X = (x_i,\ i = 1, \ldots, 16)$$

$$U = (u_j,\ j = 1, \ldots, 24)$$

$$X_1 = (x_6, x_7, x_9, x_{10}, x_{11}, x_{15})$$

$$U_1 = (u_9, u_{10}, u_{19}, u_{20}, u_{17})$$

let

$$d(u_j) = j, \quad j = 1, \ldots, 24$$

Note that the vertices $x_6, x_{10}$ and $x_{11}$ are 1-sources and the vertices $x_7, x_9$ and $x_{15}$ are 1-sinks. The transportation problem to be solved has the following cost matrix $c_{ij}$:

| *sources* \ *sinks* | $x_7$ | $x_9$ | $x_{15}$ |
|---|---|---|---|
| $x_6$ | 13 | 96 | 80 |
| $x_{10}$ | 22 | 53 | 89 |
| $x_{11}$ | 42 | 73 | 63 |

It is made up of the shortest distances from the sources $x_i$ to the sinks $x_j$ in the graph G.

The optimal solution of this transportation problem consists in matching $x_6$ with $x_7$, $x_{10}$ with $x_9$ and $x_{10}$ with $x_{15}$. Therefore, the graph $G_E(X_E, U_E)$ in Figure 17.4 is obtained. Its total length can be shown to be

$$\underbrace{13 + 53 + 63}_{\text{added arcs}} + \underbrace{\sum_{u_j \varepsilon U_1} d(u_j)}_{\text{length of } G_1} = 129 + 72 = 201$$

A tour covering $G_1$ is easily obtained by joining the vertices labeled 9,10,6,7,11,12,16,15,11,10,14,13,9 in that order.

Example 2

Here an Euler-cycle (a connected tour that covers all of the edges of a graph exactly once) is constructed. The graph is as shown in Figure 17.5. The following phases end in the desired cycle:

| | *red string* | *white string* |
|---|---|---|
| 1. | $u_1, u_6, u_{10}$ | |
| 2. | | $u_{10}$ |
| 3. | $u_1, u_6, u_5, u_9, u_8$ | |
| 4. | | $u_{10}, u_8$ |
| 5. | $u_1, u_6, u_5, u_9,$ $u_2, u_7, u_4, u_3$ | |
| 6. | | $u_{10}, u_8, u_3, u_4, u_7,$ $u_2, u_9, u_5, u_6, u_1$ |

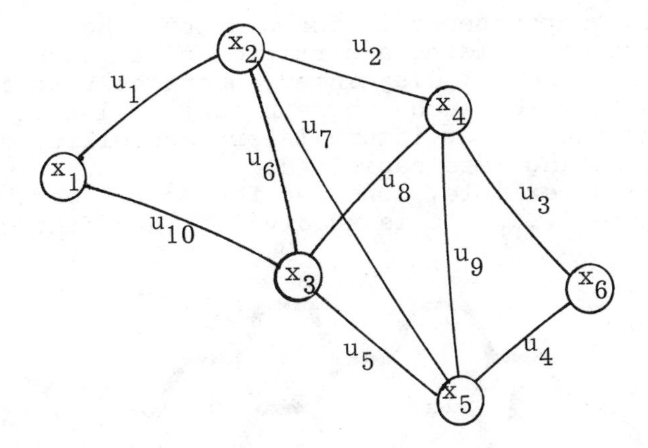


*Figure 17.5. Graph for Euler-cycle.*

*Discussion*

Routing problems for street sweeping, salting or snow removing vehicles can be solved using the above model: the lengths will be replaced by suitable time-units or others describing the range or capacity of the vehicles. The streets to be swept in a given tour make up the graph $G (X ,U )$. If they are not connectable into a closed tour, the graph $G_E(X_E,U_E)$ shows the extra street segments that must be added to make such a tour possible.

The same problem formulated for *undirected* graphs when scheduling tours on side walks is somewhat more difficult and is transformed into a *matching problem*. The existing methods for solving this problem are not nearly as swift. A further complication is introduced if the graph $G_1$ is not connected. It can then be shown that the resulting optimization problem is equivalent to a traveling salesman problem, with all the inherent difficulties.

## III. A ROUGH-PLANNING MODEL

Depending on the capacity of the cleaning vehicles, the working hours of the crew, and the cleaning frequencies, the city was divided into a number of geographically disjointed regions. The aim was to create as few regions as possible, while enabling fine-planning of the daily tours of one cleaning crew per region.

In graph-theoretic formulation, the problem consists of dividing a partition of a graph into a minimum number of disjointed subgraphs that cover it, so that for each subgraph certain linear inequalities (that insure later feasibility of fine-planning) are satisfied.

As an example, consider the "city" portrayed in Figure 17.6. It is subdivided into fine cells

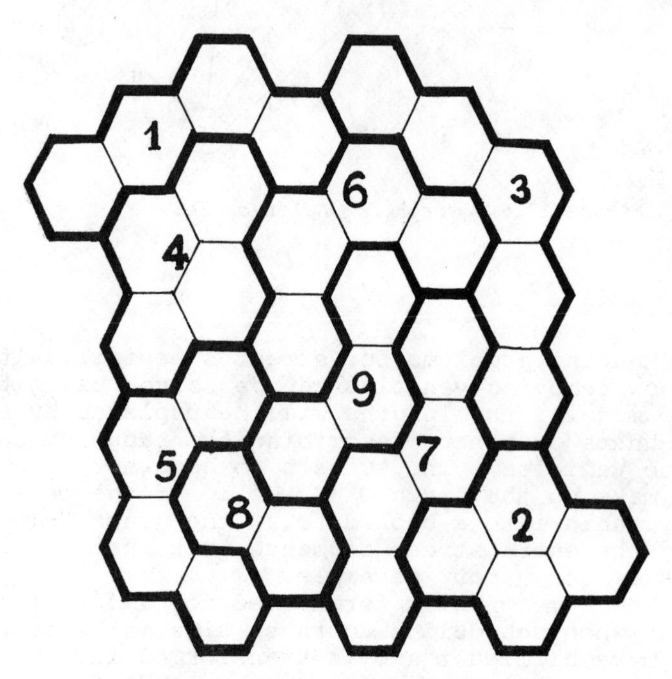


*Figure 17.6. City subdivided into cells.*

that are grouped to form regions as found by an adequate algorithm. Figure 17.7 shows the corresponding graph and Figure 17.8 the subgraphs found by the algorithm. Each vertex of the graph corresponds to one of the cells and the vertices that correspond to geographically adjacent cells. For each of the cells, the total length of streets to be cleaned once, twice and five times a week was computed. Two linear inequalities using these numbers check the feasibility of joining some of the cells into one "cleaning region." The major mathematical difficulty of this problem was that in the process of creating regions from the cells the former not split into disjointed parts.

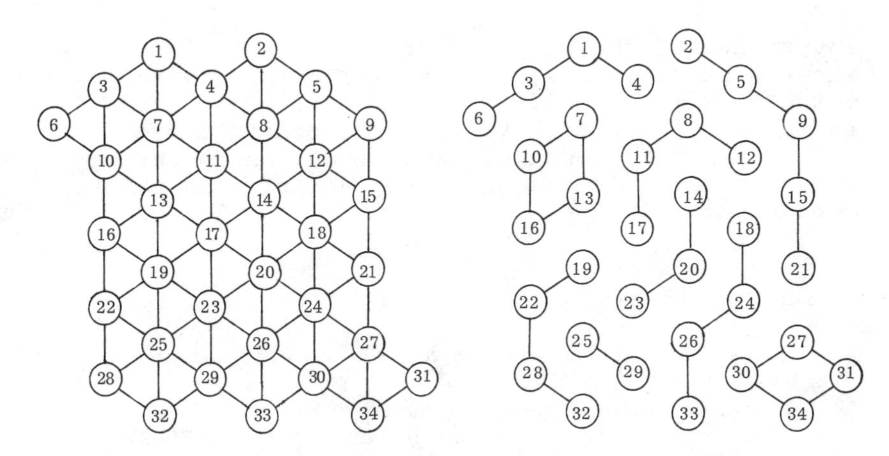


*Figure 17.7. Graphs for city in Figure 17.6.*

*Figure 17.8. Subgraphs for city in Figure 17.6.*

## IV. DISCUSSION

Problems of the type discussed above are difficult to solve exactly. There are some special cases in which the exact solution can be found without much effort, but as a rule one must be content with a good approximately. Such approximations may be determined by algorithms such as those proposed by Liebling[3] or Mayer.[4] The solution approach is a heuristic one. It consists of allowing the computer to apply "local" criteria systematically, as one would do if the problem were solved by hand, while ruling out nonsensical combinations and correcting the shape of the regions. Of course this requires some amount of computation, but problems of considerable size with the amount of data of Zurich were solved satisfactorily.

## V. IMPLEMENTATION

It may finally be worth mentioning, that contrarily to the experience often made when similar methods are first used, the City Authorities of Zurich were not only very cooperative during the development of the models, but were also able to make extensive use of their results. Thus they were able to give a sound basis to their long range policy concerning the purchase of cleaning equipment and zoning of the city. Furtheron, last winter the streets of Zurich were salted following the routes computed with the aid of the models. Increased reliability, streamlined work and considerable savings in the amount of salt needed as compared with the way this task used to be performed before were reported.

REFERENCES

1. Berge, C. *Théorie des graphes et ses applications.* (Paris: Dunod, 1967).
2. Dantzig, G. B. *Linear Programming and Extensions.* (Princeton, N.J.: Princeton University Press, 1963).
3. Liebling, Th. M. *Graphentheorie in Planungs- und Tourenproblemen am Biespiel des städtischen Strassendienstes.* (Berlin: Springer Verlag, 1970).
4. Mayer, J. *Studie für den Winterdienst der Stadt Zürich* (Zürich: IFOR-ETHZ Bericht 700119, 1971).

# PART V

# NOISE CONTROL

MODELS FOR ENVIRONMENTAL POLLUTION CONTROL


# CHAPTER 18

# ENVIRONMENTAL NOISE MANAGEMENT

Daniel P. Loucks*

## I. INTRODUCTION

Some sounds add to the quality of the environment while others do not. Sounds that are unwanted or annoying result in what is popularly called noise pollution. This paper is focused primarily on the management of environmental noise, a particular kind of noise pollution that emanates from sources that are not easily controlled by those who hear it. Environmental noise includes, but is not restricted to, what is also termed outdoor noise and community noise. Those who are subjected to environmental noise, whether they be outside or inside a building, must either endure the noise and the resulting annoyance or damage or incur a cost of protecting themselves from at least a portion of it. For example, outdoor noise entering a room through an open window may interfere with the activities of those in the room. Having to close the window, thereby reducing the circulation of air on a hot day, just to decrease the noise damage is an example of social cost. An economic cost is incurred if an air conditioner is then needed solely because of the closed window. The overall objective of any environmental noise management program is to minimize the sum of the damages resulting from noise and the social and economic costs of any measures taken to reduce noise in the environment.

---

*Daniel P. Loucks is an Associate Professor in the Department of Environmental Engineering, Cornell University, Ithaca, New York, 14850, U.S.A.

The approach to environmental noise management proposed in this paper is very similar to the approaches proposed and applied to the management of other waste residuals. Regional air quality management, water quality management and solid waste management are relatively popular and well-known terms usually implying an economically efficient or cost effective approach for planning and maintaining the quality of our airshed, waterways, and land, and for controlling the residuals that can reduce the value of those natural resources. Among the terms denoting particular aspects of the overall problem of environmental quality management should also be included the term environmental noise management: the management of a residual that in many rural and urban areas reduces the quality of the environment far more than other mass or energy residuals.

Probably most in need of environmental noise management programs are many urban areas, as evidenced by the increasing number of public awareness programs and noise abatement and control laws being enacted by the major cities of the United States and by the continued enforcement of noise regulations in many European and Asian cities. While this paper will concentrate on the urban noise management problem, the general rationale and approach to noise management extends to all environmental noise problems. The reason for the concentration on urban noise stems not only from the fact that environmental noise and its damage are often the greatest in urban areas, but also that the problem of noise management can be the most complex there. The identification of major noise sources and their characteristics, the prediction of noise attenuation from each noise source to each receptor, and the estimation of the effectiveness of noise reduction alternatives at the source and receptor sites and in the transmitting medium between source and receptor sites is made no easier by the existence of complex configurations of buildings, surfaced with various materials, and forming irregular street canyons (which tend to reduce sound attenuation) and barriers (which often increase sound attenuation).

The objective of this paper is not to recommend a set of specific environmental noise levels for various urban, suburban or rural settings, or to suggest a specific means of reducing noise in any given situation. Rather it is to discuss the development and use of quantitative procedures and techniques that can assist in the establishment of specific environmental noise abatement plans and policies. Of interest is an approach to environmental noise management that can define and evaluate numerous noise abatement alternatives based on one or more management objectives. This requires the identification of all significant noise sources and receptors and the alternative means, and costs, of reducing the noise at each source, at each receptor, and in the paths between each source and receptor. It requires the prediction of the effectiveness of various noise reduction alternatives in terms of the total decrease (or increase) in the intensity or duration of noise experienced at various locations within the affected area and a knowledge of the effects of the remaining noise on each receptor. The simultaneous consideration of all of these factors, together with information on how changes in various data and assumptions might alter the time distribution and intensity of the noise at various sites within an area, should lead to more effective environmental noise management policies.

This paper will first discuss in qualitative terms the proposed approach for defining and evaluating alternative environmental noise management policies. Following a short description of the major urban sources and typical characteristics of environmental noise, various noise reduction alternatives will be reviewed in the context of a hypothetical urban area. Finally a procedure will be outlined for defining and evaluating alternative management policies that satisfy environmental noise standards. These standards specify the maximum noise levels allowed at numerous sites within the urban area.

## II. ENVIRONMENTAL NOISE CHARACTERISTICS

It is neither necessary nor practical to identify and characterize all sources of environmental noise for the simple reason that all noise sources do not contribute in an equal manner to the total noise level. It is generally acknowledged that the major contributors of environmental noise in most urban environments are transportation, construction, and certain commercial and industrial activities. In addition, noise from heating and air conditioning systems can be locally significant. These are activities over which the typical man on the street has very little control, and yet the noise they create may affect him psychologically, sociologically, physiologically and economically. The effect of each of these types of noise sources will depend on the characteristics of the aggregate noise at various locations as well as on the personal characteristics and activities of the individuals hearing the noise at those locations.

The annoyance associated with environmental noise depends, in part, on a number of its characteristics. Perhaps the most obvious characteristic is intensity, or amplitude. Sound intensity is usually stated in units called decibels, abbreviated dB. Individuals with unimpaired hearing can detect sounds close to zero decibels. Jet aircraft, on the other hand, can generate over 125 decibels, an intensity that can temporarily if not permanently decrease one's hearing ability and is painful to all but a very few individuals. Obviously most environmental noises have intensities between these two values. Before describing in any more detail the various noise sources and the intensities associated with urban settings, additional properties of urban noise should be discussed.

The loudness of any noise and the annoyance it may cause is also a function of the frequency of the noise. A higher frequency noise of a given intensity is usually considered more annoying, up to a point, than a lower frequency noise of the same intensity. For this reason most measurements of environmental noise combine in a specific way both frequency and intensity. The net result is called an A-weighted decibel scale, abbreviated dB(A). Two sounds of equal intensity but of differrent frequencies will have different dB(A) levels, the lower frequency sound having the lower dB(A) level. This method of noise measurement appears to be satisfactory for most environmental noises containing many frequencies, *i.e.*, broadband noise. For noise containing strong pure frequency tones (*e.g.*, the whine of jet engines) other more complex measurement procedures have been devised.

It is also necessary to account for the temporal pattern of the changes in the dB(A) levels at various locations. A steady noise is usually less annoying than one with varying levels, especially if the variations are random. Also, a short duration noise is less annoying than a longer duration one.

Most urban noise environments can be characterized by three features. The first is that the aggregate noise level often varies with time over a range of more than 30 dB(A), which is usually perceived as an eight-fold range of noisiness. This eight-fold range is based on the observation that a change of approximately 10 dB represents a doubling or halving of perceived loudness or noisiness of a sound. Hence a 30 dB(A) increase represents three doublings (2x2x2) or eight times the original noisiness.

The second feature of most urban noise environments is that a relatively steady lower noise level exists upon which higher levels originating from clearly identifiable sources are superimposed. The fairly constant lower level is termed the residual or background noise level coming from sources that are not easily identifiable. Distinct sounds superimposed on the residual noise level are usually the most annoying. It is these intrusive noises that a noise management policy can attempt to reduce. If these intrusive levels are reduced, one can also expect some reduction in the background noise levels as well. However, the intrusive noise sources are the primary sources of annoyance, and, in general, the greater the difference between the intrusive noise level and the background or residual level, the greater will be the annoyance.

The third feature of these noise environments is the difference in the resulting noise level-time patterns due to the various identifiable noise sources. For example, noise levels resulting from aircraft overflights are heard for a relatively long time but occur less frequently than noise levels resulting from the passing motor vehicles on the thoroughfare or local streets. Clearly it is the rapidity of occurrence as well as the duration of the frequency weighted noise levels that contributes to the degree of annoyance.

The noise-duration patterns of outdoor environments are likely to change during different periods of the day or night and also during different times of the year. Also of importance is the noise environment of individuals who travel from one place to another during the day. The objective of any noise management plan is to control this environment, which can be done by altering the noise levels heard at specific locations.

## III. AN APPROACH TO ENVIRONMENTAL NOISE MANAGEMENT

As outlined above, at least one of the goals of any environmental noise management policy is to minimize the sum of the damages resulting from the effects of environmental noise and the net costs incurred by any measures taken to reduce the noise either at the noise source sites, between the source sites and the receptor sites, or at the receptor sites. However, like other mass and energy residuals released into the environment, the effects of noise are difficult to quantify; therefore the approach suggested here for the management of noise will be similar to that developed and applied to the management of other residuals, namely one that efficiently meets predefined environmental quality standards. Numerous noise abatement alternatives exist that can be used to satisfy standards specifying the maximum allowable duration-frequency weighted noise levels at various locations in an urban area. A method for defining these alternatives and evaluating them based on their costs will be discussed in this section of the paper. This discussion will continue to be qualitative but will introduce items to be discussed in more quantitative terms later in this paper.

Decisions regarding the expenditure of funds for the maintenance or improvement of environmental quality are often made on the basis of a mostly qualitative integration of numerous economic, political, social and technological objectives. The explicit trade-offs between each of these partially complementary and conflicting objectives have not always been clear and therefore the selection and implementation of environmental quality management plans too often have failed to meet these objectives to the extent originally envisioned. In many cases, the current approach to environmental noise management promises to be no exception.

The environmental management decision-making process is further complicated by the uncertainty of the outcome of any decision due to factors outside the control of the decision maker, and by the limited ability of decision makers to simultaneously consider all relevant information pertinent to the decision even if there were no uncertainty. Even when supplemented with computers, the capacity to plan and implement efficient environmental quality management policy is limited. To decrease the extent of this limitation with regard to the management of air, water and solid wastes, planners and decision makers have constructed simplified models of their management problem. These models enable them to process more effectively the information available in order to predict and evaluate the possible outcomes of various policies. These simplified models can range from those wholly conceptual and contained within the mind of the planner or decision maker to those specified by sets of algebraic equations designed to be solved on high speed computers. It is this latter approach applied to the management of environmental noise that is proposed in this paper. The use of this quantitative modeling approach can be of considerable value as an aid, but not as a substitute, to the responsible decision-making process.

Very briefly, the quantitative modeling approach to be discussed involves:

1. the identification of all intrusive noise sources and the measurement of the intensity and duration of each resulting noise
2. the identification of alternative methods for reducing the noise intensity and/or duration at each noise source and the net cost of this reduction as a function of the amount reduced
3. the identification of a variety of receptor sites, *i.e.*, sites where the control or management of environmental noise is desired and which are located in such a manner that the environmental noise in areas between the specific receptor sites will also be adequately controlled
4. the identification of alternatives that can be implemented at each receptor site for reducing both the intensity and/or duration of the noise heard at those sites, and the net cost of these control measures as a function of their effectiveness
5. the identification of all significant paths that sound can travel between each source and each receptor
6. the identification of measures that can be used to modify the attenuation of sound as it travels along each path from source to receptor, and the net cost and the effectiveness of each of these alternatives as a function of some appropriate scale of these alternatives, and
7. the use of both optimization and simulation techniques to aid in the economic evaluation of all combinations of noise abatement alternatives and to assist in defining the particular combinations of noise reduction measures that satisfy various noise standards and management objectives.

This systematic approach for defining and evaluating noise management alternatives obviously involves a great deal of work. However, in the case of thermal energy and gaseous, liquid and solid mass residuals, the economic benefits (costs saved) achieved from the use of such an approach have proven to be substantially more than the costs of data collection and analysis, model development, and computation. There is no reason to believe that the application of these techniques to the management of environmental noise will be any less beneficial.

Three of the steps mentioned above warrant further discussion, namely, the types of alternatives available for reducing noise intensities and durations, the prediction of noise attenuation in urban areas, and the structure of models to be used in the defining and evaluating of efficient noise management policies.

### *Noise Reduction Alternatives*

Noise can be reduced at most source sites by improving the design of the object emitting acoustical energy, by shielding the object, or by a combination of design changes and shielding. It may also be possible to relocate the noise source or change its time and/or duration of operation. In some cases a change in the method of operating certain kinds of equipment may substantially reduce the resulting noise level. Examples of these alternatives for controlling noise at its source include shielding and improving the muffling of combustion engines and air compressors, restricting the operation of noise producing equipment to certain hours of the day, reducing motor vehicle speed and the number of stops required along city streets, and rerouting both surface and air traffic. Of considerable impact in many areas would be the effective enforcement of noise abatement procedures already required by law.

The construction of solid barriers, the planting of belts of trees and other vegetation, and the modification of the material and shape of building exteriors are examples of measures that tend to modify the propagation of sound from its source to any receptor. The effectiveness of these control alternatives for increasing the attenuation of sound will depend on the location and characteristics of the noise sources as well as on the location of the receptor sites.

At the receptor sites, the insulation of buildings and the wearing of ear protectors are two obvious means of reducing the noise that is heard, but each has its serious disadvantages. Noise reduction measures applied at receptor sites are usually considered only in special circumstances, but where the social cost of other noise reduction alternatives is excessive, these measures are available and should be considered.

### *Noise Propagation Prediction*

The prediction of noise propagation and attenuation in an urban area is complicated by the existence of numerous obstructions of different shapes, sizes and materials. Nevertheless, it is possible to estimate the attenuation of noise from each noise source to each receptor site along each noise path by the application of a series of appropriate rules for the various situations encountered along each path. These situations include free field or unobstructed conditions, barriers of buildings or trees, surfaces that absorb or reflect sound to various extents, street canyons, intersections, and so on. Local meteorological conditions also affect the attenuation of noise. The noise that each identifiable source contributes to each receptor site can be derived from a knowledge of the intensity of the noise source and the total attenuation along each path from that source to the receptor site. Once this is known, it is relatively simple to predict the effect of any noise abatement measure, applied at any noise source, within the transmission medium, or at the receptor site, on the aggregate noise level heard at any particular receptor site.

### *Cost Effective Noise Management Policies*

While there are numerous noise management objectives that can and should be identified, perhaps one of the more significant objectives is cost minimization. This single objective can be used as a means of illustrating the application of quantitative modeling techniques to the definition and evaluation of noise management alternatives. Through the use of such models one can estimate how much reduction in noise intensity and/or duration is needed and the manner in which it can be achieved, in order to satisfy a set of noise standards (incorporating duration, frequency and intensity) at each receptor site at a minimum total cost.

An estimate of the solution to this cost minimization problem can be obtained by using both mathematical optimization and computer simulation techniques. As sophisticated as these models and solution techniques may be, they still are considerable simplifications of reality, hence the emphasis in this paper on their use as guides rather than substitutes for the planning and implementation of effective noise management policy.

## IV. PRELIMINARY NOISE MANAGEMENT MODELS

By defining some notation two models for the preliminary screening of alternative environmental noise management policies can be structured. The first model to be developed will include only alternatives that reduce the frequency weighted dB(A) noise levels. This model will then be modified to include noise duration as well. Of interest in this example will be the identification of cost-effective policies, *i.e.*, noise abatement alternatives that minimize the total cost of meeting standards specifying the maximum allowable noise levels at various receptor sites. Before constructing the management models, it may be useful to review in quantitative terms some of the more elementary properties of environmental noise and how it is measured.

A Bel is simply the logarithm (to the base 10) of a ratio of two quantities Q and $Q_O$. One-tenth of a Bel is a decibel, hence the decibel level resulting from a quantity Q is

$$dB = 10 \log \frac{Q}{Q_o} \qquad (1)$$

The quantity $Q_O$ is a predetermined reference quantity, and to be precise decibels should be stated with reference to this quantity. One of three ratios used to measure the intensity of sound is the ratio of the square of the sound pressures, $p^2/p_O^2$, where the reference pressure, $p_O$, equals 20 micronewtons per square meter, $20\mu N/m^2$. Thus a sound pressure of $p = 20\mu N/m^2$ is equivalent to 0 dB re $20\mu N/m^2$, which at 1000 cycles per second is the threshold of hearing for normal young ears.

Note that 0 dB is not the absence of sound; indeed some individuals can hear sound levels less than 0 dB.

Denoting $dB_i$ as the sound pressure level one unit distance from a source at site i, Equation 1 can be rewritten

$$dB_i = 10 \log 10^{dB_i/10} \qquad (2)$$

since $p^2/p_0^2$ equals $10^{dB_i/10}$. In free unobstructed fields, the square of the sound pressure decreases with the square of the distance. Letting $d_{ij}$ be the number of unit distances from source site i to a receptor site j, the sound pressure level heard at site j due to source i equals

$$dB_{ij} = 10 \log \frac{10^{dB_i/10}}{d_{ij}^2} \qquad (3)$$

Of course in most environments free field conditions do not usually exist and a variety of other equations are required to estimate $dB_{ij}$.[1]

Once each source site i and receptor site j have been identified together with the contributing sound pressure levels $dB_{ij}$ for each pair of source and receptor sites, the total sound pressure levels, $dB_j$, at each receptor site j can be estimated by summing over all source sites i the square of the corresponding sound pressure ratios

$$dB_j = 10 \log \sum_i 10^{dB_{ij}/10} \qquad (4)$$

The application of the last two equations can be illustrated by considering two noise sources, each producing 80 dB as measured one unit distance away, hence $dB_i = 80$ for $i = 1,2$. From Equation 3, the sound pressure level at a receptor site 10 unit distances away is 60 dB if just one of the two sound sources is operating.

$$dB_{ij} = 10 \log 10^{80/10} - 10 \log 10^2 = 60$$

If both sources are heard simultaneously, Equation 4 can be used to calculate the total sound pressure level at site j.

$$dB_j = 10 \log [10^{60/10} + 10^{60/10}] = 10 \log 2(10^{60/10})$$

$$= 10 \log 2 + 10 \log 10^{60/10} = 63$$

Note that this 3 dB increase when two equal sound levels are heard simultaneously will apply regardless of the original levels. For example, 0 dB + 0 dB = 3 dB.

When frequency is included in the calculations by using the dB(A) weighted decibel measure, Equation 4 may be regarded only as an approximation. Recall from the previous qualitative discussion that the dB(A) scale is simply a summation over various bands of frequencies of the corrected squares of the sound pressure ratios associated with those frequency bands. Defining the correction factor associated with the A weighting scale as $C_f$ for each frequency band f,

$$dB(A) = 10 \log \sum_f 10^{(dB_f + C_f)/10} \tag{5}$$

To accurately define $dB(A)_j$ it would be necessary to identify the contributing sound pressure levels for each frequency band f, $dB_{ijf}$ for use in Equation 6.

$$dB(A)_j = 10 \log \sum_f \sum_i 10^{(dB_{ijf} + C_f)/10} \tag{6}$$

For the purposes of preliminary screening prior to a more detailed analysis of noise abatement policies, the following approximation of Equation 6 can be used:

$$dB(A)_j \cong 10 \log \sum_i 10^{dB(A)_{ij}/10} \tag{7}$$

At this point one can begin to structure a management model for estimating the dB(A) reductions at each source and receptor site, and the scale of structures placed between various source and receptor sites that minimize the total cost of meeting a set of environmental noise standards, $dB(A)_j^{max}$. These standards specify the maximum allowable dB(A) level at each receptor site j.

$$dB(A)_j \leq dB(A)_j^{max} \tag{8}$$

Let the variables $X_i^S$ and $X_j^R$ be the dB(A) reductions at each source site i and receptor site j costing $C_i^S(X_i^S)$ and $C_j^R(X_j^R)$ respectively. The scale of any alternative h used to modify the noise attenuation between a source and receptor site can be defined by the variable $S_h$ which costs $C_h(S_h)$ and results in an attenuation of $A^h_{ij}(S_h)$ between source i and receptor j.

Combining these costs into an objective function and incorporating the decision variables $X^S_i$, $X^R_j$ and $S_h$ into Equations 7 and 8 completes the nonlinear but separable cost effective noise management model.

Model I:

$$\text{Minimize} \quad \sum_i C^S_i(X^S_i) + \sum_j C^R_j(X^R_j) + \sum_h C_h(S_h) \qquad (9)$$

subject to environmental noise standards

$$10 \log \{ \sum_i 10^{[dB(A)_{ij} - X^S_i - \sum_h A^h_{ij}(S_h)]/10} \}$$

$$-X^R_j \leqslant dB(A)^{max}_j \qquad \forall j \qquad (10)$$

and nonnegativity conditions:

$$X^S_i, X^R_j, S_h > 0 \qquad \forall i,j,h \qquad (11)$$

The above model can be transformed for solution by separable programming algorithms which are generally available at most scientific computing facilities. Letting the parameter $e_{ij}$ equal the known constant $10^{dB(A)_{ij}/10}$, Equation 10 can be expressed as

$$\sum_i e_{ij}\ 10^{-(X^S_i + X^R_j + \sum_h A^h_{ij}(S_h))/10} \leqslant 10^{dB(A)^{max}_j/10}$$

$$\forall j \qquad (12)$$

If the variable $P_{ij}$ equals $10^{-(S^S_i + X^R_j + \sum_h A^h_{ij}(S_h))/10}$, then the environmental noise standard, Equations 10 or 12, can be written,

$$\sum_i e_{ij}\ P_{ij} \leq 10^{dB(A)^{max}_j/10} \qquad \forall j \qquad (13)$$

Since $0 < P_{ij} \leqslant 1$, the logarithm of $P_{ij}$ exists. Hence

$$-10 \log P_{ij} = X_i^S + X_j^R + \sum_h A_{ij}^h (S_h) \qquad \forall i,j \qquad (14)$$

and by definition

$$P_{ij} = 10^{\log P_{ij}} \qquad \forall i,j \qquad (15)$$

Each variable "$\log P_{ij}$" can be approximated by piecewise linear segments k having slopes $L_k$ as illustrated below.

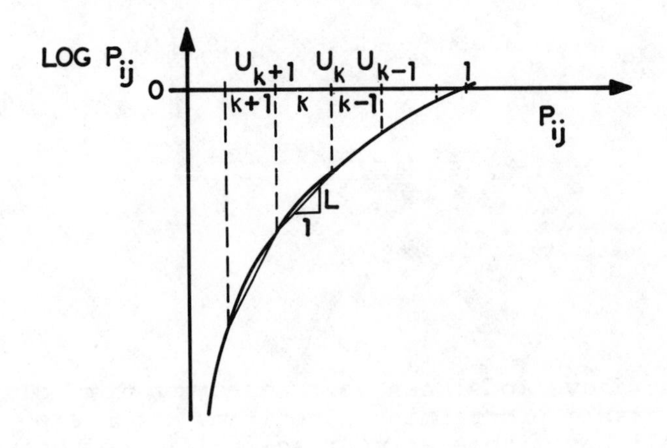


*Figure 18.1. Piecewise linear approximation of logarithmic function.*

Defining variables $P_{ijk}$ such that

$$1 - \sum_k P_{ijk}/L_k = P_{ij} \qquad \forall i,j \qquad (16)$$

$$0 \leqslant P_{ijk} \leqslant (U_k - U_{k+1})L_k \qquad \forall i,j,k \qquad (17)$$

permits a linear approximation of Equation 15

$$\log P_{ij} \cong - \sum_k P_{ijk} \qquad \forall i,j,k \qquad (18)$$

Model I, in a form suitable for solution by separable programming algorithms, can now be summarized. The objective is to estimate the minimum cost combination of dB(A) reductions $X_i^S$ and $X_j^R$ at source sites i and receptor sites j, and the scale $S_h$ of alternatives h for modifying the attenuation of sound between source and receptor sites, required to satisfy environmental noise standards expressed as a function of the dB(A) levels at each receptor site.

Model I:

$$\text{Minimize} \quad \sum_i C_i^S(X_i^S) + \sum_j C_j^R(X_j^R) + \sum_h C_h(S_h) \tag{9}$$

subject to:

$$\sum_i e_{ij} \sum_k P_{ijk}/L_k \geqslant \sum_i e_{ij} - 10^{dB(A)_j^{max}/10} \qquad \forall j \qquad (13,16)$$

$$10 \sum_k P_{ijk} = X_i^S + X_j^R + \sum_h A_{ij}^h(S_h) \qquad \forall i,j \qquad (14,18)$$

$$0 \leqslant P_{ijk} \leqslant (U_k - U_{k+1})L_k \qquad \forall i,j,k \qquad (17)$$

Experience in solving this model indicates that global optimality (in the mathematical sense) can be achieved if the slopes $L_k$ and segment bounds $U_k$ are the same for each $P_{ij}$, if the attenuation functions $A_{ij}^h(S_h)$ are concave and if the cost functions in Equation 9 are convex, as they appear to be based on the limited data currently available.[2]

The above model can be modified to include the duration of noise as well as its intensity and frequency. In fact, both occupational and non-occupational noise standards have been expressed in a manner that includes the duration of noise. The Walsh-Healy Act, which applies to individuals working in activities involved in interstate commerce, specifies the maximum time that any sound pressure level above 90 dB(A) can be heard in an 8-hour span if the remaining levels are less than 90 dB(A). These times are listed in Table 18.1.[3]

*Table 18.1*
*Walsh-Healy Noise Level Standards*

| *Sound Level Index s* | *Sound Level $dB(A)_s$* | *Permissible Duration Hours $T_s$* |
|---|---|---|
| 1 | 90 | 8.0 |
| 2 | 92 | 6.0 |
| 3 | 95 | 4.0 |
| 4 | 97 | 3.0 |
| 5 | 100 | 2.0 |
| 6 | 102 | 1.5 |
| 7 | 105 | 1.0 |
| 8 | 110 | 0.5 |
| 9 | 115 | 0.25 |

A similar set of noise levels and maximum permissible durations have been recommended for nonoccupational environments for a 16-24 hour day.[4] Both sets of standards state that for all discrete sound levels $dB(A)_s$, the actual duration of those levels, $t_s$, must be such that the sum of the ratios $t_s/T_s$, where $T_s$ is the maximum permissible duration, cannot exceed unity.

$$\sum_s \frac{t_s}{T_s} \leqslant 1 \tag{19}$$

This form of environmental noise standard can be incorporated into Model I by first noting that for both the occupational and nonoccupational noise standards just discussed the product of the square of the pressure ratios, $10^{dB(A)_s/10}$, raised to the power 0.602, times the maximum permissible duration, $T_s$, is a constant. For the occupational standards this constant equals $T_s\,(10^{dB(A)_s/10})^{0.602} = 2.095 \times 10^6$. The corresponding constant for the recommended nonoccupational standards equals $0.262 \times 10^6$.

Next, define $10^{dB(A)_{sj}/10}$ as the square of the sound pressure ratio s heard at a receptor site j. Let the constant $\alpha_j$ be the appropriate weighting exponent for that receptor site (*e.g.*, $\alpha_j = 0.602$). Multiplying each term of the numerator and denominator of Equation 19 by $(10^{dB(A)_{sj}/10})^{\alpha_j}$ does not change its value.

$$\sum_s \frac{t_s (10^{dB(A)_{sj}/10})^{\alpha_j}}{T_s (10^{dB(A)_{sj}/10})^{\alpha_j}} \leqslant 1 \tag{20}$$

If each term in Equation 20 has a constant denominator, as it does for the occupational and nonoccupational standards just described, that constant can be labeled $E_j^{max}$, the maximum allowable sound duration-pressure ratio permitted at site j. Multiplying each term in Equation 20 by $E_j^{max}$ yields an environmental noise standard equivalent to the original standard expressed by Equation 19.

$$\sum_s t_s \, (10^{dB(A)_{sj}/10})^{\alpha_j} \leqslant E_j^{max} \qquad \forall j \tag{21}$$

Finally, consider a set $N_s$ of simultaneously operating noise sources i, each individually contributing a sound pressure level of $dB(A)_{ij}$ at site j for a duration of $t_s$ hours. The total squared sound pressure ratio heard at each receptor site j during this period of $t_s$ hours follows from portions of Equations 7, 12 and 13.

$$10^{dB(A)_{sj}/10} = \sum_{i \in N_s} 10^{[dB(A)_{ij} - X_i^S - X_j^R - \sum_h A_{ij}^h (S_h)]/10}$$

$$= \sum_{i \in N_s} e_{ij} \, P_{ij} \qquad \forall j,s \tag{22}$$

Substituting Equation 22 into Equation 21 yields the environmental noise standard that, in addition to frequency and intensity, includes duration.

$$\sum_s [\sum_{i \in N_s} e_{ij} \, P_{ij}]^{\alpha_j} t_s \leqslant E_j^{max} \qquad \forall j \tag{23}$$

This standard can replace the previous standard, Equation 13, that considered only intensity and frequency.

Equation 23 can be rendered amenable to solution by separable programming algorithms by using more logarithmic transformations. Let the variable

$$Q_{js} = \sum_{i \in N_s} e_{ij} P_{ij} \qquad \forall j,s \quad (24)$$

Since $0 < Q_{js} \leqslant \sum_{i \in N_s} e_{ij}$, the logarithm of $Q_{js}$ can be defined and approximated by a piecewise linear function

$$\log Q_{js} \cong \log \sum_{i \in N_s} e_{ij} - \sum_k Q_{jsk} \qquad \forall j,s \quad (25)$$

where

$$\sum_{i \in N_s} e_{ij} - \sum_k Q_{jsk}/L'_k = Q_{js} \qquad \forall j,s \quad (26)$$

$$0 \leqslant Q_{jsk} \leqslant (U'_k - U'_{k+1})L'_k \qquad \forall j,s,k \quad (27)$$

Next let

$$R_{js} = (Q_{js})^{\alpha_j} t_s \qquad \forall j,s \quad (28)$$

Since $Q_{js}$, $\alpha_j$ and $t_s > 0$, $0 < R_{js} \leqslant t_s[\sum_{i \in N_s} e_{ij}]^{\alpha_j}$ and its logarithm exists,

$$\log R_{js} = \log t_s + \alpha_j \log Q_{js} \qquad \forall j,s \quad (29)$$

Again approximating $\log R_{js}$ by a piecewise linear function,

$$\log R_{js} \cong \log\{ [\sum_{i \in N_s} e_{ij}]^{\alpha_j} t_s\} - \sum_k R_{jsk} \qquad \forall j,s \quad (30)$$

where

$$t_s \left[\sum_{i \in N_s} e_{ij}\right]^{\alpha_j} - \sum_k R_{jsk}/L''_k = R_{js} \qquad \forall j,s \quad (31)$$

$$0 \leqslant R_{jsk} \leqslant (U''_k - U''_{k+1})L''_k \qquad \forall j,k,s \quad (32)$$

The environmental noise standard, Equation 23, can now be written in a linear form:

$$\sum_s R_{js} \leqslant E_j^{max} \qquad \forall j \qquad (33)$$

To summarize, the problem is to estimate the minimum cost combination of dB(A) reductions, $X_i^S$ and $X_j^R$, at each noise source and receptor site, and the scale $S_h$ of alternatives h that will modify the noise abatement between each source and receptor site, required to meet environmental noise standards expressed as a function of dB(A) levels and durations.

Model II:

$$\text{Minimize} \qquad \sum_i C_i^s(X_i^S) + \sum_j C_j^R(X_j^R) + \sum_h C_h(S_h)$$

subject to environmental noise standards:

$$\sum_s \sum_k R_{jsk}/L''_k \geqslant \sum_s t_s \left[\sum_{i \varepsilon N_s} e_{ij}\right]^{\alpha_j} - E_j^{max}$$

$$\forall j \qquad (31,33)$$

and definitions relating the decision variables $\overline{X}$ (dB(A) reductions) and $\overline{S}$ (scale of alternatives for modifying noise transmission) to the variables $R_{jsk}$:

$$10 \sum_k P_{ijk} = X_i^S + X_j^R + \sum_h A_{ij}^h(S_h) \qquad \forall i,j \ (14,18)$$

$$0 \leqslant P_{ijk} \leqslant (U_k - U_{k+1})L_k \qquad \forall i,j,k \qquad (17)$$

$$\sum_k Q_{jsk}/L'_k = \sum_{i \varepsilon N_s} e_{ij} \left\{\sum_k P_{ijk}/L_k\right\} \qquad \forall j,s \ (16,24,26)$$

$$0 \leqslant Q_{jsk} \leqslant (U'_k - U'_{k+1})L'_k \qquad \forall j,s,k \qquad (27)$$

$$\alpha_j \sum_k Q_{jsk} = \sum_k R_{jsk} \qquad \forall j,s \quad (25,29,30)$$

$$0 \leq R_{jsk} \leq (U''_k - U''_{k+1})L''_k \qquad \forall j,s,k \quad (32)$$

Additional piecewise linear functions may be needed to define the costs in Equation 9 and the attenuation functions $A^h_{ij}(S_h)$ in Equation 14, 18 if they are nonlinear. Also note that the upper bounds $U_1$ for the variables $P_{ij1}$, $Q_{js1}$ and $R_{js1}$ are the maximum values of $P_{ij}$, $Q_{js}$ and $R_{js}$, namely 1, $\sum_{i \in N_s} e_{ij}$ and $[\sum_{i \in N_s} e_{ij}]^{\alpha_j} t_s$ respectively, and hence the segment index k will begin at different values for these three sets of variables. As illustrated in Figure 18.1, the segments are ordered so that $U_k > U_{k+1}$ for each segment k.

It is clear that some noise control measures could modify the duration of the noise with or without a concurrent dB(A) reduction. In these situations the variables $t_s$ would be unknowns. While the algebra involved to include changes in duration among the decision variables is relatively straightforward, it is not a practical exercise at this point for a variety of reasons. One reason is that limited data exist on the costs as well as the effects of noise duration control measures. In addition, very often the duration of noise generated from each source is stochastic, and this important feature has not been included in either the above models.

*Some Numerical Examples*

This section ends with some simple numerical examples to illustrate the application of these models. Consider an area having five identifiable noise sources and two receptor sites. Table 18.2 indicates the sets of sources, $N_s$, that operate simultaneously for the times, $t_s$, specified and also the A-weighted sound pressure levels, $dB(A)_{ij}$, that each source i contributes to each of two receptor sites j.

Given the information presented in Table 18.2, each parameter $e_{ij} = 10^{dB(A)_{ij}/10}$ can be computed.

*Table 18.2*

*Durations, Combinations and Levels of Noise Sources*

| *Source Number* $i$ | *Source Level* $dB(A)_i$ | *Combinations of Sources Operating Source Set Index s* 1 | 2 | 3 | 4 | 5 | $dB(A)_{ij}$ *Levels Receptor Site j* 1 | 2 |
|---|---|---|---|---|---|---|---|---|
| | | *Source Sets* $N_s$ | | | | | | |
| 1 | 80 | * | * | | * | * | 40 | 60 |
| 2 | 85 | * | * | * | | * | 73 | 60 |
| 3 | 72 | * | * | * | * | * | 40 | 60 |
| 4 | 68 | | * | | * | * | 41 | 60 |
| 5 | 110 | | | | * | * | 97 | 60 |
| Expected duration, $t_s$ hours: | | 3 | 2 | 1 | 1 | 1 | | |

A variety of solutions were obtained using Model I in which the duration of each noise level was not considered. Also assumed for simplicity were linear cost functions and no alternatives for increasing noise abatement in the transmitting medium. Some of these solutions are included in Table 18.3 as an illustration of how the model works in situations where the optimal solution is relatively obvious. The example problems and solutions shown in Table 18.3 are clearly not intended to be realistic with respect to costs or standards, but they are a better illustration than more realistic and complex examples of the types of solutions to be expected and the reasonableness of the solutions resulting from these simplified problems.

Without any noise abatement measures, the total sound pressure level at receptor sites 1 and 2 with all sources operating simultaneously, as they do for 1 hour, equals 97 and 67 dB(A) respectively. Noise standards in the first two problems constrain the sound level at receptor site 2 to be no greater than 55 dB(A), requiring a reduction of 12 dB(A) at each source site, or a similar reduction at the receptor site, or a combination of equivalent reductions at various source and receptor sites. In the first problem the total cost is less if each source is reduced by 12 dB(A), and conversely in the second problem it is cheaper to reduce the sound level heard at the receptor site by 12 dB(A). Problems

*Table 18.3*

*Solutions from Model I*

| *Problem No.* | *Standards* $dB(A)_j^{max}$ *Receptor Site j* 1 | 2 | | *Unit Cost and dB(A) Reduction (Solution) Vectors* *Source Sites i* 1 | 2 | 3 | 4 | 5 | *Receptor Sites j* 1 | 2 |
|---|---|---|---|---|---|---|---|---|---|---|
| 1 | 100 | 55 | C: | 1 | 1 | 1 | 1 | 1 | 0 | 6 |
| | | | X: | 12 | 12 | 12 | 12 | 12 | 0 | 0 |
| 2 | 100 | 55 | C: | 1 | 1 | 1 | 1 | 1 | 0 | 4 |
| | | | X: | 0 | 0 | 0 | 0 | 0 | 0 | 12 |
| 3 | 70 | 70 | C: | 1 | 1 | 1 | 1 | 1 | 3 | 0 |
| | | | X: | 0 | 6 | 0 | 0 | 30 | 0 | 0 |
| 4 | 35 | 70 | C: | 1 | 1 | 1 | 1 | 1 | 3 | 0 |
| | | | X: | 0 | 25 | 0 | 0 | 55 | 15 | 0 |
| 5 | 35 | 70 | C: | 1 | 1 | 1 | 1 | 1 | 6 | 0 |
| | | | X: | 12 | 45 | 12 | 13 | 69 | 0 | 0 |
| 6 | 35 | 35 | C: | 1 | 1 | 1 | 1 | 1 | 2 | 1 |
| | | | X: | 0 | 11 | 0 | 0 | 40 | 29 | 30 |

3, 4 and 5 include standards that are binding for receptor site 1, and the solutions of the model are again what would be expected given the relative costs of abatement at each source as compared to the abatement costs at the receptor site. Note in Problems 3 and 4 that as the environmental noise standard is lowered so that more noise sources are affected, abatement at the receptor site becomes more attractive even though the relative abatement costs have not changed. Increasing the receptor costs, Problem 5, again shifts more of the burden of abatement to the source sites. There are other rather obvious solutions.

Problem 6 requires noise level reductions to meet the standards at both receptor sites, and in this case, as simple as it is, the solution is not as obvious as it is for the other five problems. Nevertheless as in the first five examples the combination of source and receptor reductions indicated in the solution of the model does in fact minimize the total cost of meeting the environmental noise standards.

Finally an example that includes the duration of noise can be illustrated using the information from Table 18.2 and Model II. The sets of noise sources, $N_s$, that operate simultaneously for the specified number of hours, $t_s$, are indicated by the columns of asterisks in Table 18.2. For example, $N_1=\{i=1,2,3\}$ and $N_4=\{i=1,3,4,5\}$. Accepting the recommended nonoccupational environmental noise standard of $E_j^{max} = 0.262 \times 10^6$ for receptor site $j = 1$ and a more strict standard of $E_j^{max} = 0.050 \times 10^6$ for receptor site $j = 2$, Table 18.4 summarizes the solutions obtained from Model II. In these example problems the exponent $\alpha_j$ equals 0.602 for each receptor site.

*Table 18.4*

*Example Solutions from Model II*

| *Solution No.* | *Standards $E_j^{max}$ Receptor Site j* 1 | 2 | | *Unit Cost and dB(A) Reduction (Solution) Vectors* — *Source Sites i* 1 | 2 | 3 | 4 | 5 | *Receptor Sites j* 1 | 2 |
|---|---|---|---|---|---|---|---|---|---|---|
| 1 | $0.262 \times 10^6$ | None | C: | 1 | 1 | 1 | 1 | 1 | 3 | 0 |
| | | | X: | 0 | 3 | 0 | 0 | 12 | 0 | 0 |
| 2 | None | $0.05 \times 10^6$ | C: | 1 | 1 | 1 | 1 | 1 | 0 | 2 |
| | | | X: | 0 | 0 | 0 | 0 | 0 | 0 | 3 |
| 3 | $0.262 \times 10^6$ | $0.05 \times 10^6$ | C: | 1 | 1 | 1 | 1 | 1 | 3 | 2 |
| | | | X: | 0 | 3 | 0 | 0 | 12 | 0 | 2 |

| | *Receptor Site 1* — *Noise Level $dB(A)_1$* | *Duration Hours* | *Receptor Site 2* — *Noise Level $dB(A)_2$* | *Duration Hours* |
|---|---|---|---|---|
| 1 | 70 | 6 | 66 | 3 |
| | 85 | 2 | 65 | 1 |
| | | | $\leq$ 64 | 4 |
| 2 | 73 | 6 | 64 | 1 |
| | 97 | 2 | 63 | 3 |
| | | | $\leq$ 62 | 4 |
| 3 | 70 | 6 | 64 | 3 |
| | 85 | 2 | 63 | 1 |
| | | | $\leq$ 62 | 4 |

## V. CONCLUSIONS

This paper has presented only an outline of an approach to urban noise management. Considerable research is required before quantitive methods similar to those proposed and illustrated in this study can be used with confidence. Particularly necessary is the development of improved techniques for predicting and verifying noise attenuation in urban areas, for estimating costs of noise abatement as a function of the reduction achieved, and for defining and evaluating noise control alternatives based on various economic and social criteria. Without this information and methodology, attempts to achieve desired environmental noise levels may not only be ineffective but needlessly costly.

## REFERENCES

1. Beranek, L. L. *Noise and Vibration Control* (New York: McGraw-Hill, 1971).
2. National Bureau of Standards. "The Economic Impact of Noise," Report NT1D300.14 of Environmental Protection Agency (December, 1971).
3. *Federal Register*, Volume 34, No. 96, Part 2, pp. 7948-7949, May 20, 1969; also in Volume 36, No. 105, paragraph 1910.95, May 29, 1971.
4. Cohen, A. *et al.* "Sciocusis-Hearing Loss from Non-occupational Noise Exposure," *Sound and Vibration*, p. 12 (November, 1970).

# PART VI

# TOTAL ENVIRONMENTAL MODELS

MODELS FOR ENVIRONMENTAL POLLUTION CONTROL


# CHAPTER 19

# TOTAL ENVIRONMENTAL QUALITY MANAGEMENT MODELS

Walter O. Spofford, Jr.*

## I. INTRODUCTION

Traditionally, gaseous, liquid, and solid waste management problems have been dealt with separately with little or no concern for the other environmental media--atmosphere, water courses, and land--into which residuals are discharged. This fragmented approach to waste (or residuals) management has very often resulted in the imposition of unanticipated, undesirable side effects upon society. A good example of this in recent years is the regulations imposed on emissions from apartment house incinerators in New York City by the air pollution control agency. The result: tons of additional unanticipated solid "wastes" which could not be handled adequately by the city's Sanitation Department.

Most material residuals are readily transformed from one form to another (for example: liquids to "solids" as in municipal sewage to sewage sludge; liquids to gases as in ammonia stripping; solids to gases as in combustion of municipal refuse and sewage sludge) and among forms of residual-bearing streams (for example, gaseous to liquid as in the wet scrubbing of gaseous effluent streams containing particulate matter; gases to solids as in the electrostatic precipitation of particulates) in production, consumption, and residuals modification activities. By the application of appropriate equipment and energy, virtually all undesirable substances can be removed

*Dr. Walter O. Spofford, Jr. is an environmental engineer with Resources for the Future, Inc., 1755 Massachusetts Avenue, N.W., Washington, D.C. 20036, U.S.A.

from air and water streams. Though the characteristics of material residuals may be changed, the materials do not disappear. Their total weight, which must be dealt with in one form or another, is still with us.

Because management alternatives that improve water quality can cause a degradation in air quality, because alternatives that reduce gaseous emissions can result in increased liquid emissions,[1] and finally because with present strategies all these alternatives often result in the generation of significant quantities of solid residuals, in principle comprehensive regional residuals management models should include relevant energy residuals, all three forms of material residuals, and the three receiving environmental media within a single computational algorithm.

Even when only one form of material residual and one environmental medium has been considered, classical residuals management models have been deficient. They have traditionally neglected important management alternatives for reducing the quantities of residuals initially generated, and for modifying residuals after generation other than by waste treatment. For example, water quality management models have dealt primarily with wastewater treatment activities and with low flow augmentation,[2] instream aeration,[3,4] and by-pass piping schemes.[5,6] Rarely have they considered the impact on residuals generation of changes in basic production processes, of changes in raw material inputs, or of changes in product specifications. For example, a switch from batch to continuous diffusion in beet sugar refineries resulted in a reduction of 17.9 pounds of BOD per ton of beets sliced.[7] Reducing the brightness specification for newsprint from the 66-70 range presently used to about 50 (where bleaching of both groundwood and kraft pulps would not be necessary) results in a reduction of dissolved inorganic solids and dissolved organic solids of about 75 and 30 per cent, respectively.[8] In addition, possibilities frequently exist for recirculation of residual-bearing streams to enable materials recovery and by-product production.

Recent studies indicate that for at least some industrial activities, nontreatment industry alternatives are frequently less expensive than wastewater treatment, and that the savings involved may be substantial.[7-9] Clearly, if the aim is the accurate representation of the alternatives available to society, nontreatment activities should be included in our management models just as treatment facilities are included at present.

The purpose of this paper is to delineate the entire regional residuals management system, including the range of management options available to society for improving environmental quality, and to indicate how all the components of the system may be expressed and ultimately linked together within the same computational framework. The regional residuals management model is discussed qualitatively in the next section. In the following section, the management model is formulated quantitatively, mathematical forms of the various environmental models are discussed, and computational schemes for selecting an optimal management strategy are suggested.

Before proceeding with the next section, however, we must define some terms. The term *total environmental quality management models* means *regional residuals-environmental quality management models*, where these include only the impacts on environmental quality of residuals discharges to the air, water, and land environments. No implication that the residuals sector is the only important sector relating to environmental quality is made since we prefer not to enter the debate defining environmental quality. It is clear that residuals comprise a very important, and at present obvious, component of the environmental quality problem. In addition, they comprise a sector about which decisions must be made (now), and on which coherent research is possible, including model building and analysis.

The term *residual* is used rather than *waste* for two reasons: first, the material (or energy) which is commonly called a waste stems from the residuals or leftovers of man's production and consumption activities; and second, the term residual is an economically neutral term and, therefore, does not imply value (or lack of value) to this "waste" material. Very often these leftovers are not wastes at all, but valuable resources which can be recovered and reused as inputs to production processes. The term *form of residual* is used loosely to describe both the state of the residual material itself and the state of the residual-bearing stream, whichever is applicable. Thus, suspended organic material (BOD) might be considered a solid, but as long as it is being transported in water it is referred to as a liquid residual. In addition, we will refer to particulate matter in a gaseous effluent stream as a gaseous residual, even though the particles themselves are in a solid state.

The use of the term *management model* is restricted in this paper to those situations which involve the ranking of sets of management options according to a given economic criterion. Optimization (or programming) models come under this category, as do simulation models when the costs and benefits associated with the various exogeneously determined management alternatives are delineated, compared, and ranked. The term *environmental model* is reserved for those models which describe the impact on the natural world of exogeneous inputs, such as residuals discharges, but where an economic criterion and subsequent ranking of management alternatives is not implied. Environmental quality models are a necessary part of the more encompassing residuals-environmental quality management models.

Other terms used in this paper will be defined or clarified as necessary as they are introduced.

## II. THE REGIONAL RESIDUALS MANAGEMENT SYSTEM

A regional residuals management model that includes all the relevant management options available to society for improving environmental quality consists of the following components:

1. *residuals generation and discharge models:* describe the factors influencing the generation, modification (*e.g.*, treatment), and final discharge to the environment of residuals--gaseous, liquid and solid--from individual production and consumption activities, including relevant costs
2. *environmental modification models:* describe the options available for improving the assimilative capacity of the environment, including their costs
3. *environmental quality models:* translate the time and spatial patterns of residuals discharges into time and spatial patterns of the resulting states of the natural environment (described by ambient residuals concentrations and population sizes of biological species of interest)
4. *damage functions:* relate time and spatial patterns of ambient residuals concentrations to the resulting impacts on receptors--man, animals, plants, and structures--in physical, biological, and economic terms
5. *management strategies:* consist of alternative sets of measures that affect one or more points in the management system, together with the costs and benefits associated with each strategy.

A conceptual framework for regional residuals-environmental quality management is depicted in Figure 19.1. Some examples of management options available for improving environmental quality are presented as Table 19.1.[10]

One way of specifying a management strategy, or set of management strategies, is through the use of mathematical programming (optimization) techniques. We will return to a discussion of this in the next section. Before that, though, we will discuss those components of the system that warrant further elaboration.

### *Residuals Generation and Discharge Models*

Residuals generation and discharge models are used to relate inputs--material and energy--with product and nonproduct outputs at specified locations throughout a region. Nonproduct outputs are defined as the difference (in weight, or energy content) between inputs and desired product outputs--intermediate or final. The nonproduct outputs may be reduced in quantity through recycling or by-product production. What remains to be handled, modified and disposed of in some manner are the residuals.

Incorporated in these models are the management options available for reducing, modifying, handling, and transporting solid, liquid, and gaseous residuals up to the point of discharge into the environment. These models permit choices among production processes, raw material input mix, recycle of residuals, by-product production, and in-plant adjustments and improvement, all of which can reduce the total quantity of residuals for disposal. In addition, they allow for choices among residuals modification ("treatment") processes and thus among the possible forms of the residuals to be disposed of in the natural environment and the environmental media into which they are disposed, as well as for choices among discharge locations. Landfill operations for disposing of solid residuals are included in these models. But a description of the impact of this disposal activity on the quality of the underlying groundwater[11] would fall within the environmental quality model category to be discussed below.

Residuals generation models encompass all types of economic activities in the region within which residuals are generated and subsequently discharged

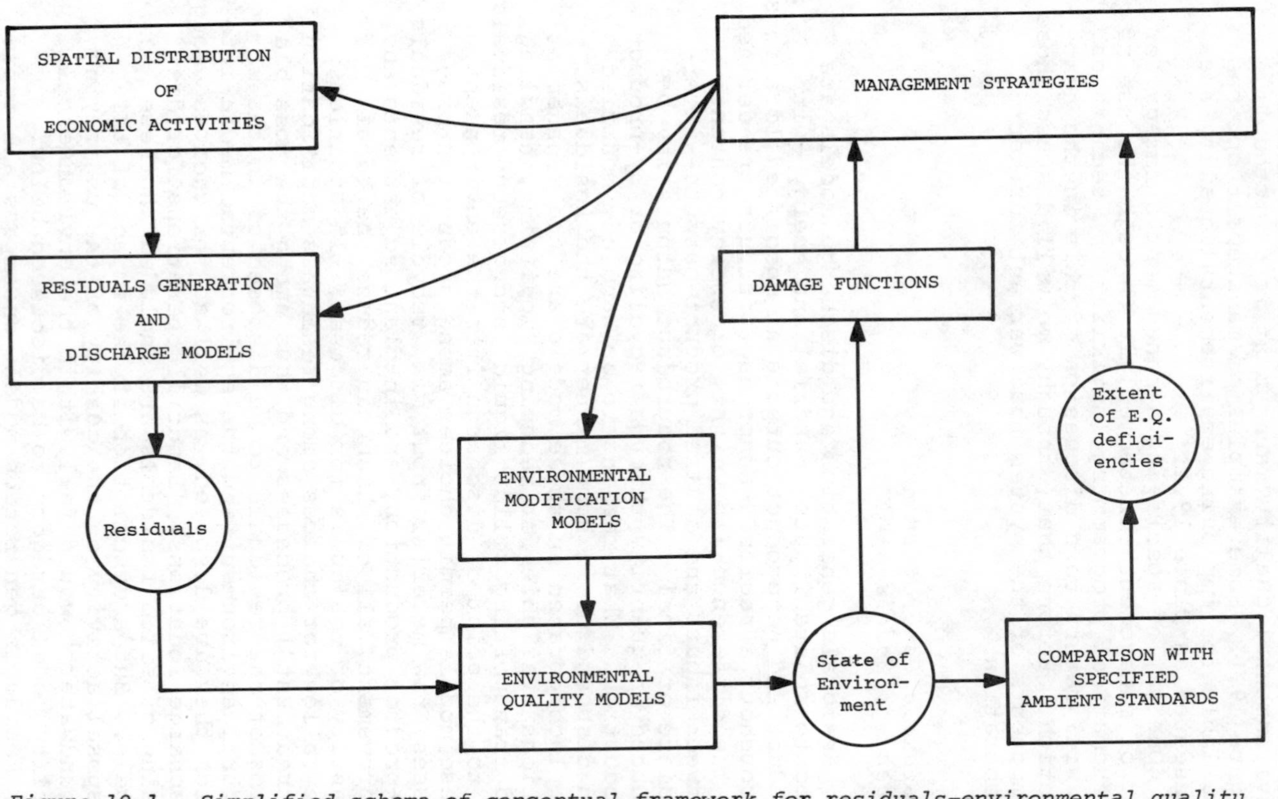


*Figure 19.1. Simplified schema of conceptual framework for residuals-environmental quality management.*

*Table 19.1*

*Classification of Current Measures for Improving Environmental Quality**

| *Category of Measure* | *Examples* |
|---|---|
| 1. Reduce residuals generation | a. Increase longevity of goods.<br>b. Change type of raw material inputs. Examples: from high to low sulfur fuels; concentrated rather than raw ore.<br>c. Change production process, including mode of transport. $H_2SO_4$ to HCl for pickling steel; internal to external combustion engines for vehicles; modify jet engine to reduce particulates, noise.<br>d. Change nature of final product outputs, that is, modify final demand. Prohibit plastic containers made from PVC; change brightness, color specifications on paper products; use short-lived, specific pesticides instead of long-lived general pesticides. |
| 2. Modify residuals after generation in on-site or collective facilities | a. Materials recovery (direct recycle).<br>b. By-product production (indirect recycle). Examples: pet foods; briquettes; pressed logs; particle board; fly ash concrete, bricks, building materials.<br>c. Modification of residuals streams ("waste treatment" activities). Examples: precipitation, clarification, biological oxidation, filtration, incineration, scrubbers, sludge disposal, compression of solid residuals. |
| 3. Make better use of, or modify, assimilative capacity of environment. | a. Build higher effluent stacks.<br>b. Change location of residuals discharge.<br>c. Artificial aeration of streams, lakes.<br>d. Flow augmentation.<br>e. Build up topography with solid residuals.<br>f. Prohibit modification of topography, such as filling of bays, which may reduce assimilative capacity.<br>g. Regulate land use, with respect to either or both residuals generators and receptors.<br>h. Regulate time pattern of activities by production scheduling, staggering office hours, regulating time of residuals discharges. |

*Adapted from B. T. Bower, "Integrated Residuals Management," Resources for the Future, Inc., unpublished (1972).

to the environment. Examples of activities within this category are (1) industrial production, both of intermediate and final goods, (2) mining activities, (3) agricultural activities, (4) forest (management) activities, (5) urban and suburban runoff, and (6) consumption activities (both commercial and residential residuals generators) including municipal waste treatment facilities. Models of these activities range from comparatively simple statistical (regression) functions relating outputs and inputs to relatively sophisticated linear programming models.[12] Substantial progress has been made in the last decade in developing models that relate residuals discharges with products, production levels, and other relevant factors of production, especially for industrial activities.[13]

### *Environmental Modification Models*

Environmental modification models are used to describe management options available for improving the assimilative capacity of the natural environment at various points in space and time. Provision of additional river flow through low flow augmentation and the addition of dissolved oxygen to water bodies in instream aeration are examples in this category.

### *Environmental Quality Models*

Environmental quality models--air and water dispersion, chemical reaction, and models of terrestrial and aquatic ecosystems--are used to describe the impact on the environment of energy and material residuals discharged from the various production and consumption activities in the region. These models are employed to predict the ambient concentrations of residuals and population sizes of species at various points in space throughout the environment. They must be given a set of residuals discharge rates from the regional residuals generation models and a set of values for the environmental parameters, including stream flow and velocity, wind speed and direction, atmospheric stability, and atmospheric mixing depth. We will return to a discussion of environmental models in the next section where we suggest possibilities for incorporating them, especially the nonlinear, simulation model variety, within an optimizing framework.

*Management Strategies*

A management strategy consists of a set of measures or activities affecting one or more of the residuals generators, the natural environment, or the receptors.[14] These strategies can be evaluated in the context of different objective functions and different sets of constraints, including both effluent and ambient environmental quality standards. Basically there are three approaches to seeking an optimal strategy for any given objective and set of constraints: (1) response surface sampling using simulation, (2) optimization (mathematical programming), and (3) a combination of (1) and (2). An example of the latter is exogenous treatment of various levels of low flow augmentation in a water quality optimization model. However, the costs (and damages if they occur) of providing the various augmented flows are included in the overall ranking of the various management alternatives.

Each approach has its advantages and disadvantages. Simulation models, in general, are able to provide a more realistic representation of real world conditions and their outputs are generally easier to obtain than optimization models are to solve. Their major disadvantage is the general difficulty of selecting *a priori* that combination of raw material inputs, production processes, recycling and by-product production opportunities, and residuals modification activities and levels that optimizes a given objective function. Exhaustive sampling of a finite number of combinations can be used. But because the total number of combinations is usually extremely large, random sampling techniques appear to be a more reasonable approach.

The major advantage of optimization models is the direct determination of the activity levels that optimize a given objective function. Their major disadvantages, given the magnitude of the regional residuals management problem, are that they are generally difficult to construct and then to solve, even when formulated as linear programming problems. Furthermore, they may not be good representations of the actual (real world) situation. For some cases, a combination of simulation and optimization techniques provide the logical approach to residuals management problems.[15] The use of one technique or the other or a combination would depend upon each individual situation.

Because of the extremely large number of management options available in a given region for improving environmental quality, optimization models are useful. Unfortunately, as pointed out earlier, they are more difficult to construct and subsequently solve than simulation models. This is particularly true when complex, nonlinear models of the natural world (for example, of aquatic ecosystems) are to be incorporated within the optimization framework. In the next section, we suggest an approach to regional residuals management modeling which deals explicitly with: (1) production, consumption, and residuals treatment activities, (2) all forms of material residuals simultaneously, and (3) complex models of the natural world.

## III. MANAGEMENT MODEL FORMULATION AND OPTIMIZATION SCHEME

In this section, we (1) present a formal mathematical description of the regional residuals management model, (2) indicate a method of handling certain kinds of constraints that are difficult or impossible to deal with in the traditional manner, and finally, (3) suggest procedures for optimizing a nonlinear objective function subject to a set of linear constraints.

The management model presented here is deterministic and steady state. Only one season, which could represent either the low flow season or an entire year, is considered at a time. Also, from an economic point of view, the model is static. Later on, it may be desirable to extend this model to incorporate dynamic aspects. This would depend, however, on the objectives of the analysis, and on the relative importance of both data and model components in specifying an optimal regional management plan. In any case, further refinement will have to wait because the regional residuals management modeling task is difficult enough under the simplifying assumptions stated above.

A quantitative framework for residuals management decisions, including both a conceptual framework for analysis and a didactic application to a small hypothetical region, has been presented elsewhere.[16] In this application, appropriate demand functions and economic damage functions associated with ambient residuals concentrations at various locations throughout the region were assumed to exist, the environmental models--air and water quality--were assumed to be linear, and the objective function was one of net regional benefits. In a follow-up but still didactic application, the model was expanded to provide information on the sociogeographic distribution of costs and benefits associated with meeting different levels of environmental quality.[17] This model was applied to an hypothetical region similar to the one mentioned above, and linear environmental models were again employed.

In a more recent effort, the beginnings of an application of this modeling approach to an actual region--the Delaware Valley region of New Jersey, Pennsylvania, and Delaware--was presented.[18] In this application, a nonlinear simulation model of an aquatic ecosystem was included within the optimization algorithm, and some computational problems associated with scaling up from relatively small hypothetical regions to significantly larger actual regions were discussed.

### *Model Formulation*

The type of objective function that is appropriate for regional residuals environmental quality management depends upon the particular situation, data availability, and the objectives of the analysis. We do not intend to examine this question here except to point out that various possibilities exist, among them, to maximize total regional net benefits and to minimize the total costs of meeting exogeneously determined ambient environmental quality standards. For purposes of exposition, we assume that we must meet specified ambient environmental quality standards and that we must supply to the region a fixed bill of goods and services. Regional imports and exports are allowed. In addition, we assume that the objective function is expressed in the form of a net benefit function, and, hence, the objective is to maximize. The positive elements in this function include gross revenues from the sale of various products. The negative elements include: all the opportunity costs of traditional production inputs, all liquid and gaseous residuals treatment (modification) costs, and all collection, transport, and landfill costs associated with the disposal of solid residuals.

In this discussion, we will concentrate on overall model formulation. Individual components and submodels will be discussed only when relevant to the overall model formulation. We will carefully examine the mathematical form of the various environmental models, as this is central to the issue of how they are handled within an optimization framework. For this development, we assume that the residuals generation and discharge models either exist or can be constructed in one of the available linear programming forms--standard LP,[12] separable programming,[19-21] or mixed integer programming.[19,20,22]

Let us now state the residuals management problem formally.

$$\max \quad \{F = f(X, R)\} \tag{1}$$

$$\text{s.t.} \quad g_i(X) = 0 \qquad i = 1, \ldots, m < n-q \tag{2}$$

$$g_i(X) \geqslant 0 \qquad i = m+1, \ldots, p \tag{3}$$

$$h_i(X) = R_i \qquad i = 1, \ldots, q \tag{4}$$

$$R_i \leqslant S_i \qquad i = 1, \ldots, q \tag{5}$$

$$R_i \geqslant 0 \qquad i = 1, \ldots, q \tag{6}$$

$$X_i \geqslant 0 \qquad i = 1, \ldots, n \tag{7}$$

where $f(X, R)$ is, in general, a nonlinear objective function; $g_i(X) = 0$, $i = 1, \ldots, m$, is a set of linear equality constraints; $g_i(X) \geqslant 0$, $i = m + 1, \ldots, p$, is a set of linear inequality constraints; $h_i(X) = R_i$, $i = 1, \ldots, q$, represents a set of environmental functions which relate ambient concentrations of residuals and residuals discharges; $X_i$, $i = 1, \ldots, n$, is a vector of decision variables, including residuals discharges; $R_i$, $i = 1, \ldots, q$, is a vector of ambient concentration levels of residuals and population sizes of species; and $S_i$, $i = 1, \ldots, q$, is a vector of ambient environmental quality standards (*e.g.*, sulfur dioxide and particulates in the atmosphere, and algae, fish, and dissolved oxygen in the water). Note that the environmental relationships could have been written directly as

$$h_i(X) \leqslant S_i \qquad i = 1, \ldots, q$$

However, we choose to deal explicitly with the variables $R_i$, $i = 1, \ldots, q$, here as they will be useful to us in a later development.

As we mentioned earlier and will demonstrate below, some of the necessary environmental functions [$h_i(X) = R_i$] are available in linear form (*e.g.*, most of the available air dispersion relationships and the Streeter-Phelps-type dissolved oxygen models). Others are only available in nonlinear analytical form, while still others are available in various other forms. As the discussion of nonlinear aquatic ecosystem models will show, no analytical expressions for them, either linear or nonlinear, of the form $h(X) = R$ are available. The variables $R_i$, $i = 1, \ldots, q$ are expressible only as a set of implicit nonlinear functions and, hence, simulation and other iterative techniques must be used to compute their values. In general, the environmental constraint set, Equation 4, represents a variety of functional forms, many of which are difficult or even impossible to deal with using traditional mathematical programmin techniques.

Because it is generally easier to deal with a linear constraint set than with a nonlinear one, and because one of the optimization schemes we suggest below requires that all the constraints be linear, we remove the environmental relationships from the constraint set and deal with them in the objective function. This modification of the problem requires the use of penalty functions for exceeding environmental quality standards. The penalty function approach for eliminating constraints is not new. It is a well-known technique and is in frequent use in one form or another under a variety of names, for example, penalty and barrier methods[23] and exterior and interior point methods[24] depending upon whether an optimum is approached from outside or from within the original feasible region.

The new optimization problem may be stated formally as

$$\max \quad \{F = f(X) - P(X)\} \qquad (8)$$

$$\text{s.t.} \quad g_i(X) = 0 \qquad i = 1, \ldots, m \qquad (2)$$

$$g_i(X) \geqslant 0 \qquad i = m + 1, \ldots, p \qquad (3)$$

$$x_i \geqslant 0 \qquad i = 1, \ldots, n \qquad (7)$$

where

$$P(X) \equiv \sum_{i=1}^{q} p_i \, [S_i, R_i = h_i(X)] \qquad (9)$$

and where $p_i (S_i, R_i)$, $i = 1, \ldots, q$ are the penalty functions associated with exceeding the environmental standards, $S_i$, i = 1, ..., q. Note that the q environmental model relationships, $h_i(X) = R_i$, i = 1, ..., q, have been used to eliminate the q element vector of ambient concentrations, $R_i$, i=1, ..., q, from the original objective function, Equation 1.

Because many nonlinear optimization schemes require that the gradient, $\nabla F$, be evaluated at each step in the ascent (descent) procedure, we require that the objective function, Equation 8, be continuous and have continuous first derivatives. A quadratic penalty function of the following form satisfies these requirements.[23,24]

$$p_i(X) = \frac{1}{r_i}\{\max [(h_i(X) - S_i), 0]^2\} \quad i=1,\ldots,q \quad (10)$$

where $r_i$ is a penalty function parameter. Note that the second derivative of this function is also defined and that it is positive for $h(X) > S$. Note also that for situations where $h_i(X) > S_i$, the penalty, $p_i(X)$, can be made as large (or as small) as desired by merely changing the value of the penalty function parameter, $r_i$.

For computer applications, Equation 10 may be written more conveniently as

$$p_i(X) = \frac{1}{r_i}\left[\frac{(h(X) - S) + |(h(X) - S)|}{2}\right]^2 \quad i=1,\ldots,q \quad (11)$$

a form which gives $p_i(X) = 0$ when $h_i(X) < S_i$.

The new optimization problem, Equations 8, 2, 3, 7 and 9 may be solved using one of the available nonlinear programming algorithms. Throughout the optimization procedure the vector of standards, S, Equation 5, is allowed to be violated, but only at some penalty to the value of the objective function. The object of the approach is to make this penalty severe enough so that at the optimum the standards will be satisfied, within some tolerance. Whatever optimization technique is employed, we must be able to evaluate total penalties, P(X), for various sets of values for the elements of the residuals discharge vector, X. This requires solving the relevant environmental models for a given discharge vector, X, in order to determine the resulting state of the natural world, and then comparing this state with the ambient environmental quality standards. In addition, gradient methods require that the vector of marginal penalties, $\partial P(X)/\partial X$, be evaluated for each state of the natural world which has been computed. Regardless of the optimization scheme employed, this analysis requires a side computation involving the environmental models, and this computation must be made at each step in the ascent (descent) procedure.

We discuss below a nonlinear programming algorithm which uses a standard linear program for selecting better and better positions (in terms of the objective function) along a response surface. Although this optimization scheme requires only that we remove those constraints (environmental relationships) that are not of the linear form $R = A \cdot X$, we note from the formulation of the new problem, Equations 8, 2, 3 and 7, that even the linear environmental models have apparently been removed (as constraints). This is optional and depends upon the model formulation and its size. If model size (the number of rows and columns) is of no consequence and if the entire management model is contained within a single linear program (LP), it is more efficient to keep the linear environmental relationships as part of the constraint set. If size is a problem and it appears desirable to divide the management model up into a number of smaller LP's to be solved sequentially, disposition of the linear environmental models is not as straightforward.

In order to separate a large LP problem into a series of smaller ones, either the smaller ones must be completely separable in the sense that none of the original constraints relate variables which are now located in different LP's, or one of the available decomposition techniques must be employed.[19,25,26] No matter how the larger LP is subdivided, the environmental relationships, which involve all the liquid and gaseous residuals discharges throughout the region, invariably link the smaller LP's. Removing these relationships from the constraint set and placing them in the objective function is similar to the decomposition method known as price coordination (Lagrange Method) or dual feasible decomposition.[27,28] It should be noted that for this particular case, even after the linear environmental models have been removed, some form of decomposition may still be necessary for dealing with the traditional resource constraints such as those on raw materials, capital, or labor.

Before proceeding with a discussion of an optimization procedure for the regional residuals management model, it might be useful to examine the mathematical form of the various environmental models that we are apt to encounter.

### *Environmental Models*

Within an optimization framework some environmental models are easier to deal with than others. It depends, in general, upon the mathematical structure of the model. In terms of the complexity involved, we find it useful to distinguish among four broad categories: (1) linear relationships where ambient concentrations are expressed as explicit functions of residuals discharges, *i.e.*, $R = AX$, (2) linear, implicit functions, *i.e.*, $X = AR$, (3) nonlinear, explicit functions, $R = f(X)$, and (4) nonlinear, implicit functions, *i.e.*, $X = f(R)$.

From a computational point of view, the first is the easiest to deal with and the last the most difficult to incorporate within an optimization framework. For illustrative purposes, we discuss two environmental models which represent these extremes. The first, a linear atmospheric dispersion model, is used to predict ambient concentration levels throughout the region of sulfur dioxide and airborne particulates. The second, a nonlinear aquatic ecosystem model, is used to predict ambient concentrations of various materials, and population sizes of certain species. This model was developed at Resources for the Future as part of a residuals management study of the Lower Delaware Valley.[29] Inclusion of these environmental models can greatly increase the usefulness of the overall management model for aiding in decisions on public policy matters, but it poses several computational problems also. A discussion of each model will consider some of the important issues and reveal some of the problems involved.

#### Atmospheric Dispersion Model

The atmospheric quality model is the air dispersion model from the U.S. Federal Government's Air Quality Implementation Planning Program.[30] It may be used to predict mean seasonal ambient concentration levels of sulfur dioxide and suspended particulates. This model is based upon a diffusion model developed by Martin and Tikvart[31,32] that evaluates concentrations downwind from a set of point and area sources on the basis of the Pasquill[33] point source Gaussian plume formulation. This particular formulation, which is based on mass continuity considerations, may be used to estimate ambient concentrations under deterministic, steady state conditions.

For a given source-receptor pair, production process and abatement device, and specified meteorologic conditions, the Gaussian plume formulation reduces to a simple linear expression for relating the resulting ambient concentration with the gaseous discharge rate. Expressed mathematically,

$$R_i = a_{ij}\, x_j \tag{12}$$

where $R_i$ is the ambient concentration at location i, $x_j$ is the gaseous discharge rate from source j, and $a_{ij}$ is commonly known as the transfer coefficient.

For n sources in the region, the ambient concentration at location i may be expressed as

$$R_i = a_{i1}\, x_1 + a_{i2}\, x_2 + \ldots + a_{in}\, x_n \tag{13}$$

For a multireceptor, multisource region, the contribution to ambient concentrations of all identified sources may be expressed as

$$R_i = \sum_{j=1}^{n} a_{ij}\, x_j \qquad i = 1, \ldots, m \tag{14}$$

or in matrix notation as

$$R = AX \tag{15}$$

where X is a vector of gaseous discharge rates (for example, sulfur dioxide and particulates), R is a vector of ambient concentrations due to these discharges (sulfur dioxide and suspended particulates) and A is a matrix of transfer coefficients that specify, for each source-receptor pair in the region and specified set of physical and meteorologic conditions, the contribution to ambient concentrations associated with a residual discharge rate of unity.

Because in any region there invariably are sources of gaseous residuals other than those specifically identified, Equation 15 is modified to incorporate background concentration levels, B.

$$R = AX + B \tag{16}$$

The important thing to note from Equation 16 is that the state of the natural world (R) is expressible directly in terms of linear, explicit algebraic functions. This particular mathematical form is relatively easy to deal with in an optimization framework. In fact, by rearranging terms slightly, Equation 16 may be incorporated directly within the constraint set of a standard linear program when one of the management objectives is to constrain ambient concentrations of residuals.

One of the requirements of all gradient type optimization schemes, when environmental models are dealt with in the objective functions, is the availability of an *environmental response matrix*, $(\partial R_i/\partial x_j)$, $i = 1, \ldots, m$; $j = 1, \ldots, n$; where m is the total number of environmental quality measures at all the designated receptor locations in the region, and n is the total number of residuals discharges in the region. This matrix, which turns out to be the matrix of transfer coefficients, A, may be obtained by differentiating Equation 16 with respect to all the residuals discharges in the region. That is,

$$\left(\frac{\partial R}{\partial X}\right) = A \tag{17}$$

It should also be noted that not all atmospheric quality models are as easy to deal with as the physical dispersion models expressed in linear, explicit analytical form. Chemical reaction models, such as for photochemical smog,[34,35] for example, would be significantly more difficult to handle within an optimization framework. The kinds of problems we would face with them are revealed in the discussion of a nonlinear aquatic ecosystem model which follows.

## Aquatic Ecosystem Model

A variety of indicators are commonly used to describe the quality of a body of water: pathogenic bacterial counts (or counts of an indicator thereof), algal densities, taste, odor, pH, turbidity, suspended and dissolved solids, dissolved oxygen, temperature, and population sizes of certain plant and animal species. Streeter-Phelps-type dissolved oxygen models have been used for many years to predict water quality, in terms of DO, as a result of discharges of organic material, most notably, sanitary sewage.[36] Given certain assumptions about the natural environment, these dissolved oxygen (DO) models can be expressed as a set of linear algebraic relationships. When the Streeter-Phelps-type differential equation set is solved analytically (as was done by Streeter and Phelps in their original formulation), the matrix of transfer coefficients (response matrix) is directly obtainable from the resulting expressions. However, when the differential equation set is solved using finite difference techniques, the matrix of transfer coefficients is obtainable only by inverting the matrix of linear coefficients associated with the original implicit algebraic function set. That is, given the differential equation set that is characteristic of all linear mass continuity models, $\dot{R} = AR - X$, where R is a vector of endogenous variables, $\dot{R}$ represents their time rates of change, and X is vector representing the influence on R of all exogeneous parameters considered (including residuals discharges), for steady state conditions, where $0 = AR - X$, the state of the natural world may be expressed as $R = A^{-1}X$ where $A^{-1}$ is the response matrix. Note that A is a square matrix because one differential equation is needed to describe the time rate of change of each endogeneous variable considered in the model.

From a computational point of view, the Streeter-Phelps-type models are relatively easy to deal with and are in widespread use today. A typical example is given for the Delaware Estuary by Thomann.[37]

The aquatic ecosystem model presented here[29] is substantially more difficult to deal with, computationally, than the more traditional, linear Streeter-Phelps-type models. These computational difficulties arise mainly because this model consists of a nonlinear system of equations that must, in general, be solved simultaneously. It is not the purpose of this paper to justify the use of complex ecosystem models in the management of environmental resources. Rather it is to indicate how the problem might be approached if it appears desirable to include them within an optimization model.

This aquatic ecosystem model is based on a trophic level approach.[38,39] The components of the ecosystem are grouped in classes (compartments) according to their function, and each class is represented in the model by an endogeneous or state variable. Eleven compartments are designated in this model. The endogeneous variables representing these 11 compartments are nitrogen, phosphorus, turbidity (suspended solids), organic material, algae, bacteria, fish, zooplankton, dissolved oxygen, toxics, and heat (temperature). Carbon is assumed not to be limiting and, hence, is not considered as either an endogeneous or exogeneous variable.

The inputs to the estuary model from the residuals discharges in the region are organic material measured by its BOD, total nitrogen, phosphorus, phenols (toxics), and heat. The outputs of this model of concern to policy makers are densities of fish biomass, algal densities, and dissolved oxygen levels. The levels of these outputs are constrained; that is, environmental standards are imposed.

The time rate of change of material in each compartment is expressed in terms of the sum of the transfers among other compartments, and between adjacent sections of the estuary (since the material is distributed spatially as well as temporally). In order to insure mass continuity of the materials considered, material entering and leaving a compartment is accounted for explicitly. The mathematical description of material transfers among compartments is based on the theoretical-empirical formulations given by Odum.[40]

Each compartment requires a separate differential equation to describe mass continuity, and in general these equations must be solved simultaneously. In this particular case, the differential equations are ordinary ones of the first order nonlinear variety. A set of similar differential equations is required for each reach of the estuary.

Finite difference forms of the more general partial differential equation set for describing mass continuity are used for the distance (space) variable. This is why we are able to write a separate set of differential equations for each reach (section) of the estuary. However, within each reach, time is expressed continuously (thus, the set of total differential equations rather than algebraic equations). When these differential equations are solved with analog computers, the concentrations of materials are continuous in time. When these equations are solved using digital computers, time must also be expressed in the finite difference form, and in the process they reduce to a simultaneous set of nonlinear algebraic equations.

The general form of the differential equation set for the *k*th reach may be expressed as

$$\left(\frac{dR}{dt}\right)^{(k)} = f\ [R(t)^{(k-1)},\ R(t)^{(k)},\ R(t)^{(k+1)},\ X(t)^{(k)}] \qquad (18)$$

where $(dR/dt)^{(k)}$ is a vector of time rates of change of the endogeneous variables in the *k*th reach, $R^{(k)}$; and $X^{(k)}$ is a vector of residuals discharges into the *k*th reach.

There are two problems associated with this ecosystem model formulation in particular which we wish to discuss in more detail: (1) that of obtaining a steady state solution, and (2) that of obtaining an environmental response matrix.

Solution Methods. In its present form, Equation 18 represents a set of ordinary nonlinear differential equations, one equation for each compartment and one set of compartmental equations for each estuary reach, that must be solved simultaneously. If we are interested in the transient (or nonsteady) states of the system, simulation techniques, *i.e.*, numerical integration (simulating over space and then time) provide us with a readily available means of solution. However, for inclusion within the optimization model, we are interested only in the steady state solution.

For determining steady state solutions, there are two possibilities (or a combination thereof), neither of which guarantees finding a stable point equilibrium: (1) simultaneous simulation of a nonlinear differential equation set, and (2) simultaneous solution of a set of nonlinear algebraic equations. If we neglect inputs to, and outflows from, each reach due to longitudinal dispersion, the system can be dealt with first over time and then space, starting with the uppermost reach and progressing systematically down the estuary. If longitudinal dispersion is neglected, even in an estuary, this is not as unreasonable as it first appears. Finite difference techniques for solving these differential equations introduce a numerical diffusion effect into the model. Inputs are immediately mixed in the volume, not because of any physical effects, but solely because of the numerical procedure.[41]

In this case, Equation 18 for the *k*th reach would reach to

$$\left(\frac{dR}{dt}\right)^{(k)} = f\,[R(t)^{(k-1)}, R(t)^{(k)}, X(t)^{(k)}] \qquad (19)$$

Now, only the 11 compartmental equations within each reach must be solved simultaneously. The state of the system within a particular reach depends only upon the inputs from upstream, $R(t)^{(k-1)}$ and the residuals discharges to the *k*th reach, $X(t)^{(k)}$, both of which may now be treated as exogeneous inputs. In addition, if the resulting steady state solution, $R^{*(k)}$, is independent of the time paths of rates of inputs, $R(t)^{(k-1)}$ and $X(t)^{(k)}$, Equation 19 reduces to

$$\left(\frac{dR}{dt}\right)^{(k)} = f\,[R^{*(k-1)}, X^{*(k)}, R(t)^{(k)}] \qquad (20)$$

Usually, ecological models are solved by simulation.[42] Simulation of the differential equation set (a set of equations similar to Equations 18 through 20) poses no particular problem, but the steady state solution, if one exists at all, may take considerable time. Oscillations can, and do, occur, and solutions may be otherwise unstable (become infinitely large). However, May[43] has demonstrated for a set of reasonable assumptions and a similar predator-prey nonlinear model that these systems possess either a stable point equilibrium or a stable limit cycle.

At steady state, $(dR/dt) = 0$, and thus the differential equation set above, Equation 20 reduces to a set of nonlinear algebraic equations of the following form:

$$0 = f\,[R^{*(k-1)}, X^{*(k)}, R^{*(k)}] \qquad (21)$$

The endogeneous variables, $R^{*(k)}$, are implicitly expressed in this formulation.

Techniques such as Gauss-Seidel and Newton's method have been used with success on nonlinear algebraic equation sets, but each has its faults (see, for example, Mankin and Brooks[44]). Gauss-Seidel (also known as "Successive Approximations") has slow convergence properties, but it is relatively stable. Newton's method has more rapid convergence properties but is sensitive to initial conditions and it is often unstable.

Determination of steady state values for the endogeneous variables in a nonlinear ecosystem model is difficult due to the nonuniqueness and complexity of the solution. Even the stability characteristics of the steady state solution cannot be determined prior to its solution. With linear models, we can solve for the eigenvalues (or characteristic values) of the differential equation set. These will tell us whether or not the time independent solution converges to a finite set of values or diverges to infinity, or even if oscillations are involved--stable, diverging, or converging. For the nonlinear differential equation set, the best we can do is linearize the system at some point and examine the eigenvalues of the resulting linear form. But this only tells us what is happening locally.

A combination of Newton's method and simulation appears to be a reasonable approach for obtaining a steady state solution to this problem.

The Environmental Response Matrix. To include such a nonlinear aquatic ecosystem model within a residuals management (optimization) model, in addition to determining a set of steady-state values, one must evaluate the response throughout the ecosystem to changes in the rates of the residuals discharges. It is necessary to know, for example, the effect on algae in reach 17 of an additional BOD load discharged into reach 8, and so on. This requirement results in a considerable number of additional computations, but this, in conjunction with the penalty functions mentioned above and to be discussed again below, is the key to utilizing these complex ecosystem models within the optimization framework.

The environmental response matrix we wish to compute may be expressed in matrix notation as $(\partial R/\partial X)$, where R is a vector describing the state of the system throughout the entire length of the estuary, and X is a vector of residuals discharges throughout the region. Using Equation 21 for each reach of the estuary and the relationship

$$Z^{(k)} = q^{(k)} R^{(k-1)} + X^{(k)} \tag{22}$$

where $Z^{(k)}$ is a vector of inputs to the $k$th reach, $R^{(k-1)}$ is a vector of concentrations of materials in reach (k-1), $q^{(k)}$ is the estuary advective flow rate into the *k*th reach, and $X^{(k)}$ is a vector of residuals discharges to the *k*th reach, a section of the system response matrix may be computed accordingly:

$$\frac{\partial R^{(i)}}{\partial X^{(j)}} = \frac{\partial R^{(i)}}{\partial Z^{(i)}} \cdot \frac{\partial Z^{(i)}}{\partial R^{(i-1)}} \cdot \frac{\partial R^{(i-1)}}{\partial Z^{(i-1)}} \cdots \frac{\partial R^{(j+1)}}{\partial Z^{(j+1)}} \cdot \frac{\partial Z^{(j+1)}}{\partial R^{(j)}} \cdot \frac{\partial R^{(j)}}{\partial Z^{(j)}} \cdot \frac{\partial Z^{(j)}}{\partial X^{(j)}} \quad (23)$$

From Equation 22 we note that

$$\left(\frac{\partial Z^{(k)}}{\partial R^{(k-1)}}\right) = q^{(k)} I \quad (24)$$

and

$$\left(\frac{\partial Z^{(k)}}{\partial X^{(k)}}\right) = I \quad (25)$$

where I is the identity matrix. Thus, the $\partial Z^{(k)}/\partial R^{(k-1)}$ terms are known *a priori* and are exogeneous parameters in the ecosystem model.

The other terms, $\partial R^{(k)}/\partial Z^{(k)}$, are evaluated from Equation 21 according to the rules for differentiating implicit functions.[45] That is,

$$\left(\frac{\partial f}{\partial R}\right) \cdot \left(\frac{\partial R}{\partial Z}\right) = -\left(\frac{\partial f}{\partial Z}\right)$$

or

$$\left(\frac{\partial R}{\partial Z}\right) = -\left(\frac{\partial f}{\partial R}\right)^{-1} \left(\frac{\partial f}{\partial Z}\right)$$

This operation involves the inversion of the Jacobian matrix ($\partial f/\partial R$). In addition, because the system of equations is nonlinear, the Jacobian matrix ($\partial f/\partial R$) must be recomputed for each state of the natural world.

### *Optimization Procedures*

Thus far we have presented a formal mathematical statement of the regional residuals management problem, discussed a penalty function method for eliminating all but the linear constraints, restated the problem in terms of a nonlinear objective function and linear constraint set, and discussed the mathematical forms of, and problems associated with, various types of environmental models. Very little, however, has been said about an appropriate nonlinear optimization procedure. That is the purpose of this section.

A variety of nonlinear programming techniques exist for solving problems similar to the regional residuals management problem stated in Equations 1 through 7 above. In the last decade considerable progress has been made in the development of general nonlinear programming algorithms that can handle a nonlinear objective function and nonlinear constraints of both the equality and inequality type.[24] For problems where the objective function is nonlinear and the constraints are linear, Rosen's gradient projection method has been used with considerable success.[46] To ensure the proper form for Rosen's linear constraint method, the nonlinear constraints could be eliminated using one of the penalty or barrier methods discussed above.

For purposes of exposition, we will present still another technique. A discussion of the method will reveal the types of computations necessary for any of these nonlinear procedures. It is not the purpose of this paper to compare the relative efficiencies of the different optimization techniques but only to point out that a variety of methods do exist, and that some may be better than others, depending upon the particular situation.

Two inherent characteristics of the regional residuals management problem which could affect the selection of an appropriate optimization scheme should be noted. First of all, the regional residuals management problem, in any detail at all, is likely to be extremely large. Even if the problem were cast as a completely linear one, decomposition techniques, for dealing with sheer model size, might be necessary. A relevant question, then, is what size problems can the various nonlinear algorithms handle?

Second, because of all the different submodels--residuals generation and discharge models as well as environmental models--required by this approach, the analyst will inevitably be faced with adapting, and subsequently employing, those models which are available to but not necessarily developed by him. The analyst is likely to be faced with attempting to link models together which may or may not have been developed for that purpose. This situation can impose severe restrictions on the choice of an appropriate optimization procedure. Out of necessity, efficiency of the optimization scheme by itself may become a secondary consideration to the efficiency of the model building and model operation stages taken jointly. For example, many of the existing residuals generation models are linear, but proportionately they contain very many equality constraints. Although each equality constraint could be used to eliminate a variable, this is an error-prone and time-consuming procedure once the models are built. In addition, some of the eliminated variables are likely to provide useful information for management decisions and, hence, would have to be computed anyway after the optimization phase of the analysis was complete. In these cases, one may decide to retain the larger models, which would reduce the model building (or model adaptation) time and also provide additional information, rather than attempting to reduce the size solely for the purpose of improving the efficiency of the optimization procedure.

## An Heuristic Approach

A formal presentation of this nonlinear programming algorithm which is used to optimize a nonlinear objective function subject to a set of linear constraints has been presented elsewhere.[16] Only the essence of the scheme is repeated here. Relevant equations and expressions used for this procedure are restated, and the objective function, Equation 8, is modified accordingly.

This optimization scheme is analogous to the gradient method of nonlinear programming.[47,48] The technique consists of linearizing the response surface in the vicinity of a feasible point, $X^{(k)}$. To do this, we construct a tangent plane at this point by employing the first two terms of a Taylor's series expansion (up to first partial derivatives). This linear approximation to the nonlinear response surface will, in general, be most accurate in the vicinity of the point $X^{(k)}$ and less accurate as one moves farther away from this point. Because of this, a set of "artificial" bounds (constraints) is imposed on the system to restrict the selection of the next position along the response surface to that portion of the surface most closely approximated by the newly created linear surface. The selection of the appropriate set of artificial bounds is analogous to choosing a step size in other gradient methods of nonlinear programming.

Because the newly created subproblem is in a linear form, we are able to make use of standard linear programming techniques for finding a new optimal point, $X^{(k+1)}$. This point locates the maximum value of the linearized objective function within the artificially confined area of the response surface. Because, in general, the linearized surface will not match the original nonlinear surface, the original nonlinear objective function must be evaluated at this point to determine whether or not this new point, $X^{(k+1)}$, is in fact a better position than the previously determined one, $X^{(k)}$. That is, the following condition must be satisfied:

$$F[X^{(k+1)}] > F[X^{(k)}].$$

If this condition is satisfied, a new tangent plane is constructed at the point $X^{(k+1)}$ and a new set of artificial bounds is placed around this point. As before, a linear programming code is employed to find a new position, $X^{(k+2)}$, which maximizes the linearized objective function, and so on until a local optimum is reached. This procedure, like all gradient methods, finds only the local optimum. If the response surface contains more than one optimum, the problem becomes one of finding the *global* optimum. One way of approaching this is to start the procedure at different points within the feasible region, where the starting points may be chosen at random. Techniques on random starts within a feasible region defined by a linear constraint set have been presented by Rogers.[49]

This optimization procedure requires that we linearize the objective function at a point $X^{(k)}$. We do this according to the following formulation:

$$F[X^{(k+1)}] = \nabla F[X^{(k)}] \cdot X^{(k+1)} + \gamma \qquad (27)$$

where $\gamma$ is a constant. Expressing our revised objective function, Equation 8, in terms of Equation 27 results in

$$\begin{aligned} F &= \nabla f(X) \cdot X - \nabla P(X) \cdot X + \gamma \\ &= \left(\frac{\partial f(X)}{\partial X}\right) \cdot X - \left(\frac{\partial P(X)}{\partial X}\right) \cdot X + \gamma \end{aligned} \qquad (28)$$

In our residuals management problem $(\partial f(X)/\partial X)$ is a vector of linear cost coefficients associated with traditional production inputs and residuals handling, modification, and disposal activities; and $(\partial P(X)/\partial X)$ is a vector of marginal penalties associated with the discharge into the environment of each residual.

Given that $R = h(X)$, Equation 4, we see from Equation 10 that

$$\frac{\partial P(X)}{\partial x_j} = \sum_{i=1}^{q} \frac{2}{r_i} (\max[(R_i - S_i), 0]) \frac{\partial R_i}{\partial x_j} \qquad j=1, \ldots, n \qquad (29)$$

The term $\{2/r_i(\max[(R_i - S_i), 0])\}$ represents the slope, $dp_i/dR_i$ of the *i*th penalty function evaluated at the point $R_i$. The term $\partial R_i/\partial x_j$ represents the marginal response of the *i*th descriptor of the natural world (or ecosystem) to changes in the discharge of the *j*th residual. It is an element of the environmental response matrix. Equation 29 may be expressed more generally as

$$\frac{\partial P(X)}{\partial x_j} = \sum_{i=1}^{q} \frac{dp_i}{dR_i} \cdot \frac{\partial R_i}{\partial x_j} \qquad j=1, \ldots, n \qquad (30)$$

or in matrix notation as

$$\frac{\partial P(X)}{\partial X} = \left(\frac{\partial R}{\partial X}\right)^T \cdot \frac{dp}{dR} \qquad (31)$$

For linear environmental systems $(\partial R/\partial X)$ is the matrix of transfer coefficients, A, when the environmental functions are expressed, linearly, as

$$R = h(x) = A \cdot X \qquad (32)$$

Hence, for the case of linear environmental systems, the marginal penalties, Equation 31, may be expressed in matrix notation as

$$\left(\frac{\partial P}{\partial X}\right) = A^T \cdot \left(\frac{dp}{dR}\right) \qquad (33)$$

For the case of nonlinear environmental models, the situation is similar except that evaluation of the environmental response matrix $(\partial R/\partial X)$, is somewhat more involved, as demonstrated by Equation 23, and in addition because the response is nonlinear it must be recomputed for each state of the natural world.

A schematic diagram of this approach to regional residuals management modeling is delineated in Figure 19.2.

### The Linearized Subproblem

Now that we have presented the essence of this optimization scheme, including a discussion of the LP subproblem which must be both constructed and solved at each step along the ascent procedure, we can restate the regional residuals management problem in these terms:

$$\max \{F[X^{(k+1)}] = [\nabla F[X^k] - \nabla P[X^k]] \cdot X^{(k+1)} + \gamma\} \quad (34)$$

$$\text{s.t.} \quad g_i(X) = 0 \qquad i = 1, \ldots, m \quad (2)$$

$$g_i(X) \geqslant 0 \qquad i = m + 1, \ldots, p \quad (3)$$

$$x_i \geqslant 0 \qquad i = 1, \ldots, n \quad (7)$$

$$x_j \leqslant \beta_j \qquad j = 1, \ldots, s \quad (35)$$

$$x_j \geqslant \alpha_j \qquad j = 1, \ldots, s \quad (36)$$

where $\beta_j$ and $\alpha_j$ are, respectively, upper and lower bounds on the $s$ discharge variables at the (k+1)th iteration. As mentioned previously, setting these bounds is similar to selecting a step size with other gradient methods of nonlinear programming. The efficiency of the optimization scheme depends directly on the selection of these bounds. This is one of the areas that needs to be explored in more detail if this method is to be given serious consideration for use on full scale regional applications.

## IV. SUMMARY AND CONCLUDING REMARKS

There is no question that the regional residuals management model suggested here is more complex than the classical management model approach and that it requires a substantial amount of computer resources. Two questions we face but hope to answer as a result of research efforts on large-scale regional residuals management models are:

1. Is it necessary to include all forms of material residuals within a single computational framework?

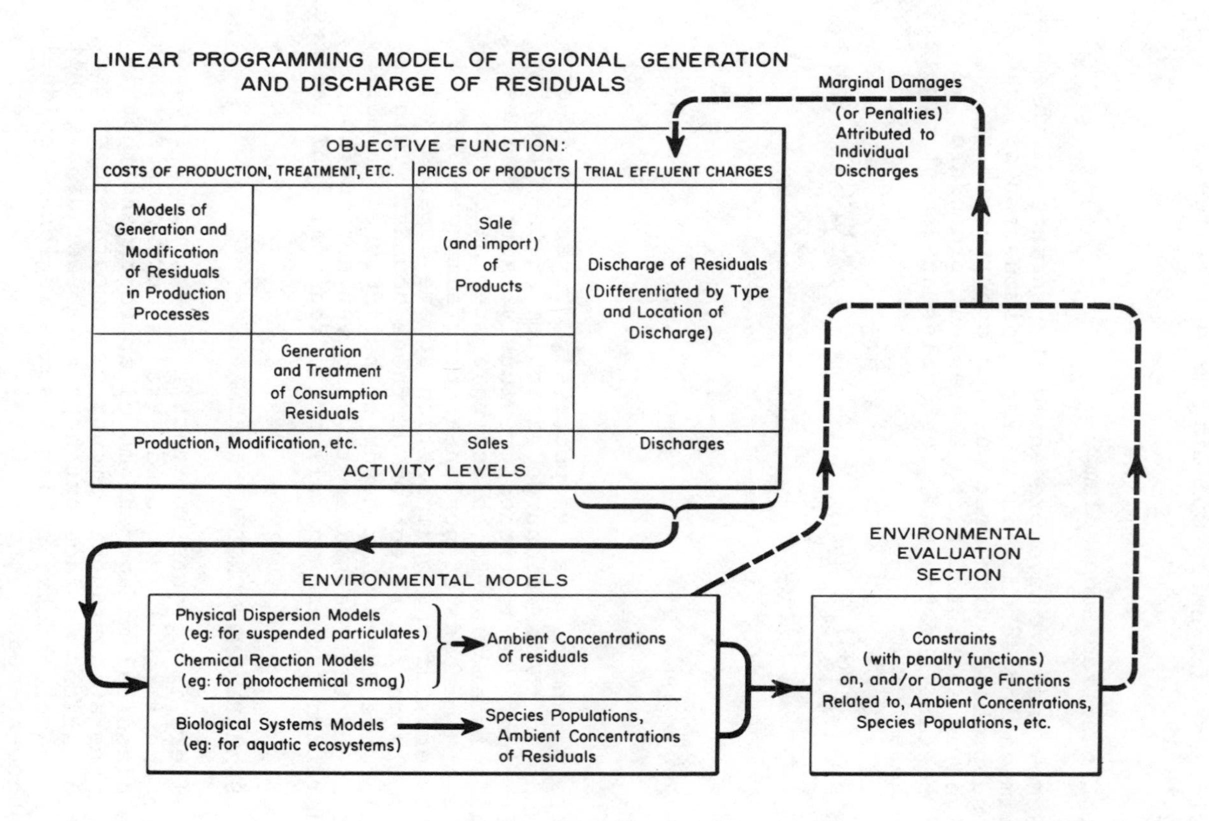


*Figure 19.2. Schematic diagram of the regional residuals management model.*

In principle, it is. But in practice does the additional information on the interactions among forms of residuals warrant the added effort and expense?

2. Nonlinear environmental models, such as for photochemical smog and of aquatic ecosystems, can provide additional useful information for making public policy on environmental resources, but can we incorporate these models within an optimizing framework without completely expending the computer budget of a regional residuals management agency?

Although we have been criticized on the first point, none of us will know the answer to this question until some one (or group) tries an integrated approach to residuals management and compares its output with that of the classical residuals management models.

ACKNOWLEDGMENTS

I am grateful to my colleagues, Blair T. Bower and Clifford S. Russell, for their helpful comments on earlier versions of this paper.

REFERENCES

1. Schuman, J. "The Solution of Air Pollution can Cause Water Pollution," *Water Waste Eng 7(11)* (1970).
2. Loucks, D. P. and H. D. Jacoby. "Flow Regulation for Water Quality Management," in *Models for Managing Regional Water Quality*, R. Dorfman, H. Jacoby, and H. A. Thomas, Jr., Eds. (Cambridge, Mass.: Harvard University Press, 1972).
3. Whipple, W., *et al.* *Instream Aeration of Polluted Rivers*, Water Resources Research Institute at Rutgers University, New Brunswick, New Jersey (1969). Also available as PB 192 637 from Clearinghouse for Federal Scientific and Technical Information, Springfield, Virginia.
4. Ortolano, L. "Artificial Aeration as a Substitute for Wastewater Treatment," in *Models for Managing Regional Water Quality*, R. Dorfman, H. Jacoby, and H. A. Thomas, Jr., Eds. (Cambridge, Mass.: Harvard University Press, 1972).
5. Graves, G. H., G. B. Hatfield, and A. Whinston. "Water Pollution Control Using By-Pass Piping," *Water Resourc. Res. 5(1)*:13 (1969).
6. Graves, G. H., G. B. Hatfield, and A. Whinston. "Mathematical Programming for Regional Water Quality Management," *Water Resourc. Res. 8(2)*:273 (1972).

7. Löf, G. O. G. and A. V. Kneese. *The Economics of Water Utilization in the Beet Sugar Industry.* (Baltimore, Md.: The Johns Hopkins University Press, 1968), p. 22.
8. Bower, B. T., G. O. G. Löf, and W. M. Hearon. "Residuals Management in the Pulp and Paper Industry," *Nat. Res. J. 11(4)*:605 (1971). Also available as Reprint No. 100, Resources for the Future, Inc., Washington, D.C., January 1972.
9. Russell, C. S. *Residuals Management in Industry: A Case Study of Petroleum Refining.* (Baltimore, Md.: The Johns Hopkins University Press, 1973).
10. Bower, B. T. "Integrated Residuals Management," Resources for the Future, Inc., unpublished (1972).
11. Freeze, R. A. "Subsurface Hydrology at Waste Disposal Sites," *IBM J. Res. Devel. 16(2)*: 117 (1972).
12. Russell, C. S. "Models for the Investigation of Industrial Response to Residuals Management Action," *The Swedish Journal of Economics 73(1)* (1971). Also available as Reprint No. 95, Resources for the Future, Inc., June, 1971.
13. Bower, B. T. "Studies of Residuals Management in Industry," Resources for the Future, Inc., presented at *Conference on Economics of the Environment*, sponsored jointly by Universities-National Bureau Committee for Economic Research and Resources for the Future, Chicago, Illinois (November 10-11, 1972).
14. Bower, B. T. and W. R. D. Sewell. "Selecting Strategies for Air-Quality Management," Resource Paper No. 1, Policy Research and Coordination Branch, Department of Energy, Mines and Resources, Information Canada, Ottawa (1971).
15. Jacoby, H. and D. P. Loucks. "The Combined Use of Optimization and Simulation Models in River Basin Planning," *Water Resourc. Res. 8 (12)* (1972).
16. Russell, C. S. and W. O. Spofford, Jr. "A Quantitative Framework for Residuals Management Decisions," in *Environmental Quality Analysis: Theory and Method in the Social Sciences,* A. V. Kneese and B. T. Bower, Eds. (Baltimore, Md.: The Johns Hopkins University Press, 1972).
17. Russell, C. S., W. O. Spofford, Jr., and E. T. Haefele. "Residuals Management in Metropolitan Areas," Resources for the Future, Inc., presented at the *International Economics Association Conference on Urbanization and the Environment,* Copenhagen (June 19-24, 1972).
18. Spofford, W. O., Jr., C. S. Russell, and R. A. Kelly. "Operational Problems in Large Scale Residuals Management Models," Resources for the Future, Inc., presented at *Conference on Economics of the Environment,* sponsored jointly by Universities-National Bureau Committee for Economic Research and Resources for the Future, Chicago, Illinois (November 10-11, 1972).

19. Orchard-Hays, W. *Advanced Linear-Programming Computing Techniques* (New York: McGraw-Hill Book Co., 1968).
20. Driebeek, N. J. *Applied Linear Programming* (Reading, Mass.: Addison-Wesley Publishing Co., 1969).
21. International Business Machines Corporation. "Mathematical Programming System Extended (MPSX): Linear and Separable Programming, Program Description," Program No. 5734-XM4, IBM Corporation, White Plains, New York (February, 1971).
22. International Business Machines Corporation. "Mathematical Programming System Extended (MPSX): Mixed Integer Programming (MIP), Program Description," Program No. 5734-XM4, IBM Corporation, White Plains, New York (February, 1971).
23. Zangwill, W. I. *Nonlinear Programming: A Unified Approach* (Englewood Cliffs, N.J.: Prentice-Hall, Inc., 1969).
24. Fiacco, A. V. and G. P. McCormick. *Nonlinear-Programming: Sequential Unconstrained Minimization Techniques* (New York: John Wiley and Sons, Inc., 1968).
25. Hadley, G. H. *Linear Programming.* (Reading, Mass.: Addison-Wesley Publishing Co., 1962).
26. Dantzig, G. B. *Linear Programming and Extensions.* (Princeton, N.J.: Princeton University Press, 1963).
27. Silverman, G. J. "Primal Decomposition of Mathematical Programs by Resource Allocation: I - Basic Theory and a Direction-Finding Procedure," *Operations Research 20(1):* 58 (1972).
28. Lasdon, L. S. *Optimization Theory for Large Scale Systems* (New York: MacMillan, 1970).
29. Kelly, R. A. "Conceptual Ecological Model of Delaware Estuary," Resources for the Future, Inc., Washington, D.C., (1973). To be published in *Systems Analysis and Simulation in Ecology, Vol. III,* B. C. Patten, Ed. (New York: Academic Press, 1974).
30. TRW, Inc. "Air Quality Implementation Planning Program," Environmental Protection Agency, Vols. I and II (November, 1970). Also available from National Technical Information Service, Springfield, Virginia, 22151, accession numbers PB 198 299 and PB 198 300 respectively.
31. Martin, D. O. and J. A. Tikvart. "A General Atmospheric Diffusion Model for Estimating the Effects on Air Quality of One or More Sources," *APCA Paper 68-148*, presented at the Annual Meeting of the Air Pollution Control Association, St. Paul, Minn. (June, 1968).
32. Martin, D. O. "An Urban Diffusion Model for Estimating Long Term Average Values of Air Quality," *J. Air Poll. Cont. Assoc. 21(1):*16 (1971).
33. Pasquill, F. *Atmospheric Diffusion* (London: D. Van Nostrand Co., 1962).
34. Friedlander, S. K., and J. H. Seinfeld. "A Dynamic Model of Photochemical Smog," *Environ. Sci. Technol. 3(11):*1175 (1969).

35. Eschenroeder, A. Q. and J. R. Martinez. "Concepts and Applications of Photochemical Smog Models," Tech. Memo. 1516, General Research Corporation, Santa Barbara, California (June, 1971).
36. Streeter, H. W. and E. B. Phelps. "A Study of the Pollution and Natural Purification of the Ohio River," Public Health Bulletin No. 146 (Washington, D.C.: U.S. Public Health Service, 1925).
37. Thomann, R. V. *Systems Analysis and Water Quality Management* (New York: Environmental Science Services Division of Environmental Research and Applications, Inc. 1972).
38. Williams, R. B. "Computer Simulation of Energy Flow in Cedar Bog Lake, Minnesota, based on the classical studies of Lindeman," in *Systems Analysis and Simulation in Ecology, Vol. I*, B. C. Patten, Ed. (New York: Academic Press, 1971), p. 543.
39. Odum, H. T. *Environment, Power and Society*. (New York: Wiley-Interscience, 1971).
40. Odum, H. T. "An Energy Circuit Language for Ecological and Social Systems: Its Physical Basis," In *Systems Analysis and Simulation in Ecology, Vol. II*, B. C. Patten, Ed. (New York: Academic Press, 1972), p. 139.
41. O'Connor, D. J. and R. V. Thomann. "Water Quality Models: Chemical, Physical, and Biological Constituents," *Estuarine Modeling: An Assessment*, Environmental Protection Agency (Washington, D.C.: U. S. Government Printing Office, February, 1971), Stock No. 5501-0129, Chapter III, pp. 102-173.
42. Patten, B. C., Ed. *Systems Analysis and Simulation in Ecology, Vol. I*. (New York: Academic Press, 1971).
43. May, R. M. "Limit Cycles in Predator-Prey Communities," *Science 177*:900 (1972).
44. Mankin, J. B. and A. A. Brooks. "Numerical Methods for Ecosystem Analysis," Oak Ridge National Laboratory, Oak Ridge, Tennessee, ORNL-IBP-71-1 (June, 1971).
45. Sokolnikoff, I. S. and R. M. Redheffer. *Mathematics of Physics and Modern Engineering*. (New York: McGraw-Hill Book Co., 1958).
46. Rosen, J. B. "The Gradient Projection Method for Nonlinear Programming, Part I: Linear Constraints," *J. Soc. Ind. Appl. Math. 8(1)*:181 (1960).
47. Saaty, T. L. and J. Bram. *Nonlinear Mathematics*. (New York: McGraw-Hill Book Co., 1964).
48. Hadley, G. H. *Nonlinear and Dynamic Programming*. (Reading, Mass.: Addison-Wesley Publishing Co., 1964).
49. Rogers, P. P. *Random Methods for Non-Convex Programming*. Ph.D. dissertation, Harvard University, Cambridge, Mass. (June, 1966).

MODELS FOR ENVIRONMENTAL POLLUTION CONTROL


# CHAPTER 20

# ENVIRONMENTAL ACCOUNTABILITY PARALLELS BETWEEN NUCLEAR SAFEGUARDS AND POLLUTION CONTROL

W. Häfele

This paper reflects on the connections between the field of the peaceful application of nuclear energy and that of pollution control. And indeed there are many connections and parallels. But there is one main area that may be significant for pollution control--the safeguards of nuclear material.

In 1966/1967 safeguards was a very challenging subject, when it became one of the hard core problems of the Treaty of the Non-Proliferation of Nuclear Weapons. This challenge was of a universal nature because the Non-Proliferation Treaty was designed to cover truly the whole globe. And there was also a tremendous lack of experience. How to meet such a challenge? All this is more or less also true for pollution control. It is of a global nature, and we have little or no experience in meeting such a challenge. The international character of the problem is also obvious. So it does make sense to describe in the first portion of this paper how international safeguards of nuclear material were developed.

Let us recall: After the first round of the nuclear arms race during the fifties, when nuclear weapons were considered more of a sword than a shield, the increasing number of more sophisticated and powerful atomic weapons and the growing awareness of the strategic complexities of atomic warfare made nuclear weapons more of a shield and the conventional

---

*Prof. Dr. W. Häfele is with the Institut für Angewandte Systemtechnik und Reaktorphysik, Kernforschungszentrum Karlsruhe, 75 Karlsruhe, West Germany.

forces more of a sword. In the early sixties Jerome Wiesner in particular developed the scheme of second strike capability and bipolar stability. This led to a configuration where the further proliferation of nuclear weapons was considered to be a main threat that could well be detrimental to the delicate bipolar stability. It was then only natural that the Non-Proliferation Treaty (NPT) was conceived. In the course of the basic thinking and international debate that accompanied it, it became clear that the underlying problem of the NPT consisted of three equally large sectors:

1. arms control and disarmament
2. structure of alliances
3. the peaceful uses of atomic energy as an independent dimension.

The first sector is self-explanatory since arms control was the genuine meaning of the NPT. The second sector refers to the internal structure, particularly of NATO, as an alliance between nuclear weapon states and non-nuclear weapon states. But also EURATOM and other alliance structures came out to be a problem. For many involved in the Non-Proliferation Treaty the third sector and its equality to the others was not obvious from the beginning. The peaceful applications of nuclear energy had passed the threshold to commercial competitivity only in the United States (that was the order for the nuclear power station Oyster Creek on purely commercial grounds). Many were under the impression that such successful implementation would be somewhat impeded in non-nuclear weapon states. But the commercial orders for the State and Würgassen plants in Germany proved the opposite. So safeguards came out to be a problem. If the non-nuclear weapon states were accepting inequality in the military domain, they had to insist that equality in the civilian domain was not only not impeded but even fostered.

The challenges of today are population control, energy shortage, pollution control, shortage of material resources, and waste. They are all in the civilian domain and are already beginning to overshadow the challenges in the military domain. In retrospect it is obvious that the problem of international safeguards had to come up, but it was not so obvious in 1967.

In 1967 the debate on international safeguards was heated. One group maintained that international safeguards had to be total and absolute. This group wanted to achieve unlimited objectives with limited resources. Another group said that they wanted no safeguards at all. Commercial competition, they argued, is so sharp that there must be enterprise which is absolutely free. It should be remembered that safeguards at this time were foreseen only for the non-nuclear weapon states. Most of the debate at that time was highly emotional and highly moral, with a great deal of vested interest. This compares nicely with today's debate on pollution control.

How did the question of international safeguards appear at that time? First there were no clear objectives: Was it protection of nuclear material or was it the detection of the diversion of nuclear material that was at stake? What did the inspectors look for, the mg amounts in the plate-out in pipes, g amounts, or kg's? Also the subject of safeguards was not well-defined. Was it the nuclear facility as such or was it the material involved or the equipment? Was a pump or a valve or even a large digital computer subject to safeguards?

The Non-Proliferation Treaty was intended to be a committing international treaty with a duration of 25 years. In view of the seriousness of the problem of non-proliferation of nuclear weapons the problem of safeguards had to be solved because security was at stake. Also at stake is now the problem of pollution control in order to maintain clean air and water.

How was the challenge of having meaningful international safeguards met? A few groups in the world in connection with the International Atomic Energy Agency in Vienna began a process of research and development, realizing that meaningful international safeguards had to embrace politically inconsistent groups. Up to that point international safeguards had been applied only punctually or within politically consistent groups such as EURATOM. Further, the demand for equality of nations in the civilian domain had to be met. Modern international safeguards had to be rational, objective and formalized. They had to be rational because the resources to operate such safeguards were and continue to be limited. They had to be objective to make the cooperation between politically inconsistent groups possible. And they had to be formalized to cut the open-endedness inherent in most inspection or safeguard processes.

The objective of the safeguards was easily identifiable: because the link between peaceful and explosive application of nuclear energy is only the nuclear material, only the nuclear material can be subject to safeguards. Further, due to the international character of the safeguards in question, it must be the detection of diversion of nuclear material that is more specifically the objective of safeguards. After some time it became clear that international safeguards of nuclear material had three principal components:

1. accountability of nuclear material
2. containment
3. surveillance.

By defining the physical place of the flow, the flow can then be measured. The nuclear material flows through the nuclear fuel cycle: fuel fabrication, nuclear reactor, chemical reprocessing, and again fuel fabrication. Most of the relevant facilities have a strong containment. Numbers, which can be made objective, are produced by accounting for the flow of the nuclear material. Preestablished requirements for measurement accuracies cut the open-endedness with surveillance filling in for the remaining aspects.

Accountability accomplishes the task of quantification. Every systems analyst knows how principal the step of quantification is. After it is solved the tremendous tools of existing mathematics can be applied. By making accountability the most important component of modern international safeguards it was possible to make the problem tractable, and then systems analysis could be applied.

The important thing to be seen from this example is that once the quantifiable part of the problem became tractable, the remainder eventually became tractable also. More general systems analysis is not confined to the quantifiable aspects only. The application of systems analysis forces the introduction of relevant parameters and various interconnections. In particular, it helps to clarify the objective more precisely.[1] It is better to have an exact problem with an approximate solution than an approximate problem with an exact solution. And this leads to the problem of systems analysis of pollution control. Consider therefore the following ideas:

-- Pollution control is truly of a global nature, as mentioned earlier.
-- Pollution control makes it mandatory for cooperation between politically inconsistent groups. The relevant

inconsistency is not between the West and the East, but between the North and the South. Look, for example, at the Mediterranean where there are highly developed, industrialized countries and also developing countries. It is obviously not possible for the industrialized countries to tell the developing countries to stay underdeveloped in order to limit pollution.

-- Pollution control makes cooperation between the operator of an industrial plant and environmental protection agencies necessary in the same sense as there is now cooperation between the operator of a nuclear facility and the inspector.
-- Pollution control will also somehow limit the freedom of entrepreneurship in the commercial domain.
-- Pollution control must have a clearly defined objective. It is not sufficient to say that clean air and clean water are desired.
-- Pollution control faces the problem of open-endedness. Modern groups of intervenors like to demand no pollution at all. But anything that man does somehow pollutes nature. What is nature in an absolute sense?
-- Pollution control faces the problem of public acceptance, the Gofman-Tamplin debate in the U.S. being indicative.

The "Institut für Angewandte Systemtechnik und Reaktorphysik" of the Kernforschungszentrum Karlsruhe has tried to look deeper into the problem of environmental accountability, believing that it is the mean by which systems analysis will be applied to pollution control. Figure 20.1 gives an outline of this. The first approximation of the objective of environmental accountability may be at hand today. This indicates the type of information on emission and immission one must seek. On the one hand, one is led into the modelling of the physics of emission/immission. It might be possible to conclude from that how to establish and optimize a set of measurement-points in the water or in the air. On the other hand, one is led into the problem of an accountability of an industrial plant, for instance an oil refinery. If there is full knowledge on the path of pollutions within such an industrial plant and the necessity of these pollutions, one is induced to minimize pollution and, more generally, to have good housekeeping. So-called integral experiments may help to establish such an in-house accountability, as was the case for nuclear facilities. The two

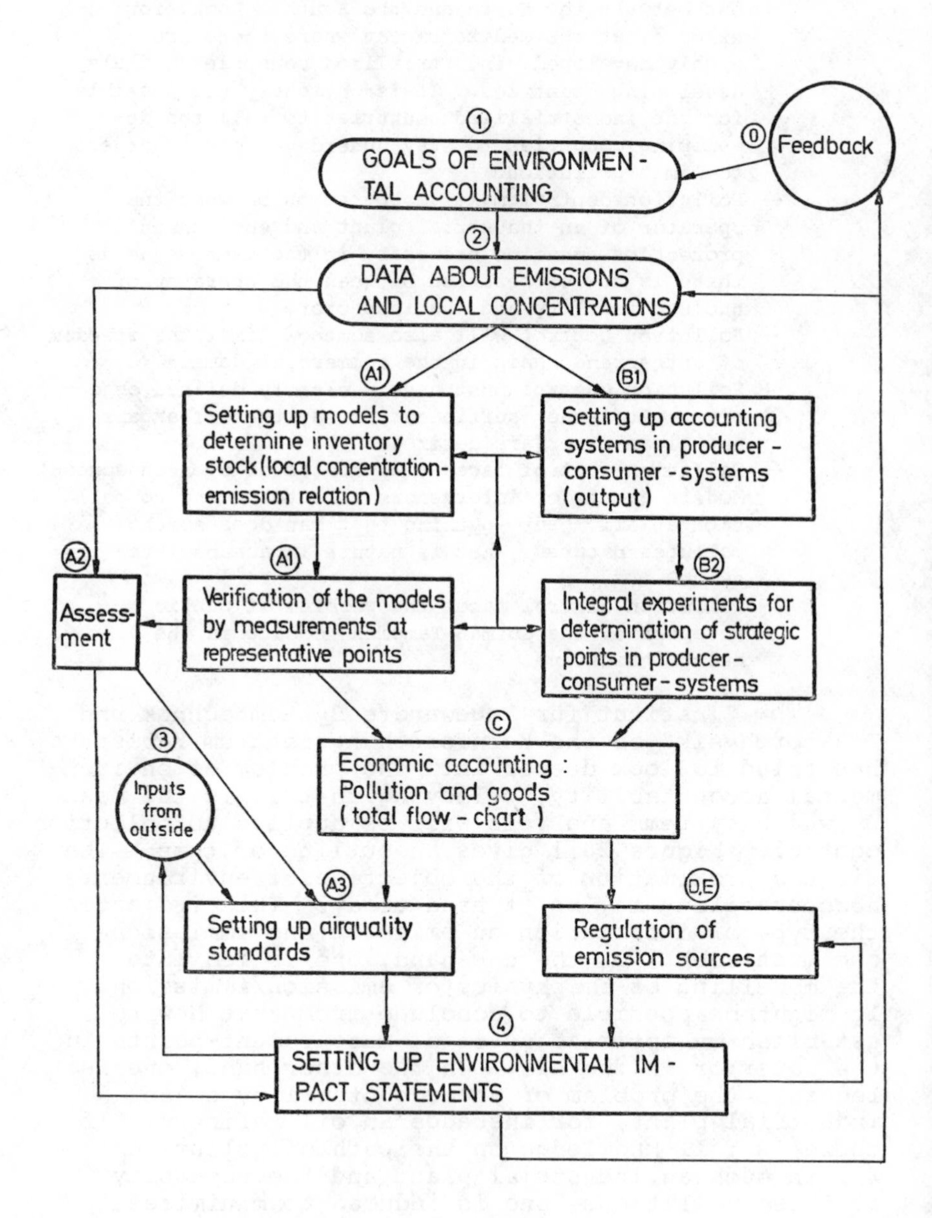


*Figure 20.1. Flow-chart for realization of an environmental accounting system.*

branches combine and eventually make it possible to establish a correlation between goods and immissions. This correlation can be influenced by technological developments (such as filters) and the alteration of consumers' behavior; the yardsticks for all that are environmental standards. The establishment of such standards requires an input in terms of toxicology and of public acceptance, as well as in terms of the correlation between goods and immissions. But such establishment is also a political process (as indicated by the round circle in the figure). The result of this is that all are environmental input statements. It is to be expected that these input statements will make it necessary to go through the whole procedure again. One must envisage, then, a permanent iteration process.

## REFERENCE

1. Häfele, W. "Systems Analysis in Safeguards of Nuclear Material," IV Geneva Conference 1971, P/771. *Proceedings of the Fourth International Conference, Geneva,* September 6-16, 1971. United Nations and the International Atomic Energy Agency (1972).

## INDEX